OCEAN WAVE
DYNAMICS

Other Related Titles from World Scientific

Theory and Applications of Ocean Surface Waves
(In 2 Volumes)
Third Edition
by Chiang C Mei, Michael Aharon Stiassnie and Dick K-P Yue
ISBN: 978-981-3147-17-1 (set)
ISBN: 978-981-3147-18-8 (set pbk)
ISBN: 978-981-3147-21-8 (vol. 1)
ISBN: 978-981-3147-22-5 (vol. 1 pbk)
ISBN: 978-981-3147-23-2 (vol. 2)
ISBN: 978-981-3147-24-9 (vol. 2 pbk)

Ocean Surface Waves: Their Physics and Prediction
Third Edition
by Stanisław R Massel
ISBN: 978-981-3228-37-5
ISBN: 978-981-3230-14-9 (pbk)

Computational Wave Dynamics
by Hitoshi Gotoh, Akio Okayasu and Yasunori Watanabe
ISBN: 978-981-4449-70-0

OCEAN WAVE DYNAMICS

edited by

Ian Young
Alexander Babanin
University of Melbourne, Australia

NEW JERSEY • LONDON • SINGAPORE • BEIJING • SHANGHAI • HONG KONG • TAIPEI • CHENNAI • TOKYO

Published by

World Scientific Publishing Co. Pte. Ltd.
5 Toh Tuck Link, Singapore 596224
USA office: 27 Warren Street, Suite 401-402, Hackensack, NJ 07601
UK office: 57 Shelton Street, Covent Garden, London WC2H 9HE

Library of Congress Cataloging-in-Publication Data
Names: Young, Ian R. (Ian Robert), 1932–   editor. |
 Babanin, Alexander V., 1960–   editor.
Title: Ocean wave dynamics / edited by Ian Young, Alexander Babanin,
 University of Melbourne, Australia.
Description: Singapore ; Hackensack, New Jersey : World Scientific, 2020. |
 Includes bibliographical references and index.
Identifiers: LCCN 2019027696 | ISBN 9789811208669 (hardcover) |
 ISBN 9789811208676 (ebook)
Subjects: LCSH: Ocean waves.
Classification: LCC GC211.2 .O238 2020 | DDC 551.46/3--dc23
LC record available at https://lccn.loc.gov/2019027696

British Library Cataloguing-in-Publication Data
A catalogue record for this book is available from the British Library.

Copyright © 2020 by World Scientific Publishing Co. Pte. Ltd.

All rights reserved. This book, or parts thereof, may not be reproduced in any form or by any means, electronic or mechanical, including photocopying, recording or any information storage and retrieval system now known or to be invented, without written permission from the publisher.

For photocopying of material in this volume, please pay a copying fee through the Copyright Clearance Center, Inc., 222 Rosewood Drive, Danvers, MA 01923, USA. In this case permission to photocopy is not required from the publisher.

For any available supplementary material, please visit
https://www.worldscientific.com/worldscibooks/10.1142/11509#t=suppl

Typeset by Stallion Press
Email: enquiries@stallionpress.com

Dedication

This book is dedicated to Mark Donelan of the University of Miami who sadly passed away in 2018 while this book was being written. Mark was a giant of ocean wave research. If one peruses the table of contents for this book, then you will have a summary of the areas in which Mark has made a profound impact. He could have written every chapter. Air–sea interaction, nonlinear interactions, wave breaking, ocean wave modeling, extreme conditions and remote sensing are all areas where Mark left a major legacy. The authors of this book have all been shaped by Mark Donelan. He was a wonderful scientist, teacher and human being. We are all richer for having been guided by him as a scientist and having known him as a person.

Preface

Surface waves are one of the most obvious processes in the ocean and in physics. They are immediately obvious to the casual observer, and one does not have to be a scientist or an engineer to have an opinion about them. They are also important for many reasons. The Allianz insurance company estimated that in 2017 a total of 94 large ships were lost at sea. In ocean engineering, they impose the main loading on offshore structures and are the main factor in design. Closer to shore, we know that waves play a critical role in coastal erosion and flooding. More than 800 million people live in 570 cities exposed to the combined threats of coastal erosion and inundation. In 2010, 290 million people worldwide lived below the 100-year flood level and US$9,600 billion of assets were exposed to inundation. Over the last century, more than 1 million lives have been lost due to coastal flooding. Ocean waves also play an important part in our climate system. They have a role in determining the aerodynamic roughness of the air–water interface and hence impact winds, fluxes of heat, energy and gas through the interface and the rate of breakup of ice sheets. There is also evidence that ocean waves are changing due to climate change, potentially exacerbating the risks already identified above.

    The physics of ocean waves are complex. Winds clearly generate ocean waves, but a full understanding of the exact mechanisms by which energy is transferred to the wave field, then nonlinearly transformed and finally lost due to random wave breaking have been actively investigated for more than seven decades. There are also conflicts of scale when considering ocean waves. Waves grow at the scale of thousands of wave periods under continuous wind action and lose it in a faction of the period in sporadic wave breaking events. The waves themselves have wavelengths from a few centimeters to hundreds of meters. They are generated by weather systems many hundreds of kilometers across and propagate as swell across oceanic basins thousands of kilometers in extent. Finally, they respond to climate change on multi-decadal timescales. Hence, even the computational resources of modern super computers are tested by both the complexity of the physics and the conflicting scales.

In this book, we have brought together a multi-disciplinary group of the world's leading ocean wave researchers to provide a state-of-the-art overview of ocean wave dynamics. Following an initial introduction to ocean wave mechanics, we provide an overview of the three major physical processes active in ocean wave evolution: atmospheric input, nonlinear interactions and wave breaking. We then provide a comprehensive overview of the two most common approaches to wave modeling: phase-resolving and phase-averaging models. A description of the statistical estimation of extreme wave conditions is also provided. Finally, we look at the role satellite oceanography has played in providing a description of global wave (and wind) climate and how it is changing.

About the Authors

Ian Young is Kernot Professor of Engineering at the University of Melbourne. Prior to this appointment, he held the administrative roles of Vice-Chancellor of the Australian National University and Vice-Chancellor of the Swinburne University of Technology. Prior to joining Swinburne, he was Executive Dean, Faculty of Engineering, Computer & Mathematical Sciences and Pro-Vice-Chancellor (International) at the University of Adelaide. His research interests concern wind-generated ocean waves. He has an extensive publication record in areas, such as the physics of air–sea interaction, the numerical modeling of waves, finite depth waves, satellite remote sensing and ocean wind and wave climate. In recent years, he has conducted a range of studies aimed at understanding global ocean wind and wave climate. These studies use large satellite databases which he has compiled. The studies investigate seasonal wind and wave climate, extremes and long-term trends in wind speed and wave height. He has published extensively in aspects of remote sensing of the oceans with applications to both engineering design and climate. He is the author of more than 150 refereed papers and two major research monographs in the field. He is also a consultant to offshore industries in Australia, United States and Asia, as well as an advisor to the US Navy on ocean wave physics.

Alexander V. Babanin is Professor in Ocean Engineering at the University of Melbourne, Australia. He completed his BSc in Physics, MSc in Physical Oceanography from the Lomonosov Moscow State University, Russia, and PhD in Physical Oceanography from the Marine Hydrophysical Institute, Sebastopol, Russia. He worked as a Research Scientist in the Marine Hydrophysical Institute, as an academic in the University of New South Wales, ADFA, Canberra, The University of Adelaide, South Australia, Swinburne University of Technology, Melbourne. Areas of expertise, research and teaching include wind-generated waves, maritime and coastal engineering, air–sea interactions, ocean turbulence and ocean dynamics, climate, environmental instrumentation and remote sensing of the ocean. These include extreme Metocean conditions from tropical cyclones to Arctic and Antarctic environments. He has authored 280+ publications in total in his career.

Contents

Preface   vii
About the Authors   ix
List of Symbols   xiii
List of Figures   xix
List of Tables   xxxi

Chapter 1.   Wind-Generated Waves   1
    Ian Young

Chapter 2.   Air–Sea Interaction   23
    Brian Haus

Chapter 3.   Wave Breaking and Ocean Mixing   47
    Alex Babanin

Chapter 4.   Nonlinear Processes   103
    Takuji Waseda

Chapter 5.   Phase-Averaged Wave Models   163
    W. Erick Rogers

Chapter 6.	Phase-Resolving Models *Dmitry Chalikov*	205
Chapter 7.	Extreme Conditions *Kevin Ewans and Philip Jonathan*	271
Chapter 8.	Satellite Observations and Climate *Ian Young*	321

Index   357

List of Symbols

a	wave amplitude
a_b	semi-excursion distance of water particle at bed
A	horizontal semi-axis of water particle ellipse
A	integral of the directional spectrum [Babanin and Soloviev, 1998]
A_s	wave asymmetry
b_T	breaking probability (0–1)
b_1	proportionality coefficient for wave turbulence production rate
B	vertical semi-axis of water particle ellipse
c_h	Charnock parameter
C	wave phase speed
C_d	atmospheric drag coefficient
C_f	bed friction coefficient
C_g	group velocity
C_p	phase speed of components at the spectral peak frequency
d	water depth
$D(f,\theta)$	directional spreading function
E	wave energy per unit crest length
f	frequency $= 1/T$
f^*	non-dimensional frequency $= fu_*/g$
f_c	Coriolis parameter
f_p	frequency of spectral peak
$F(f)$	one-dimensional or omni-directional wave frequency spectrum

$F(f,\theta)$	directional frequency spectrum
g	gravitational acceleration
G	quadruplet nonlinear interaction coupling coefficient
h_w	height of whitecap region of a wave
H	wave height
$H_{\max}$	maximum wave height
H_s	significant wave height
H_{rms}	root-mean-square wave height
$\bar{H}$	mean wave height
k	wave number
K	normalized directional spectrum
K_p	pressure attenuation factor
K_r	refraction coefficient
K_s	shoaling coefficient
L	wave length
L_o	Obukhov scale height
L_w	length of whitecap region of a wave
M_I (BFI)	Modulation Instability (Benjamin–Feir) Index
M_{Id}	directional modulation instability index
n	action density
N	number of waves to breaking
p	pressure
$p(H)$	probability density function of wave height
p_w	pressure exerted on water surface by whitecap
p_0	central pressure of a hurricane
$\bar{P}$	energy flux of waves
$Q(k_x, k_y)$	wave number spectrum of waves
Q_b	fraction of waves breaking
Q_s	normalization factor for directional spreading
R_b	bulk Richardson number
Re	Reynolds number
s	directional spreading exponent
S	Jeffreys' sheltering coefficient
S_k	skewness of wave record
S_b	bottom interaction source term

List of Symbols

S_{bf}	bottom friction source term
S_{brk}	depth limited breaking source term
S_{ds}	whitecap dissipation source term
S_{in}	atmospheric input source term
S_{nl}	quadruplet nonlinear interaction source term
S_{tot}	total source term
S_{tri}	triad nonlinear interaction source term
t	time
t_d	duration for which the wind blows
t^*	non-dimensional time $= tg/u_*$
T	wave period
T_a	temperature of air
T_r	length of recorded time series
T_v	mean virtual temperature of air at a height of $10\,\text{m}$
T_w	temperature of water
u	horizontal velocity (x-direction)
u_b	velocity at the bed
u_*	friction velocity
$\bar{u}$	depth-averaged velocity
U_{10}	wind speed measured at a height of $10\,\text{m}$
U_g	gradient wind speed
$U_{L/2}$	wind speed measured at a height of $L/2$
$U_{\min}$	minimum wind speed which can generate waves
U_∞	free stream wind velocity outside boundary layer
v	horizontal velocity (y-direction)
V_{fm}	velocity of forward movement of a hurricane
$V_{\max}$	maximum wind speed within a hurricane
w	vertical velocity (z-direction)
x	horizontal length coordinate
x	fetch length over which the wind blows
$X(f)$	Fourier transform
y	horizontal length coordinate
z	vertical length coordinate
z_0	surface roughness length
α	JONSWAP Phillips parameter

List of Symbols

$\hat{\alpha}$	spectral steepness parameter
β	Donelan et al. [1985] spectrum scale parameter
β	empirical parameter in Π_2
β_s	Donelan et al. [1985] directional spreading factor
β_x	growth rate of Benjamin–Feir instability
γ	energy growth rate
γ	JONSWAP peak enhancement factor
γ_d	Donelan et al. [1985] peak enhancement factor
γ_{ds}	damping coefficient for whitecapping
Γ	fractional increase in wave energy per radian
δ	boundary layer thickness
δ	non-dimensional water depth $= gd/U_{10}^2$
ε	non-dimensional energy $= \sigma^2 g^2/U_{10}^4$
ε	wave steepness ($\varepsilon = ak$)
ε_p	spectral peak wave steepness (see Eq. (3.21))
ε_k	volumetric dissipation rate of turbulence
ζ	vertical displacement of water particle
η	water surface elevation
ψ	phase angle of water wave
θ_m	mean wave direction
$\bar{\theta}$	mean directional width of spectrum
κ	von Karman constant ≈ 0.4
κ	non-dimensional peak wave number $= k_p U_{10}^2/g$
λ_r	radar wave length
λ_w	Bragg scattering wave length
Λ	atmospheric turbulence wave number
ν	non-dimensional peak frequency $= f_p U_{10}/g$
ν_a	kinematic viscosity of air
ν_w	kinematic viscosity of water
ξ	horizontal displacement of water particle
$\Pi(k,\omega)$	pressure wave number-frequency spectrum
Π_1	modulation instability index for one-dimensional spectrum
Π_2	modulation instability index for two-dimensional spectrum

ρ_a	density of air
ρ_w	density of water
σ	JONSWAP peak width parameter
σ^2	variance or total energy of the wave record
ς	non-dimensional duration $= gt/U_{10}$
τ	shear stress
$\tilde{\tau}$	non-dimensional timescale for change in wind direction
ϕ	velocity potential
ϕ_{mn}	cross-spectrum
φ	wave direction
χ	non-dimensional fetch $= gx/U_{10}^2$
ω	radian measure of frequency $= 2\pi f$
$\tilde{\omega}$	depth scaled non-dimensional frequency $= \omega^2 d/g$
$\bar{\omega}$	mean frequency
$\Omega(\omega)$	turbulent frequency pressure spectrum

List of Figures

Chapter 1

1.1 Example of a wave record, recorded at a location in the Southern Ocean. 2
1.2 The frequency spectrum of the time series as shown in Fig. 1.1. 8
1.3 Data from a number of fetch-limited studies showing the development of the non-dimensional energy ε as a function of the non-dimensional fetch χ. After Young [1999] 10
1.4 Data from a number of fetch-limited studies showing the development of the non-dimensional peak frequency ν as a function of the non-dimensional fetch χ. After Young [1999] 11
1.5 Growth law relationships given by (1.29) (top) and (1.30) (bottom). The shaded regions show the typical spread of published results and give an indication of the accuracy of these relationships. After Young [1999] 12
1.6 Comparison between proposed forms for the one-dimensional frequency spectrum. JONSWAP (1.33) (dashed line) and Donelan et al. [1985] form (1.36) (solid line). 14
1.7 JONSWAP data [Hasselmann et al., 1973] showing the relationship between α and non-dimensional peak frequency ν. The solid line shows the form (1.35). 14

1.8 JONSWAP data [Hasselmann et al., 1973] showing the relationship between γ and non-dimensional peak frequency ν. The solid line shows the mean value $\gamma = 3.3$. . . 15

1.9 JONSWAP data [Hasselmann et al., 1973] showing the relationship between σ_a, σ_b and non-dimensional peak frequency ν. The solid lines show the mean values $\sigma_a = 0.07$ and $\sigma_b = 0.09$. 15

1.10 The data of Donelan et al. [1985] showing the dependence of β on ν. The solid line is (1.38). 17

1.11 The data of Donelan et al. [1985] showing the dependence of γ on ν. The solid line is (1.39). 17

Chapter 3

3.1 Profiles of steep waves. (left) Real waves exhibiting asymmetry: Black Sea (left) and laboratory (right) — waves propagate from right to left. (right, reproduced from Babanin et al. [2007a]) Comparison of shapes for linear and nonlinear waves of the same steepness — waves propagate from left to right. 61

3.2 Spectra for evolution of steep nonlinear waves. (left) From top to bottom, spectra of wave steepness, skewness, their coherence and phase. (right) Same for skewness and asymmetry. 63

3.3 Statistics of features of the incipient breaking: skewness versus steepness (top left), asymmetry versus steepness (right), asymmetry versus skewness (bottom). 65

3.4 Segment of a time series with $IMS = 1.8\,\text{Hz}$, $IMS = 0.30$ and $U/C = 0$. The top panel shows the water surface elevation, η as a function of time in seconds. The highest wave in each group is an incipient breaker at the measurement point, breaking immediately after the wave probe. The subsequent panels show properties determined for each of the waves: steepness (second panel), skewness (third panel), asymmetry (fourth panel) and frequency (bottom panel, solid line signifies $IFM = 1.8\,\text{Hz}$). Figure is reproduced from Babanin et al. [2007a]. 68

3.5 Laboratory statistics for incipient breakers (five steepest waves). $IMF = 1.8\,\text{Hz}$, $IMS = 0.30$, $U/C = 0$. (top left) Skewness versus steepness. (right) Asymmetry versus steepness. (bottom) Frequency (inverse period) versus steepness. $IMF = 1.8\,\text{Hz}$ is shown with the solid line. Figure is reproduced from Babanin et al. [2007a]. 69

3.6 Parameterization of the breaking probability, laboratory data. Number of wavelengths to breaking versus IMS. No wind forcing: o — $IMF = 1.6\,\text{Hz}$; × — $IMF = 1.8\,\text{Hz}$; + — $IMF = 2.0\,\text{Hz}$. Filled circles represent — $IMF = 2.0\,\text{Hz}$, with wind forcing applied. The parameterization (3.10) is shown with the solid line. Figure is reproduced from Babanin et al. [2007a], where $\lambda = L$. 70

3.7 Ratio of maximal crest height to the initial wave amplitude versus mean wave steepness, in the modulational-instability scenario of evolution of nonlinear wave groups. Figure is reproduced from Hwung et al. [2005]. 73

3.8 Maximal volumetric dissipation rates max(Diss) versus wave amplitude a. The shaded area corresponds to the range and confidence limits of the laboratory experiment by Babanin and Haus [2009], obtained below the troughs. Dots are obtained by means of the wave-turbulence coupled model of Babanin and Chalikov [2012], vertical segments indicate their standard deviation. Figure is reproduced from Babanin [2011]. 91

Chapter 4

4.1 Four-wave resonant interaction experiments by McGoldrick et al. [1966] (left plate) and Longuet-Higgins and Smith [1966] (right plate). 107

4.2 Reproduced from Waseda and Tulin [1999]. Growth rate of the sidebands (seeded run). For waves 1.0 Hz, $\varepsilon = 0.175$, $\hat{\delta} = 0.343$, and for 0.514, 0.686, 0.857, 1.03, and 1.0 Hz, $\varepsilon = 0.133$, $\hat{\delta} = 0.842$. The curves are theoretical predictions from Benjamin and Feir [1967], and from (4.19) based on Krasitskii [1994]. 114

4.3 Spectral evolution and time series of the recurrent modulated wave train. Left: Reproduced from Lake et al. [1977]. Right: Reproduced from Tulin and Waseda [1999]. 116

4.4 The effect of spectral bandwidth demonstrated by Dysthe's equation. From upper left to lower right figures, the number of wave modes are 32, 16, 8, and 4. $ak = 0.1$, $df/f = 0.9$, $a/a = 0.3535$, and initial phases are $-pi/4$. Figure adopted from Waseda [1997]. 120

4.5 Classification of a temporally evolving [(a) case-SPTE] and a spatially evolving [(b) case-TPSE] unstable Stokes wave train. The wave profile and time series are given for each case. Double-headed arrows indicate the comparison in after Houtani et al. [2018]. 121

4.6 The time series (a), the difference between SPTE and TPSE (b) and enlarged view of the surface elevations, from the tank experiment (left), and from the Akhmediev solution (right). Solid gray line: SPTE; dashed black line: TPSE. Adapted from Houtani et al. [2018]. 122

4.7 A range-time schematic allowing a quantitative analysis of the range-time radar return image from a low-grazing angle radar. This is a schematic depiction presented by Tulin [1996]. 125

4.8 (Left) Range-time-intensity diagram of a 2.3-m wave group with 0.165 initial steepness, 12 m/s wind. Tracks of individual breaking waves can be identified in HH polarization. [After Fuchs et al., 1999] (Right-top) Range-time amplitude image of the range-time radar data collected in a wind–wave tank. [After Lamont-Smith et al., 2003] (Right-bottom) Image of the backscattered power for resolved waves. The set of short solid lines outlines the propagation of 3.1-s waves toward the radar. The single line shows an additional modulation propagating at the group velocity. [After Smith et al., 1996] 127

4.9 Sideband peaks and continuous downshifting. Reproduced from Hatori [1984]. 128

4.10 The $Q_p - \sigma_\theta$ diagram during two marine incidents. The timing of the Suwa-Maru accident in June 29, 2008 and Onomichi-Maru accident on December 30, 1980 are indicated by red dots. (Adapted from Tamura *et al.*, 2009; Waseda *et al.*, 2014) 134

4.11 The spectral parameters during marine accident cases. Adapted from Fujimoto *et al.* [2018]. 136

4.12 The JKEO site is indicated by the solid triangle ▲, and the typhoon center locations when the two freak waves were observed (□ and ◇). The JKEO-Narrow and JKEO-Broad freak wave time series are shown in the right diagrams. Directional spectra are shown below. 138

4.13 Kurtosis time-series of simulated JKEO-Narrow (red, left) and JKEO-Broad (blue, right) wave fields for HOSM $M = 3$ (solid lines) and $M = 2$ (dotted lines). Shaded area is the standard deviation of ensemble mean kurtosis. Observed on October 26 19:00 and 27 16:00, 2009. 139

4.14 Freak wave groups simulated by HOSM $M = 3$. Left and right panels show JKEO-Narrow and JKEO-Broad cases, respectively. Color indicates indices assigned to each freak wave group. After Fujimoto *et al.* [2018]. 142

4.15 The pdf of lifetime (left). Side views of average freak wave shapes for (middle) JKEO-Narrow and (right) JKEO-Broad of HOSM $M = 3$ calculation. After Fujimoto *et al.* [2018]. 143

4.16 Effect of narrow bandedness to the coupled NLS simulation of an unstable wave train of the carrier, and the two sidebands. After Waseda [1997]. 146

4.17 The evolution of sideband amplitudes and the carrier wave amplitude of an unstable wave train. Adapted from Melville [1982], and from Tulin and Waseda [1999]. 148

4.18 Evolution of the amplitude of the initially non-existing member of the DIA quartet numerically estimated by solving Zakharov's equation. The solid line corresponds to exact resonant interaction whereas the other lines correspond to cases with randomly perturbed resonance detuning parameters. After Waseda et al. [2015]. 152

4.19 Frequency spectra of irregular directional waves without and with opposing current, -5.2, -8.8 and $-13.8\,\mathrm{cm\,s^{-1}}$. The unperturbed spectrum is JONSWAP with a directional spread of $G(\theta) \propto \cos^{10}\theta$. Right figure is the saturated spectrum $B(f) = S(f) \times f^5$. After Waseda et al. [2015]. 152

Chapter 5

5.1 Significant wave height (shading, in meters) and mean wave direction (arrows) from a WAVEWATCH III hindcast, south of Alaska. The three contours indicate wave heights of 12, 13, and 14 meters. 166

5.2 Comparison of the frequency distributions of dissipation of wave energy by sea ice. Wadhams et al. [1988], Ardhuin et al. [2016], Doble et al. [2015], and Meylan et al. [2014] are observational studies. The "WA3" lines are based on model-data inversion using method similar to that used by Rogers et al. [2016, 2018], from Wave Array 3 [Thomson et al., 2015] of the ONR-supported "Sea State" field experiment. "SWIFT", "UK", and "NIWA" are buoy types. 172

5.3 Fetch/duration comparisons of four models. Model source term package is indicated in title above each plot. See text for explanation. 175

5.4 Results from four models, for a duration-limited case, with 10-meter wind speed of $15\,\mathrm{ms^{-1}}$ after 15 hours duration, starting from rest. $E(f)$ is energy density. $S_{\mathrm{in}}(f)$, $S_{\mathrm{ds}}(f)$, $S_{\mathrm{nl4}}(f)$ are source terms for wind input, whitecapping, and four-wave nonlinear interactions, respectively. $S_{\mathrm{tot}}(f)$ is the summation of the three source terms, and is the growth/decay rate, since advection terms are zero (unlimited fetch). 176

5.5 Histograms of wave height error from a WAVEWATCH III hindcast. Observed wave height is from buoys. Color scaling indicates air–sea temperature differences (°C) taken from buoy observations. 185

5.6 Validation of m_4 for three models: ST1/SWAN, ST4/WW3, and ST6/SWAN, using Waverider buoy CDIP 166, at Ocean Station Papa, owned and operated by APL/UW. Time period is 0400 UTC 4 November to 1200 UTC December 25, 2015. The m_4 parameter is proportional to the contribution to the mean square slope of the sea surface by frequency components up to 0.58 Hz. [These figures are taken from a poster presented by the author in 2017] 193

Chapter 6

6.1 Phase velocities of wave components as a function of wave number (solid curve) for different *rms* steepness of elevation (numbers at top right corner). Gray vertical bars correspond to *rms* of phase velocities. Dashed line is the linear dispersion relation. Dotted line is the wave spectrum (right axis). 213

6.2 Rate of growth γ (contours) as a function of δ and AK. 214

6.3 An example of development of a 2D wave spectrum $S(k_x, k_y)$; t is the time expressed in wave peak period. . . . 219

6.4 A schematic illustrating the HOS-NWT setting. In front of the domain, initial waves are assigned. At the back side of the domain, the wave absorption is formulated. 220

6.5 Distribution of spectral energy $\log_{10} \bar{S}$ ($\bar{S}$ is a spectral density normalized by its maximum) in the coordinates $(\omega_{\text{lin}}, \omega_{\text{mod}})$ where $\omega_{\text{lin}} = k^{1/2}$ and ω_{mod} is an actual phase velocity. Solid line shows the spectral density averaged over directions. 224

6.6 The wave spectra obtained upon reaching the 94th peak wave period corresponding to runs starting from the same wave spectrum but with different random sets of initial phases. Thick curve corresponds to the spectrum averaged over the ensemble; dashed curves characterize dispersion of the data. 226

6.7 Top panel corresponds to the initial Pierson–Moskowitz 2D wave spectrum $\log_{10}(S)$. Bottom panel corresponds to the final spectrum after integration over 318 peak wave periods. 227

6.8 Part (a) shows positions of local maxima in a single spectrum. Part (b) shows positions of maxima in 50 parallel runs. 228

6.9 Probability of wave surface deviations PH from a mean level, 1 — calculations for the data reproduced by a nonlinear model; 2 — calculations for an ensemble of linear waves given by formula (6.38). 234

6.10 Integral probability P_f of wave height from trough to crest: 1 and 3 — a 2D algorithm for the nonlinear and linear wave fields, respectively; 2 and 4 — the same for a 1D algorithm. 235

6.11 The imaginary part of the β-function. 238

6.12 Probability of curvilinearity $\eta_{\xi\xi}$. Thick curve calculated with a full 3D model; thin curve is the probability calculated over an ensemble of linear modes with the same spectrum. 244

6.13 Evolution of the integral characteristics of solution (a rate of evolution of the integral energy multiplied by 10^7) due to: 1 — tail dissipation D_t (Eqs. (6.41) and (6.42)); 2 — breaking dissipation D_b (Eqs. (6.43)–(6.46)); 3 — input of energy from wind I (Eqs. (6.39) and (6.40)); 4 — balance of energy $I + D_t + D_b$. Curve 5 shows evolution of wave energy $10^5 E$. Vertical bars of gray color show instantaneous values; thick curve shows a smoothed behavior. . . 247

6.14 Wave spectra $S_h(r)$ integrated over angle ψ in polar coordinates and averaged over a period of approximately 100 units of non-dimensional time t. The spectra grow and shift from right to left. 248

6.15 Sequence of 3D images of $\log_{10}(S(k,l))$ where each panel corresponds to single curve in Fig. 6.14. The left side refers to wave number $l(-M_y \leq l \leq M_y)$ while the front side corresponds to $k(-M \leq k \leq M)$. 249

6.16 A sequence of the 2D images of $\log_{10}(S(k,l))$ averaged over the consequent seven period length $\Delta t = 200$. The numbers indicate the period of averaging (first panel marked 0, refers to the initial conditions). The horizontal and vertical axes correspond to wave numbers k and l, respectively. 251

6.17 The spectrum of energy input $I(r)$ integrated over angle ψ in the polar coordinates and averaged over approximately 100 units of non-dimensional time t. 252

6.18 The tail dissipation spectra $D_t(r)$ integrated over angle ψ in the polar coordinates and averaged over approximately 100 units of non-dimensional time t. 253

6.19 Breaking dissipation spectra $D_b(r)$ integrated over angle ψ in the polar coordinates and averaged approximately 100 units of non-dimensional time t. 254

6.20 An example of energy input due to breaking $D_b(x)$. 255

6.21 Evolution of the number of wave breaking events N_b expressed as the percentage of the number of grid points $N_x \times N_y$. 255

6.22 A sequence of wave spectra $S_h(r)$ (thick curves) and non-linear input term $N(r)$ (thin curves) averaged over the consequent eight periods of length $\Delta t = 100$ starting from sixth period. 257

6.23 Time evolution of ratio E_l/E_k. 258

6.24 Time evolution of: the weighted frequency ω_w (1) Eq. (6.34); spectral peak frequency ω_p (2); full energy E (3) Eq. (6.25). Thin curves are the empirical dependence for peak wave number and energy. F is the distance passed by the spectral peak. 258

Chapter 7

7.1 Probability distributions and density functions of H_s values derived from a Weibull 2-parameter distribution.

Continuous blue line is the empirical distribution estimated from H_s values, the remaining dash and dash-dotted lines are estimates of the cumulative distribution and probability density functions of the hourly wind speeds (blue), the annual maximum hourly H_s values (red), and the 100-year maximum hourly H_s values (black). 278

7.2 10- (blue), 100- (red), and 1000-year (orange) $H_s - T_p$ contours. Crosses mark respective design point sea states, corresponding to the median maximum crest elevation for the contours. Dashed lines are lines of constant median maximum crest elevation corresponding to the respective design point sea states. 298

7.3 Storm peak value of associated T_p versus H_s^{sp} for measured NNS storm peak data [Jonathan et al., 2010]. . . . 303

7.4 Plots of storm peak H_s^{sp} (in meters, horizontal) versus associated T_P (in seconds, vertical) for the most severe (20%) storms emanating from six directional sectors (ordered, clockwise from 20°). Note that there are no data for sector [20,140). [After Jonathan et al., 2013]. . . 304

7.5 Left: Northern North Sea location and directional sectors with distinctive wave characteristics. Right: Return values of H_s^{sp} (inner circle) and conditional values of associated T_P (outer circle). Inner dashed lines (on common scale): storm peak H_s^{sp} with non-exceedance probability 0.99 (in 34 years), with (white) and without (black) directional effects. Outer solid lines (on common scale): median associated T_P with (white) and without (black) directional effects; outer dashed lines give corresponding 2.5% and 97.5% values for associated T_P. [After Ewans and Jonathan, 2013]. 305

Chapter 8

8.1 Spherical altimeter pulses radiating from the satellite antenna and areas illuminated in successive range gates. [After Young, 1999] 323

8.2	Examples of the one second average return pulse for the GEOSAT altimeter for (a) $H_s = 0.36$ m and (b) $H_s = 7.47$ m. [After Young, 1999]	324
8.3	Relationship between radar cross-section, σ_0 and wind speed, U_{10} for Ku band altimeters. For $U_{10} < 18$ ms^{-1} the relationship is given by Abdalla [2007] and for $U_{10} \geq 18$ ms^{-1} by Young [1993].	325
8.4	The scan geometry of a radiometer instrument, showing the approximately 1400 km wide swath and the fact that the instrument senses emissions at multiple frequencies. [After Hollinger et al., 1990]	326
8.5	The scan geometry of a rotating bean scatterometer, showing the swath width and a location imaged by multiple beams. As the scatterometer flies past the location, it will again image the location by the circular scan on the rear-side of the circle.	327
8.6	Satellite missions from altimeter, radiometer and scatterometer which have been operational since 1985.	329
8.7	Buoy co-location points used for satellite calibration. [After Young et al., 2017]	330
8.8	An example of a match-up calibration between buoy and (a) radiometer and (b) scatterometer. [After Young et al., 2017]	330
8.9	Difference in wind speed ΔU_{10} between buoy data and CRYOSAT-2 altimeter. A discontinuity is clear around mid-2014. In this case, separate calibration relations were applied either side of the discontinuity. [After Ribal and Young, 2019]	331
8.10	Q–Q plot between the f15 radiometer and buoy data, showing the radiometer measures higher than the buoys above approximately 18 ms^{-1}. [After Young et al., 2017]	331
8.11	Satellite–satellite validations for a number of different satellite combinations. [After Young et al., 2017]	333
8.12	Global wind speed, U_{10} during the month of January from ALT data [unit: ms^{-1}]. (a) Mean monthly wind speed (b) 90th percentile wind speed.	335

8.13 Global wind speed, U_{10} during the month of July from ALT data [unit: ms^{-1}]. (a) Mean monthly wind speed (b) 90th percentile wind speed. 336

8.14 Global significant wave height H_s during the month of January from ALT data [unit: m]. (a) Mean monthly significant wave height (b) 90th percentile significant wave height. 337

8.15 Global significant wave height, H_s during the month of July from ALT data [unit: m]. (a) Mean monthly significant wave height (b) 90th percentile significant wave height. 338

8.16 Global values of 100-year return period wind speed and significant wave height obtained from ALT data. (a) 100-year return period wind speed, U_{10}^{100} [unit: ms^{-1}], (b) 100-year return period significant wave height, H_s^{100} [unit: m]. 340

8.17 Global values of 100-year return period wind speed, U_{10}^{100} obtained from SCAT data [unit: ms^{-1}]. 343

8.18 Global trends in ALT wind speed. (a) Mean monthly wind speed, $\bar{U}_{10}$, (b) 90th percentile monthly wind speed, U_{10}^{90} [units cm/s/yr]. 346

8.19 Global trends in ALT significant wave height. (a) Mean monthly significant wave height, $\bar{H}_s$, (b) 90th percentile monthly significant wave height, H_s^{90} [units cm/yr]. 347

8.20 Global trends in ALT wind speed and wave height mode of the probability distribution. (a) Wind speed trend [units cm/s/yr]. (b) Significant wave height trend [units cm/yr]. 349

8.21 Probability distributions functions for ALT wind speed and wave height for a region west of South America. (a) Wind speed pdf. (b) Significant wave height pdf. . . . 350

8.22 Global trends in SCAT wind speed. (a) Mean monthly wind speed, $\bar{U}_{10}$, (b) 90th percentile monthly wind speed, U_{10}^{90} [units cm/s/yr]. 352

List of Tables

Chapter 1

1.1 Representative waves calculated from the Rayleigh distribution [after Young, 1999]. 6

Chapter 4

4.1 Parameters appearing in literature representing the relative significance of nonlinearity and dispersion (unidirectional). 130
4.2 Parameters appearing in literature extending the BFI to include directionality. 132
4.3 The steepness ak, frequency bandwidth Q_p, directional spreading σ_θ, the Benjamin–Feir Index (unidirectional) BFI, non-dimensional depth $k_p d$, and kurtosis κ_4, for BG 2014 [Bitner-Gregersen *et al.*, 2014], D 2015 [Dias *et al.*, 2015], F 2016 [Fedele *et al.*, 2016], T 2015 [Trulsen *et al.*, 2015] and F 2018 [Fujimoto *et al.*, 2018]. 137

Chapter 1

Wind-Generated Waves

Ian Young

University of Melbourne, Parkville VIC 3010, Australia

1.1. Introduction

As this book considers the dynamics of water waves on the interface between air and water (the ocean in this context), it is necessary to define how one can describe these waves. The simplest approach is to consider that such waves can be approximated by a two-dimensional sinusoidal form. That is, waves which have infinitely long crests with the surface elevation, η is defined by

$$\eta = a\sin(kx - \omega t), \qquad (1.1)$$

where a is the wave amplitude, $k = 2\pi/L$ is the wave number, L is the wave length, $\omega = 2\pi/T$ is the frequency, T is the wave period, and x and t are space and time, respectively. Small amplitude or linear wave theory [Airy, 1945] can be used to define relationships between quantities such as the phase speed of the waves, C, orbital motions beneath the surface and k and ω. A full description of linear wave theory is not contained here and the reader is referred to one of the many texts which cover this subject [e.g., Holthuijsen, 2007; Young, 1999].

1.2. Frequency or Omni-Directional Spectrum

Although linear wave theory provides a useful mathematical context to consider ocean surface waves, its applicability to real ocean

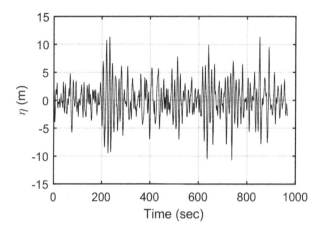

Fig. 1.1. Example of a wave record, recorded at a location in the Southern Ocean.

situations is questionable. Figure 1.1 shows a time series of water surface elevation measured at a wave buoy in the Southern Ocean. Although this is a case of extremely large waves, it is typical of wind-generated ocean waves. Although there is clearly a dominant wave period, there is marked variability in both the amplitude and period of individual waves. That is, the wave record does not conform to the sinusoidal form described by (1.1).

It is common within many areas of physics to approximate wave records such as that shown in Fig. 1.1 by the use of a spectral or Fourier model. Under this model, the water surface is represented by the linear summation of sinusoidal components of the form described by (1.1):

$$\eta(t) = \sum_{i=1}^{N} a_i \sin(\omega_i t + \phi_i), \qquad (1.2)$$

where a_i, ω_i and ϕ_i are the amplitude, frequency and phase of component i, respectively. Provided enough components are included in the summation in (1.2), almost any water surface can be approximated by the summation. The power of this approach is that each of the components in the summation will satisfy the properties of linear wave theory.

It is often more convenient to consider the energy of the wave per unit crest length E, rather than the amplitude a. From linear wave theory, this energy is as follows:

$$E = \frac{1}{8}\rho_w g H^2 L, \tag{1.3}$$

where $H = 2a$ is the wave height and ρ_w is the density of water. The energy per unit area or specific energy, $\bar{E}$ is as follows:

$$\bar{E} = \frac{1}{8}\rho_w g H^2. \tag{1.4}$$

From (1.4) and (1.2), the average energy of the wave profile is as follows:

$$\bar{E} = \frac{\rho_w g}{8N} \sum_{i=1}^{N} H_i^2 \tag{1.5}$$

or

$$\frac{\bar{E}}{\rho_w g} = \frac{1}{2N} \sum_{i=1}^{N} a_i^2 = \sigma^2, \tag{1.6}$$

where σ^2 is the variance of the water surface elevation record. It follows from (1.6) that the amplitude components a_i^2 are related to the energy of the wave record and that one could display the distribution of this energy with frequency by plotting a_i^2 versus ω_i. Such a plot is termed a discrete amplitude spectrum, being defined only at the discrete values of the Fourier summation.

In the limit, as $N \to \infty$, the amplitude spectrum can be transformed into a continuous spectrum, $F(f)$, where

$$F(f)\Delta f = \frac{a_i^2}{2}. \tag{1.7}$$

The spectrum $F(f)$ is called the frequency spectrum or omni-directional spectrum (as no direction is associated with the spectrum) or variance [as the integral of the spectrum is the variance from (1.6)] spectrum:

$$\sigma^2 = \int_0^\infty F(f)\,df. \tag{1.8}$$

1.3. Directional Spectrum

The spectral form $F(f)$ acknowledges that the water surface is composed of components across a range of frequencies; however, it does not consider that these components could propagate in a variety of directions. An extension of (1.2) to account for direction would take the form

$$\eta(x,y,t) = \sum_{i=1}^{N} a_i \sin[k_i(x\cos\theta_i + y\sin\theta_i) - \omega_i t + \phi_i], \tag{1.9}$$

where θ_i is the angle between the x-axis and the direction of propagation of component i. In a similar manner to the frequency spectrum $F(f)$, Eq. (1.9) allows a directional spectrum $F(f,\theta)$ to be defined as

$$\sigma^2 = \int_0^{2\pi} \int_0^\infty F(f,\theta)\,df\,d\theta. \tag{1.10}$$

The directional frequency spectrum $F(f,\theta)$ defines the distribution of energy as a function of frequency and direction. It follows from (1.9) that noting that wave number is a vector, the energy could also have been described in terms of a wave number spectrum $Q(k_x,k_y)$ or $Q(k,\theta)$. Then the variance in a similar fashion to (1.10) is as follows:

$$\sigma^2 = \iint Q(k_x,k_y)\,dk_x\,dk_y = \iint F(f,\theta)\,df\,d\theta. \tag{1.11}$$

Noting that $dk_x dk_y = |\mathbf{k}|dk\,d\theta$, the spectral forms can be related as

$$F(f,\theta) = Q(k_x,k_y)|\mathbf{k}|\frac{dk}{df}, \tag{1.12}$$

and the quantity dk/df can be determined from linear wave theory.

1.4. Statistical Properties of Ocean Wave Heights

In Eq. (1.2), the statistical variability of the water surface elevation is modeled using a Fourier series. Assuming the Fourier amplitudes a_i (or spectrum) are narrow banded, it can be shown that the probability density function (pdf) for individual wave heights follows a Rayleigh distribution [Rice, 1954].

$$p(H) = \frac{H}{4\sigma^2} e^{-H^2/8\sigma^2}, \qquad (1.13)$$

where the pdf must satisfy the requirement $\int p(H)dH = 1$.

Longuet-Higgins [1952] derived relationships for a number of characteristic wave heights based on (1.13):

Mean wave height:

$$\bar{H} = \int H p(H) dH = \sqrt{2\pi\sigma^2}, \qquad (1.14)$$

Root mean square (rms):

$$H_{\text{rms}}^2 = \int H^2 p(H) dH = 8\sigma^2. \qquad (1.15)$$

From (1.15) and (1.13), the Rayleigh distributions can be written as

$$p(H) = \frac{2H}{H_{\text{rms}}^2} e^{-H^2/H_{\text{rms}}^2}. \qquad (1.16)$$

The cumulative form of (1.16), i.e., the probability that H is greater than $\hat{H}$, is given by

$$P(H > \hat{H}) = \int_{\hat{H}}^{\infty} p(H) dH = e^{-H^2/H_{\text{rms}}^2}. \qquad (1.17)$$

The average height of all waves greater than $\hat{H}$ is given by

$$\bar{H}(\hat{H}) = \frac{\int_{\hat{H}}^{\infty} H^2 e^{-(H/H_{\text{rms}})^2} dH}{\int_{\hat{H}}^{\infty} H e^{-(H/H_{\text{rms}})^2} dH}. \qquad (1.18)$$

Table 1.1. Representative waves calculated from the Rayleigh distribution [after Young, 1999].

N	$H_{1/N}/\sigma$	$H_{1/N}/\bar{H}$	$H_{1/N}/H_{\rm rms}$	Comments
100	6.67	2.66	2.36	
50	6.24	2.49	2.21	
20	5.62	2.24	1.99	
10	5.09	2.03	1.80	Highest 1/10 wave
5	4.50	1.80	1.59	
3	4.00	1.60	1.42	Significant wave
2	3.55	1.42	1.26	
1	2.51	1.00	0.87	Mean wave

Rather than wanting to know the average height of all waves greater than a specified value, it is more common to want to know the average height of the highest $1/N$ waves. This can also be determined from the Rayleigh distribution [Goda, 1985]. Following Young [1999], typical values are given in Table 1.1.

From Table 1.1, it follows that the wave whose height is equal to the average of the highest 1/3 of the waves, is also equal to 4 times the variance of the record. This value is called the significant wave height H_s and is often used as a representative wave height for a record:

$$H_s = 4\sigma \approx H_{1/3}. \qquad (1.19)$$

Similarly, it follows that the highest 1% of waves is given by $H_1 \approx 1.66 H_s$ and it can also be shown that the maximum wave height is $H_{\max} \approx 2 H_s$. Therefore, the significant wave height can be calculated either from a time-domain analysis (e.g., $H_{1/3}$) or from the integral of the spectrum (1.10) (e.g., H_s).

1.5. Spectral Analysis of Recoded Data

A number of instruments can be used to measure the water surface elevation at a discrete sampling interval Δt. A good summary of approaches can be found in Tucker [1991]. Assuming the water surface elevation has been measured, following (1.2) it can be represented as

$$\eta(t) = \int_{-\infty}^{\infty} X(f) e^{i\omega t} dt, \qquad (1.20)$$

where $i = \sqrt{-1}$ and $X(f)$ is the Fourier transform of η. Noting that $e^{i\theta} = \cos\theta + i\sin\theta$, Eq. (1.20) can be expressed as

$$\eta(t) = \int_{-\infty}^{\infty} X(f)[\cos\omega t + i\sin\omega t] dt, \qquad (1.21)$$

which is the same form as in (1.9). As $\eta(t)$ is a real quantity, the integral in (1.21) must also be a real quantity. Hence, the Fourier transform $X(f)$ must, in general, be a complex quantity. It follows that

$$X(f) = \int_{-\infty}^{\infty} \eta(t) e^{-i\omega t} dt \qquad (1.22)$$

or

$$X(f) = \int_{-\infty}^{\infty} \eta(t)[\cos\omega t - i\sin\omega t] dt. \qquad (1.23)$$

The complex nature of $X(f)$ signifies that the sinusoidal spectral component has both an amplitude a_i and a phase ϕ_i. As the water surface is recorded as discrete values over a finite period of time, it is convenient to convert the integral in (1.23) to a discrete summation:

$$\underbrace{X\,(n/T_r)}_{=f} = \frac{T_r}{N} \underbrace{\sum_{j=0}^{N-1} \eta(jT_r/N)[\cos(2\pi jn/N) - i\sin(2\pi jn/N)]}_{=dt},$$
$$(1.24)$$

where N is the number of points in the discrete time series, T_r is the length of the time series and n is a counter which ranges from 0 to $N-1$. With the Fourier transform (1.24) defined, the frequency spectrum is as follows:

$$F(f) = \frac{2}{T_r} |X(f)|^2. \qquad (1.25)$$

The modulus of the Fourier transform in (1.25) means that all phase information in the Fourier approximation to the water surface is lost. Therefore, the spectrum contains no phase information. As will be outlined in Chapter 6, this means that models which predict the spectrum are often termed phase-averaged models, whereas models which represent the actual water surface elevation record are commonly termed phase-resolving models (Chapter 7).

Figure 1.2 shows the spectrum obtained from Fourier analysis of the time series shown in Fig. 1.1. Based on the integral of the spectrum, the significant wave height is given by $H_s = 12.18\,\text{m}$. It is interesting to note that the highest crest to trough height in the time series is approximately 21 m. Noting that $H_1 \approx 1.66 H_s$, this yields a value of 21.2 m in good agreement with the observed largest wave in the record.

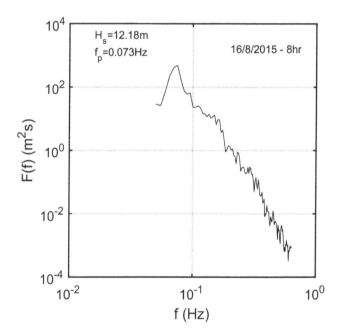

Fig. 1.2. The frequency spectrum of the time series as shown in Fig. 1.1.

1.6. Fetch-Limited Growth

A basic understanding of the important factors governing wind wave growth can be obtained from the study of one of the simplest wave evolution cases: *fetch-limited growth*. In fetch-limited growth, a constant wind U_{10} blows for a long period (such that it reaches steady state) perpendicular to an infinitely long straight coastline. The evolution of the wave spectrum is then studied with distance offshore: the fetch x. Based on a suggestion by Olson [1943], Sverdrup and Munk [1947] and Kitaigorodskii [1962, 1973] assumed that the following variables should define the situation: the fetch length x; the wind speed U_{10}; gravitational acceleration g; the variance of the water surface elevation $\sigma^2 = H_s^2/16$ (1.19); and the peak frequency of the spectrum f_p. Dimensional analysis yields three non-dimensional groupings of these quantities:

$$\varepsilon = \frac{\sigma^2 g^2}{U_{10}^4} \quad \text{(non-dimensional energy)}, \tag{1.26}$$

$$\nu = \frac{f_p U_{10}}{g} \quad \text{(non-dimensional frequency)}, \tag{1.27}$$

$$\chi = \frac{gx}{U_{10}^2} \quad \text{(non-dimensional fetch)}. \tag{1.28}$$

There have been numerous field and laboratory studies aimed at determining $\varepsilon = f_1(\chi)$ and $\nu = f_2(\chi)$, where f_1 and f_2 denote the functional dependence.

Young [1999] considers the results from a number of these studies, which include:

- the pioneering work of Sverdrup and Munk [1947] and Bretschneider [1952, 1958], who developed the so-called SMB curves;
- the studies of fully-developed or long fetch asymptotic limits by Pierson and Moskowitz [1964];
- the JONSWAP studies of Hasselmann *et al.* [1973];
- measurements in the Bothnian Sea by Kahma [1981];

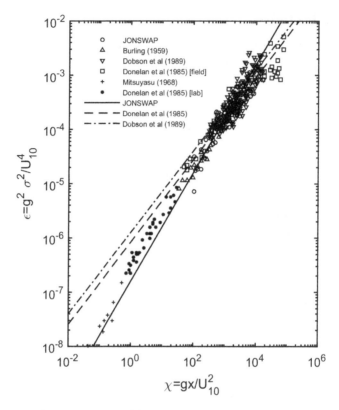

Fig. 1.3. Data from a number of fetch-limited studies showing the development of the non-dimensional energy ε as a function of the non-dimensional fetch χ. After Young [1999].

- Lake Ontario measurements by Donelan et al. [1985];
- measurements in the North Atlantic by Dobson et al. [1989].

Young [1999] combined these results and a variety of other datasets into composite diagrams relating the non-dimensional variables. Figure 1.3 shows the combined datasets for ε versus χ and Fig. 1.4 shows the corresponding results for ν versus χ.

These results show that the functions f_1 and f_2 are well described by power laws, finally asymptoting to a constant value at large

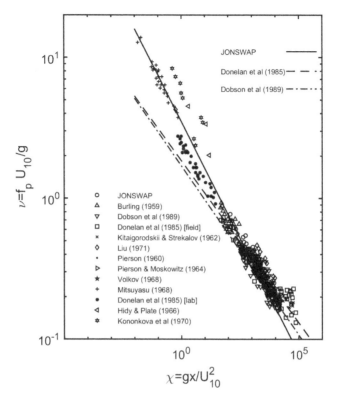

Fig. 1.4. Data from a number of fetch-limited studies showing the development of the non-dimensional peak frequency ν as a function of the non-dimensional fetch χ. After Young [1999].

non-dimensional fetch, χ. Young [1999] approximated the composite dataset by

$$\varepsilon = \max \begin{cases} (7.5 \pm 2.0) \times 10^{-7} \chi^{0.8}, \\ (3.6 \pm 0.9) \times 10^{-3}, \end{cases} \quad (1.29)$$

$$\nu = \max \begin{cases} (2.0 \pm 0.3) \chi^{-0.25}, \\ (0.13 \pm 0.02). \end{cases} \quad (1.30)$$

Equations (1.29) and (1.30) are shown in Fig. 1.5.

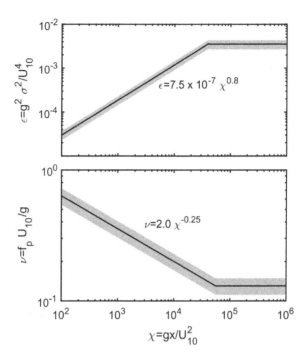

Fig. 1.5. Growth law relationships given by (1.29) (top) and (1.30) (bottom). The shaded regions show the typical spread of published results and give an indication of the accuracy of these relationships. After Young [1999].

1.7. Spectral Evolution

Although evolution of integral parameters of the spectrum, such as H_s and f_p (ε and ν), provide valuable insight, it is often more informative to understand how the full spectrum $F(f)$ evolves as a function of fetch, x.

As can be seen from Fig. 1.2, the high frequency face of the one-dimensional frequency spectrum can be approximated by a power law of the type $F(f) \propto f^{-n}$. Phillips [1958] assumed that this region was controlled by gravity such that dimensional analysis yields

$$F(f) \propto g^2 f^{-5}. \tag{1.31}$$

Toba [1973] assumed that the wind speed as characterized by the friction velocity u_* was also important, dimensional analysis yielding

$$F(f) \propto g u_* f^{-4}. \tag{1.32}$$

1.7.1. JONSWAP form: f^{-5}

Based on the high-frequency spectral form proposed by Phillips [1958], Hasselmann et al. [1973] considered data obtained from the Joint North Sea Wave Project (JONSWAP), proposing a spectral form

$$F(f) = \underbrace{\alpha g^2 (2\pi)^{-4} f^{-5} \left[\frac{-5}{4} \left(\frac{f}{f_p} \right)^{-4} \right]}_{\text{Pierson–Moskowitz spectrum}} \cdot \gamma^{\exp\left[\frac{-(f-f_p)^2}{2\sigma^2 f_p^2} \right]}, \tag{1.33}$$

where

$$\sigma = \begin{cases} \sigma_a & \text{for } f \leq f_p, \\ \sigma_b & \text{for } f > f_p. \end{cases} \tag{1.34}$$

Figure 1.6 shows a typical spectral form generated by (1.33). This relationship has five free parameters. The parameters α and f_p are scale parameters. The parameter α defines the overall scale of the spectrum and was first proposed by Phillips [1958] and f_p defines the peak of the unimodal spectral form. The remaining three parameters are shape variables. The peak enhancement parameter γ defines the ratio of the maximum spectral energy to the maximum spectral energy of the corresponding Pierson–Moskowitz [Pierson and Moskowitz, 1964] form. The parameters σ_a and σ_b are the left and right spectral width parameters, respectively. These shape parameters define the shape of the spectral peak region. They have little influence on the spectrum, as one moves away from this region. At high frequency (e.g., $f > 3f_p$), the spectrum reverts to a form $F(f) \approx f^{-5}$.

As part of the JONSWAP experiment, Hasselmann et al. [1973] parameterized the spectral parameters α, γ, σ in terms of the non-dimensional peak frequency ν (1.27). Through (1.30), these relationships could also be expressed in terms of the non-dimensional

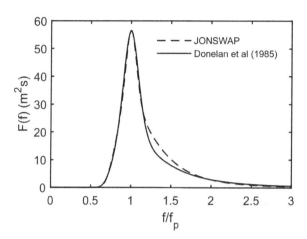

Fig. 1.6. Comparison between proposed forms for the one-dimensional frequency spectrum. JONSWAP (1.33) (dashed line) and Donelan *et al.* [1985] form (1.36) (solid line).

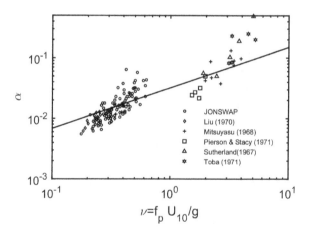

Fig. 1.7. JONSWAP data [Hasselmann *et al.*, 1973] showing the relationship between α and non-dimensional peak frequency ν. The solid line shows the form (1.35).

fetch χ (1.28). Figures 1.7–1.9 show the JONSWAP data for these parameters.

Hasselmann *et al.* [1973] found no consistent relationship within the data scatter for the shape parameters, adopting their mean

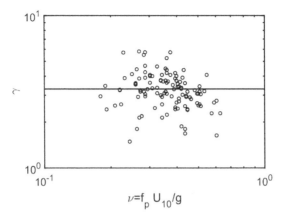

Fig. 1.8. JONSWAP data [Hasselmann et al., 1973] showing the relationship between γ and non-dimensional peak frequency ν. The solid line shows the mean value $\gamma = 3.3$.

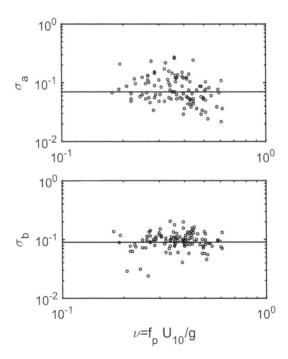

Fig. 1.9. JONSWAP data [Hasselmann et al., 1973] showing the relationship between σ_a, σ_b and non-dimensional peak frequency ν. The solid lines show the mean values $\sigma_a = 0.07$ and $\sigma_b = 0.09$.

values, $\gamma = 3.3$, $\sigma_a = 0.07$ and $\sigma_b = 0.09$. This is not surprising as these are notoriously difficult parameters to determine accurately from a curve fit to a spectrum defined at only discrete values of frequency. For α, they obtained the following relationship (Fig. 1.7):

$$\alpha = 0.033\nu^{0.67}. \tag{1.35}$$

1.7.2. The Toba form: f^{-4}

As shown by (1.32), Toba [1973] proposed an alternative high frequency spectral form proportional to f^{-4}. Utilizing this form, Donelan et al. [1985] proposed a spectral formulation similar to JONSWAP

$$F(f) = \beta g^2 (2\pi)^{-4} f_p^{-1} f^{-4} \left[-\left(\frac{f}{f_p}\right)^{-4} \right] \cdot \gamma_d^{\exp\left[\frac{-(f-f_p)^2}{2\sigma^2 f_p^2}\right]}. \tag{1.36}$$

Based on a combination of both field and laboratory data, Donelan et al. [1985] proposed the relationships for the parameters

$$\varepsilon = 6.365 \times 10^{-6} \nu^{-3.3}, \tag{1.37}$$

$$\beta = 0.0165 \nu^{0.55}, \tag{1.38}$$

$$\gamma_d = \begin{cases} 6.489 + 6 \log \nu, & \nu \geq 0.159, \\ 1.7, & \nu < 0.159, \end{cases} \tag{1.39}$$

$$\sigma = 0.08 + 1.29 \times 10^{-3} \nu^{-3}. \tag{1.40}$$

The Donelan et al. [1985] data relating β and ν are shown in Fig. 1.10 and relating γ_d and ν in Fig. 1.11. As shown by (1.37)–(1.40), Donelan et al. [1985] found consistent relationships for all parameters with much less scatter than Hasselmann et al. [1973]. This may be because (1.36) is a more appropriate spectral form than (1.33) or simply that the data quality is superior for the Donelan et al. [1985] experiment.

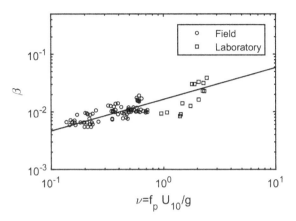

Fig. 1.10. The data of Donelan *et al.* [1985] showing the dependence of β on ν. The solid line is (1.38).

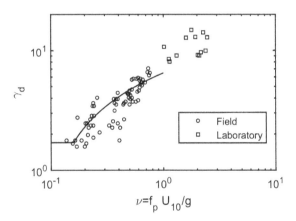

Fig. 1.11. The data of Donelan *et al.* [1985] showing the dependence of γ on ν. The solid line is (1.39).

1.8. Directional Spreading

The directional spectrum (1.10) is often represented in terms of the one-dimensional spectrum, $F(f)$ [Longuet-Higgins *et al.*, 1963], as follows:

$$F(f,\theta) = F(f)D(f,\theta). \tag{1.41}$$

The directional spreading function $D(f,\theta)$ is defined by

$$\int D(f,\theta)d\theta = 1. \tag{1.42}$$

A number of different forms have been proposed for $D(f,\theta)$. The most detailed data was compiled by Donelan et al. [1985] who proposed the form

$$D(f,\theta) = 0.5\beta_s \text{sech}^2 \beta_s[\theta - \theta_m(f)], \tag{1.43}$$

where θ_m is the mean wave direction at frequency f and β_s is given by

$$\beta_s = \begin{cases} 2.61 \left(\frac{f}{f_p}\right)^{1.3} & \text{for } 0.56 < f/f_p < 0.95, \\ 2.28 \left(\frac{f}{f_p}\right)^{-1.3} & \text{for } 0.95 < f/f_p < 1.60, \\ 10^{\{-0.4+0.8393\,\exp[-0.567\ln(f/f_p)^2]\}} & \text{for } f/f_p > 1.6. \end{cases} \tag{1.44}$$

The first two branches of the relationship (1.44) were obtained from the Donelan et al. [1985] wave staff data. The relationship for $f/f_p > 1.6$ was obtained from stereo-photographic data by Banner [1990].

Equations (1.44) define directional spreading which is narrowest at frequencies near the spectral peak and broadens at frequencies both above and below this value.

1.9. Spectral Energy Balance

The original studies of spectral shape used dimensional arguments [Phillips, 1958; Toba, 1973]. A major development in understanding of the processes responsible for the evolution of the spectrum was brought about by the work of Hasselmann [1969] and

Hasselmann *et al.* [1973]. It can be assumed that the evolution of the directional spectrum can be defined by the total derivative

$$\frac{DF(f,\theta)}{Dt} = S_{\text{tot}}, \qquad (1.45)$$

where, for deep water, the total source term S_{tot} is given by

$$S_{\text{tot}} = S_{\text{in}} + S_{\text{nl}} + S_{\text{dis}}, \qquad (1.46)$$

and

$S_{\text{in}} \equiv$ atmospheric input from the wind

$S_{\text{nl}} \equiv$ nonlinear interaction between spectral components

$S_{\text{dis}} \equiv$ dissipation due to whitecap wave breaking.

It is the balance between these terms which give rise to a spectrum of the general form shown in Fig. 1.6. Numerous field measurements have shown that wind-generated spectra (as opposed to swell) consistently follow a form similar to that shown in Fig. 1.6, i.e., a spectrum which is unimodal with a high frequency face $\approx f^{-n}$. The peak of the spectrum gradually "downshifts" to lower frequencies with increasing non-dimensional fetch (i.e., fetch and/or wind speed increasing). Although, all the source terms in (1.46) play a role in this evolution, it is the nonlinear term which is principally responsible for this consistent spectral form. Young and van Vledder [1993] showed using numerical experiments that S_{nl} has a "shape stabilizing" role, continually forcing the spectrum back to this form. Even in cases of strongly turning winds, the nonlinear term shapes the spectrum towards this standard spectral form. However, should wave components become separated too far in either frequency or direction space, the nonlinear coupling can breakdown and separate swell systems can develop, largely independent of the wind-generated waves in the sea (spectrum).

References

Airy, G.B. (1845). *Tides and Waves, Encycl. Metrop.*, London, 241–392.
Banner, M.L. (1990). Equilibrium spectra of wind waves, *J. Phys. Oceanogr.*, **20**, 966–984.
Bretschneider, C.L. (1952). Revised wave forecasting relationships, in *Proc. 2nd Conf. Coastal Engineering*, ASCE, Council on Wave Research.
Bretschneider, C.L. (1958). Revisions in wave forecasting: Deep and shallow water, in *Proc. 6nd Conf. Coastal Engineering*, ASCE, Council on Wave Research.
Burling, R.W. (1959). The spectrum of waves at short fetches, *Dtsch. Hydrogr. Z.*, **12**, 96–117.
Dobson, F., Perrie, W. and Toulany, B. (1989). On the deep-water fetch laws for wind-generated surface gravity waves, *Atmos.-Ocean*, **27**, 210–236.
Donelan, M.A., Hamilton, J. and Hui, W.H. (1985). Directional spectra of wind-generated waves, *Philos. Trans. R. Soc. Lond. A*, **315**, 509–562.
Goda, Y. (1985). *Random Seas and Design of Marine Structures*, University of Tokyo Press.
Hasselmann, K. (1969). Weak-interaction theory of ocean waves, *Basic Developments Fluid Mech.*, **2**, 117–182.
Hasselmann, K. et al. (1973). Measurements of wind–wave growth and swell decay during the Joint North Sea Wave Project (JONSWAP), *Dtsch. Hydrogr. Z.*, **8**, 1–95.
Hidy, G.M. and Plate, E.J. (1966). Wind action on water standing in a laboratory channel, *J. Fluid Mech.*, **25**, 651–687.
Holthuijsen, L.H. (2007). *Waves in Oceanic and Coastal Waters*, Cambridge University Press.
Kahma, K.K. (1981). A study of the growth of the wave spectrum with fetch, *J. Phys. Oceanogr.*, **11**, 1503–1515.
Kitaigorodskii, S.A. (1962). Applications of the theory of similarity to the analysis of wind-generated wave motion as a stochastic process, *Bull. Acad. Sci. USSR, Geophys. Ser.*, **1**, 105–117.
Kitaigorodskii, S.A. (1973). *The Physics of Air–Sea Interaction*, A. Baruch, translator, P. Greenberg, ed., Israel Program for Scientific Translations, 237 pp.
Kitaigorodskii, S.A. and Strekalov, S.S. (1962). Contribution to an analysis of the spectra of wind-caused wave action, *Izv. Akad. Nauk SSSR Geophys.*, **9**, 1221–1228.
Kononkova, G.E., Nikitina, E.A., Poborchaya, L.V. and Speranskaya, A.A. (1970). On the spectra of wind driven waves at small fetches, *Fizika Atmosfery i Okeana*, **VI**(7), 747–751 (Eng. Trans. Royal Aircraft Establishment, No. 1634).
Liu, P.C. (1970). Normalized and equilibrium spectra of wind waves in Lake Michigan, *J. Phys. Oceanogr.*, **1**, 249–257.
Longuet-Higgins, M.S. (1952). On the statistical distributions of sea waves, *J. Mar. Res.*, **11**(3), 245–265.

Longuet-Higgins, M.S., Cartwright, D.E. and Smith, N.D. (1963). Observations of the directional spectrum of sea waves using the motions of a floating buoy, in *Ocean Wave Spectra*, Prentice-Hall, pp. 111–136.

Mitsuyasu, H. (1968). On the growth of the spectrum of wind-generated waves. 1, *Rep. Res. Inst. Appl. Mech., Kyushu Univ.*, **16**, 1–459.

Olson, H.F. (1943). Dynamic Analogies, D. van Norstrand Inc., Princeton, 277p.

Phillips, O.M. (1958). The equilibrium range in the spectrum of wind-generated waves, *J. Fluid Mech.*, **4**, 426–434.

Pierson, W.J. (ed.). (1960). *The Directional Spectrum of a Wind-Generated Sea as Determined from Data Obtained by the Stereo Wave Observation Project*, New York Univ. Coll. Eng., Met. Pap. 2, 6.

Pierson, W.J. and Moskowitz, L. (1964). A proposed spectral form for fully developed wind seas based on the similarity theory of S.A. Kitaigorodskii, *J. Geophys. Res.*, **69**, 5181–5190.

Pierson, W.J. and Stacy, R.A. (1973). The elevation, slope and curvature spectra of a wind roughened sea surface, NASA Contract Rep., CR-2247, 126pp.

Rice, S.O. (1954). Mathematical analysis of random noise, in *Selected Paper on Noise and Stochastic Processes*, Dover Pub. Inc., pp. 133–294.

Sutherland, A.J. (1967). Growth of spectral components in a wind-generated wave train, *J. Fluid Mech.*, 33, 545.

Sverdrup, H.V. and Munk, W.H. (1947). Wind, sea, and swell: Theory of relation for forecasting, *Hydrographic Office, U.S. Navy*, Publ. No. 601.

Tucker, M.J. (1991). *Waves in Ocean Engineering: Measurement, Analysis, Interpretation*, Ellis Horwood.

Toba, Y. (1971). Local balance in the air-sea boundary processes: I. On the growth process of wind waves, *J. Oceanog. Spc. Japan*, 28, 109–120.

Toba, Y. (1973). Local balance in the air–sea boundary process, *J. Oceanogr. Soc. Japan*, **29**, 209–220.

Volkov, Yu. A., 1968. Analysis of the spectra of sea excitation, Izv. AN SSSR, Fizika Atkosfery i Okeana IV, 9.

Young, I.R. (1999). *Wind Generated Ocean Waves*, Elsevier Sciences Ltd.

Young, I.R. and van Vledder, G.Ph. (1993). The central role of nonlinear interactions in wind–wave evolution, *Philos. Trans. R. Soc. Lond. A*, **342**, 505–524.

Chapter 2

Air–Sea Interaction

Brian Haus

University of Miami, FL 33146, USA

2.1. Introduction

The fact that the wind blowing over a water surface causes waves to form, move and grow has likely been observed and understood at some level since before the dawn of civilization. Given the key role of seafaring to early developed and undeveloped societies alike, an empirical understanding of the nature of these interactions was a critical skill. It is therefore a testament to the fundamental complexity of the processes involved and the difficulties in making quantitative observations at the air–sea interface over open water that an academic understanding of these interactions was largely lacking until the 20th century. Because the ocean surface is a dynamic interface between two fluids of dramatically different physical properties with motions occurring over a large range of scales, a comprehensive understanding requires defining the dynamical properties of each fluid and the state of the interface between them. The state of this wavy interface is also of fundamental importance for maritime activities and coastal dynamics and the motivation for this book.

The three areas of interest: the atmospheric surface layer, wavy interface and oceanic boundary layer while inextricably linked have typically (out of necessity and disciplinary development) been approached using distinct frameworks. Given that the subject of this chapter is air–sea interaction, here we will attempt to provide an overview to each of these frameworks and more importantly to discuss the linkages between them. Upon providing this basic background constrained within the traditional disciplinary boundaries, which has been extensively covered in other published works [e.g., Janssen, 2004], it is our primary purpose to introduce the rapidly developing conceptual and computational understanding of the air–sea interface as a unified system. While enabled by the advancements in computing power that make simplifying approaches less necessary, this new paradigm seeks to fill obvious and dramatic shortcomings in the performance of simplified approaches as demonstrated by the growing body of observational evidence.

2.1.1. *History*

In the era before widely available oceanographic instrumentation and computational resources, the initial theoretical insights to the fundamental boundary layer problems were provided by analytical reasoning supported by relatively sparse empirical observations. These descriptions were limited to highly idealized conditions in general and for moderate forcing conditions. In this section, we will describe these foundational analytical treatments for the coupling between the atmosphere and ocean and the associated wave dynamics in a relatively cursory manner but hopefully one that gives some overall understanding.

2.1.1.1. *Atmospheric boundary layer*

The description of the air-flow over the ocean surface has been derived from fundamental fluid dynamics expressions developed for turbulent flow over flat solid surfaces with some inherent roughness (here we are of course interested in the waves as roughness) and thermal properties. Under the assumptions of steady state atmospheric flows over a homogeneous (in the horizontal plane) ocean with no pressure gradients, no thermal gradients and with turbulent eddy properties that can be expressed using an atmospheric eddy viscosity (ν_a), the momentum balance for the atmospheric boundary layer flow over the ocean can be expressed by the familiar "Law of the Wall" from traditional fluid dynamics $U(z) = u_*/\kappa \ \ln(z/z_0)$. This is for general purposes a reasonable assumption given the large density differences between air and water and the typically small perturbations of the ocean surface (as compared to terrestrial surfaces).

The most widely accepted theory for the atmospheric boundary layer, which is almost universally invoked for oceanographic applications was developed my Monin and Obukhov [1954]. Traditionally, the flux-gradient and flux-variance relationships in the surface layer are described by application of their Monin–Obukhov similarity theory, hereinafter referred to as MOST. MOST assumes stationarity and horizontal homogeneity of the underlying surface, including surface fluxes as well as aerodynamic and thermal roughness. MOST is derived from the Reynolds averaged Navier–Stokes equations. Buoyancy forcing is included through the Boussinesq approximation. MOST describes the atmospheric surface layer (ASL) flow as a function of a single parameter (z/L_O), where z is the distance above the interface and L_O is a length scale (known as the Obukov length) which expresses the ratio between the atmospheric shear forcing and the surface heating.

$$-\varsigma = -\frac{z}{L_O} = \frac{g}{\Theta}\overline{w'\theta'} \cdot \frac{\kappa z}{u_*^3},$$

where g is gravity, Θ is the virtual potential temperature and w' is the fluctuating vertical component of the atmospheric velocity, u_* is the friction velocity and U is the mean horizontal velocity. The same balance in a gradient form had been used by Richardson [1920]. This enabled him to define what is now known as the flux Richardson number. Invoking the assumption that the eddy conductivity (ν_κ) is equal to the eddy viscosity (ν_a) led to the gradient Richardson number:

$$R_i = \frac{g}{\Theta}\frac{\partial\Theta/\partial z}{(\partial U/\partial z)^2}.$$

By MOST, these scaling parameters can only be functions of z/L_0. The functional form of the dependencies can only be determined empirically. Through a series of experiments conducted over well-defined and quasi-uniform roughness elements (wheat field stubble of various heights), Businger et al. [1971] derived empirical functional forms for the MOST expressions. For corrected forms of these expressions in the case of both unstable ($z/L_0 < 1$) and stable ($z/L_0 > 1$) conditions, see Donelan [1990]. The combination of MOST and the aforementioned empirical expressions has proven to be a robust and useful model of the air-flow in the ASL. It has served as the backbone for air–sea interaction studies over the last seven decades and provides the fundamental framework for understanding the momentum transfer between the wind and the waves. As such it has been comprehensively described in other works [e.g., Csanady, 1974], and will only be introduced here.

Despite its wide use, it is clear that in many specific relevant applications the fundamental assumptions that were invoked to derive MOST are violated. While this may not preclude their use, it does suggest that their application should be handled with care. In the later part of this chapter, we will focus on several regimes where the underlying flows do not conform to the assumptions put forth in MOST and discuss recent efforts to define the air–sea coupling in these cases.

2.1.1.2. *Oceanic boundary layer*

For most oceanic applications, the dynamics of the upper layer of the ocean (hereinafter OBL) are typically treated in an averaged way to remove direct contributions from surface waves but retain the local and non-local wind-forced motions. Although our focus here is on the waves, the wave-averaged OBL flows are still of interest because of their feedback effects on the wave field. These effects are largely driven by horizontal shear in the mean flow and act disparately on waves of different wavelengths.

In the upper ocean boundary layer, the average velocity of the flow in an earth-fixed reference frame can be expressed as the sum of local wind-forced currents, the mean geostrophic flow which is typically produced by large-scale atmospheric forcing modified by basin-scale topography and tidal flow which can be important in coastal areas. If considering the flow in a Lagrangian frame of reference, then the wave-induced Stokes drift can be important in the OBL particularly very near the surface (within $\sim 1/2$ the surface wavelength).

These averaged flows can alter the coupling between surface waves and wind through modifying the relative velocity between the wind and the waves and in the case of horizontally sheared currents by changing the propagation direction of the waves [Kenyon, 1971]. Vertical shear and/or turning of the near-surface currents can cause different wave components to "feel" different currents because the depth to which they are sensitive is proportional to wavelength [see Stewart and Joy, 1974].

The most well-known theoretical treatment of the oceanic component of the air–sea boundary layer is that developed by Ekman [1925], which expressed the balance between wind stress at the surface and Coriolis forcing to describe the mean velocity vector profile in the upper layer.

The key result of the Ekman balance is the clockwise rotation and decrease of the mean velocity vector as the distance from the surface increases. This produces the important result for basin scale and coastal circulation that the integrated volume transport of the Ekman layer (defined by the Ekman depth parameter) to be directed to the right/left of the wind direction in the northern/southern hemisphere. It is important for the purpose of understanding wave dynamics however that the Ekman solution is considerably simplified, requiring: infinitely deep ocean with no lateral boundaries, steady state, homogeneous density and a constant vertical eddy viscosity.

Here we are interested in the dynamics of the upper layer within the region of wave impact so it is necessary that we look beyond Ekman-type dynamics. In general, given the exponential rate of wave orbital velocity decay from the surface as predicted by linear wave theory, the region of interest for wave dynamics studies is generally within a distance of $1/2$ the wavelength (L) from the surface. For typical wind waves near the spectral peak with L ranging from 20 m to 100 m conditions this is less than the Ekman depth, which in turn suggests that the assumption of a constant vertical eddy viscosity is not valid.

The most commonly applied treatment in the upper layer is simply to empirically scale the surface velocity by the wind speed. This results in the so-called 3% rule as originally defined by Wu [1983], which included both direct wind shear currents and wave-induced mass transport (Stokes drift). At that time, it was argued that the dominant waves were the major contributor to the Stokes drift component and thus had an exponential decay scale based on their wavelength. In reality, the Stokes drift velocity is composed of contributions from waves of all scales because it is dependent directly on the wave steepness instead of the wavelength. This led Tamura *et al.* [2012] and Webb and Fox-Kemper [2011] to the result that the Stokes drift velocity depends on the cube of the wave frequency and thus is most strongly influenced by high-frequency (short wavelength) waves.

It can then be argued that the 3% rule should only apply to a very thin layer near the surface which would match the conditions that were used in the Wu [1983] laboratory measurements, which considered the drift velocity of thin paper punches. Laxague *et al.* [2018] recently measured high resolution velocity profiles within the very near-surface layer under light winds using optical methods and found this rule to hold at a depth of less than 1 cm. Below this depth, the velocity decayed rapidly with depth until at a depth of 1 m it was ∼50% of the value in the uppermost layer discussed above.

This drift velocity is implicitly included in the Ekman balance for wavelengths long enough to have a vertical decay e-folding scale of the same order as the Ekman e-folding scale. However, as shown in an experimental study by Clarke *et al.* [2018] for these waves then the Stokes drift velocity is not balanced by the Coriolis force and the Stokes drift contribution to the ocean surface transport remains. They further showed that Stokes drift was an important contributor to net surface transport and was directed primarily in the local wind direction.

Interpreting the role of Stokes drift in the context of surface wave dynamical effects is less straightforward. Common treatment would suggest that in an averaged sense it is only the contributions of waves shorter than a given surface wave that would produce a net Stokes drift in a Lagrangian sense, but the longer wave would not average to zero because of the partial range of phases that would occur during a given time interval. However, since the short wave contributions to the near-surface currents remain in an averaged sense through the Stokes drift, then these could affect longer wavelength waves. Contradicting this is that the vertical e-folding scale is smaller than that of the longer waves such that if the ratio of wavelengths is large enough then the longer waves may not feel the shorter waves primarily because of the depth of influence. The limits at which these differing relative effects are important on a particular surface wave have not been resolved.

2.2. Wind–Wave Coupling

The action of the wind blowing over a water surface and waves forming is readily observed in bodies of water of all scales. The question of what physical process causes the first waves to form and then grow is one that has provoked considerable inquiry.

The pioneering work of Jeffreys [1924, 1925] suggested that pressure differences across the front and rear faces of surface waves provided the mechanism for wave generation by wind. However, his "sheltering" mechanism could not explain the initiation of waves on a flat surface nor could it provide sufficient momentum transfer to match empirical observations of growth rates so it fell out of favor for many decades. Recent field observations by Donelan et al. [2006] have shed new light on the role that this mechanism can play in wave growth.

The Phillips [1957] resonant interaction mechanism whereby some components of turbulent fluctuations at the interface are in resonance with the nascent surface waves and grow rapidly could provide for initiation of wave growth but still could not produce sufficient momentum transfer to develop full-scale ocean waves. Miles [1957, 1959] then described a mechanism for wave growth from small to large energy containing scales through resonance between wave-induced pressure fluctuations and the surface waves.

A consistent explanation for wave growth from initiation to large ocean waves was then available by combining the Phillips wave initiation mechanism with the Miles wave growth model. While this framework was intellectually satisfying, unfortunately physical reality as defined through wave observations did not match the predictions of these combined models particularly in strongly forced conditions or when the winds were not aligned with the wave direction (see Chapter 5).

2.2.1. *Stationary, spatially homogeneous conditions*

While there are still some fundamental uncertainties in the description of wind–wave coupling in stationary, homogeneous forcing, the statistics of wind–wave behavior are relatively well described through empirical studies and modern wave models do a reasonable job of describing the wave field (see Chapter 5).

In a spectral representation, considering the evolution of a wave field under the action of a steady wind, the wave energy evolution in time and space was expressed by Phillips [1977], Komen *et al.* [1994] and Cavaleri *et al.* [2007] as

$$\frac{\partial F(x,k,\theta)}{\partial t} + \frac{\partial(xF)}{\partial x} + \frac{\partial(k,F)}{\partial k} + \frac{\partial(\theta,F)}{\partial \theta} = S_{\text{in}} + S_{\text{ds}} + S_{\text{nl}},$$

where F is the wave energy variance spectrum, x is the position, k is the wave number, and θ is the directional coordinate, respectively. S_{in} is the wind input, S_{ds} is the dissipation term (typically dominated by wave breaking) and S_{nl} is the nonlinear interaction term.

Recent models for wave growth under wind forcing have invoked modified forms of the Jeffreys' sheltering hypothesis [see, e.g., Belcher and Hunt, 1994]. Donelan *et al.* [2012] developed an expression for the wind-input source term based on the Jeffrey's [1924, 1925] sheltering hypothesis and the Donelan *et al.* [2006] data as follows:

$$S_{\text{in}} = A|U_{L/2}\cos(\varphi) - C(k) - u\cos(\theta)|$$
$$\times (U_{L/2}\cos(\varphi) - C(k) - u\cos(\theta))\frac{k\omega}{g}\frac{\rho_a}{\rho_w}F(k,\theta),$$

where $U_{L/2}$ is wind speed at the height of half of the wavelength and ϕ is the angle between the wind and wave directions, thus $U_{L/2}\cos\varphi$ is the wind component in the direction of wave propagation. U is the effective Eulerian current in the OBL between the depth of $L/2$ and the surface which was discussed previously. C is the intrinsic phase speed and ω is the angular wave frequency. A is the sheltering coefficient, a non-dimensional parameter that ranges between 0.01 and 0.11 depending on the wind–wave alignment regime [Donelan et al., 2012].

2.2.1.1. Critical layer theory

While the details of the wave initiation and growth mechanism may not be completely understood, it is clear from observation that waves will continue to grow under the action of a steady wind, increasing in both amplitude and wavelength. In an unbounded basin, this growth will continue until such a point that the speed of the waves approaches the wind velocity. It is convenient to express the ratio of the two velocities to define a non-dimensional parameter (C/U_{10}) which is known as the wave age, such that young waves are those where the wind is much faster than the waves and for old waves the wave phase speed approaches the wind speed.

Given the logarithmic wind velocity profile discussed above a critical level can be defined [following Miles 1957] at which the wind velocity is equal to the wave velocity. Below this critical level, the air-flow can be perturbed by the surface waves. The importance of the critical layer for a given wave field depends strongly on the wave age, or more directly the ratio of the wave phase velocity to the friction velocity (C/u^*). Based on the Miles theory and their concept of non-separated sheltering, Belcher and Hunt [1993] suggested that there were three distinct regimes when considering the location of the inner layer where waves perturb the flow and therefore affect the turbulent stresses. For very short wavelength wind waves $C/u^* < 15$, the critical layer height is very small and it lies within the wave affected layer. For intermediate values $15 < C/u^* < 25$, the wave affected layer is thick and is approximately the same as the critical layer height. For swell waves with $C/u^* > 25$, the critical layer height is large and the wave affected layer is thin. Laboratory studies [Buckley and Veron, 2016], field experimental studies [i.e., Hristov et al., 2003; Grare et al., 2013] and numerical simulations [e.g., Chalikov, 1978; Sullivan et al., 2000, 2010] have in general confirmed this behavior.

At some point, the wind input of energy to the waves is balanced by dissipative forces (primarily wave breaking) and the waves can no longer gain energy or grow. This is known as the equilibrium condition, at which energy can be transferred to other wave numbers but the overall energy level is static. It is common to invoke this situation in wave modeling because if the input and dissipation are balanced then it is not necessary to model these sources and sinks of energy directly and the problem is substantially simplified.

2.2.1.2. *Turbulent stresses*

The coupling between the atmosphere and ocean can be expressed as the stress carried by each fluid across the interface. The total stress above the interface can be divided into three components: the viscous stress (τ_V), the turbulent stress (τ_t) and the wave coherent stress (τ_w). The viscous stress term or "skin drag" is usually neglected (as it will be here) in wave dynamics studies because it is only significant in the very thin (mm scale) viscous sublayer above the interface and is essentially unmeasurable. The τ_w term is also known as the "form drag" and is responsible for the majority of the wind–wave growth.

The generation of turbulence in an atmospheric boundary layer flow is known to be proportional to the vertical gradient of the mean flow ($\partial U/\partial z$). In turbulent flow, this proportionality can be expressed by a turbulent eddy viscosity (ν_t),

$$\tau = \rho \nu_t \frac{\partial U}{\partial z},$$

where ρ_a is the air density. In most applications related to wave modeling, the vertical gradient of the mean wind is unknown so a drag coefficient (C_D) is applied such that the stress can be expressed using the mean wind at some level (z).

$$\tau = \rho_a C_D U_z^2.$$

It is readily observed that the air exerts a force on the water below it that is in some way related to the wind velocity. Charnock [1955] postulated that as the wind speed increases then C_D should also increase. A large body of experimental observation [e.g., Smith, 1980; Large and Pond, 1981] confirmed this relationship for low-moderate wind speeds. A widely used method for calculating C_D, derived from these and many other field observations is the COARE 3.5 algorithm [Edson et al., 2013]. Included in the COARE 3.5 are three options for expressing the roughness length based on the wind speed, wind speed and wave age and wind speed, wave age and wave steepness.

Recent developments have focused on explicitly defining C_D based on local wave conditions. These efforts have included Drennan et al. [2003] which was expanded by Hogstrom et al. [2015] and focused on windsea only conditions. Zhou et al. [2015] developed a parametric model for C_D based on tower measurements in the South China Sea. It covered a range of wind speeds to hurricane force, where the limiting C_D from Donelan et al. [2004] was used and included depth limited waves. A numerical simulation based model that included wind speed and depth limited waves was also proposed by Jimenez and Dudhia [2018]. Additional experimental validation of these varied approaches is needed to determine if they provide added value over wind speed only formulations.

2.2.1.3. *Wave coherent pressure force*

The wave coherent stress is the result of the pressure force acting on the interface, this pressure force produces no stress on a lateral surface because it is then uniform in space. However, it can act on a sloping surface and it is expressed as follows: $\tau_w = \rho P_0 (\delta \eta)/\delta x$ in one dimension, where P_0 is the atmospheric pressure at the interface and $(\delta \eta)/\delta x$ is the instantaneous local slope.

The wave coherent pressure force (τ_w) defined above is also known as the pressure–slope correlation and for obvious reasons can transfer momentum from the wind to the waves in the general case of growing seas when the wind velocity (U) is greater than the wave phase velocity (C_p) or the momentum can be transferred upward from the waves to the wind in the case of fast moving swell waves and/or light winds.

The pressure–slope correlation is a very difficult measurement to make, particularly in the open ocean in energetic conditions. To precisely define it requires maintaining a pressure measurement very close to the ocean surface for sufficient time (many wave periods) to establish a stable average of the pressure–slope product. Because the pressure is not amenable to a remote measurement, this then requires some sort of adaptive measurement strategy whereby a probe is kept at a small fixed distance above the wavy interface and is therefore both non-intrusive and highly mobile to avoid disturbing the local flow. This strategy was successfully used by Donelan *et al.* [2006] to define the spectral distribution of the wind-input to surface waves.

2.3. Non-stationary and Spatially Inhomogenous Conditions

It is typical for oceanographers, meteorologists and or wave modelers to assume that MOST conditions are valid even in circumstances where they are likely violated. Often this is out of necessity due to a lack of sufficient information on the surface conditions to appropriately assess whether MOST holds. Other times, it is because the tools to treat non-stationarity and/or inhomogeneity are not well known or lacking. In this section, an attempt is made to inform the reader of progress in a variety of such conditions and to highlight existing knowledge gaps regarding them.

2.3.1. *Time-dependent winds*

Inspection of any time series record of wind velocity will immediately reveal a large degree of variability in both wind magnitude and direction, ranging in temporal scales from seconds to tens of minutes. This "gustiness" is usually treated by averaging the wind and waves over time (or space) scales such that reliable quasi-stationary statistics can be derived under the assumption that the averaging intervals are larger scale than the turbulent eddies that lead to wind gusts. While this may be sufficient for many purposes, to fully define the spectral characteristics of the wave field it is necessary to capture a range of eddy scales including larger scale fluctuations in the wind–wave field due to large-scale fluctuations.

The prevalence of rapid shifts in the wind velocity vector magnitude and direction is well known but their impact on the wave field remains relatively little studied, primarily because of the paucity of experimental information on the wind and wave fields at temporal scales less than tens of minutes and spatial scales of less than 1 m. It is a testament to the difficulty in observing the short-scale wave motions that the pioneering sun glint based study of Cox and Munk [1954] has remained the standard experimental work on this subject despite the relatively limited range of conditions tested. An exception to this was some innovative laser-based experimental work in the early 1990s by Hara *et al.* [1994] and Hwang *et al.* [1996].

Recent advances in digital imaging technology and computational capabilities has made possible measurements on temporal and spatial scales such that the response of surface waves to short timescale wind shifts can be observed. An example of application of new imaging technology is the work of Laxague *et al.* [2015] who confirmed the importance of short gravity waves for wind–wave coupling [Hwang, 2005; Phillips, 1985] in quasi-steady conditions and showed that for steadily increasing wind velocity over a range from $4\,\mathrm{ms}^{-1}$ to $11\,\mathrm{ms}^{-1}$, the gravity-capillary wave components ($1\,\mathrm{rad/m} < k < 316\,\mathrm{rad/m}$) responded most quickly to wind velocity magnitude shifts.

To fully define wind–wave coupling, wind direction shifts also need to be considered. A modeling study by Young *et al.* [1987] explored how the nonlinear terms in the WAM3G wave model handled these shifts as compared to full nonlinear interaction calculations. They showed that when the wind directional shifts were small ($<60°$) the wave field smoothly adjusted to the new conditions. This was consistent with the available field evidence at the time from the JONSWAP experiments [Hasselmann *et al.*, 1973] and from measurements made off North Carolina by Allender *et al.* [1983]. These observations were limited to waves with periods longer than about 4 seconds and therefore could not capture the gravity-capillary components. Recent work by Aijaz *et al.* [2016] considered new wave source terms in the SWAN model based on field measurements made by Young *et al.* [2005].

They found that consistent with the earlier work, small shifts in wind direction led to gradual decoupling of the windsea from the old wave spectrum. For large wind shifts ($\sim 180°$), the new wind waves developed without interaction with the existing waves.

2.3.2. *Spatially inhomogeneous conditions*

There are many circumstances in which the fundamental MOST assumption of spatial homogeneity is violated. That the distance over which the wind is blowing is an important consideration for estimating the wave conditions is immediately apparent to mariners and has been the focus of significant experimental, theoretical and numerical investigation since well before the comprehensive and highly influential JONSWAP experiments [Hasselmann *et al.*, 1973]. Swell waves impart a nonlocal scale of temporal and spatial homogeneity on local wind–wave coupling. They also can occur over large ocean regions, making their inclusion of broad importance for ocean wave modeling. Other sources of spatially inhomogeneity are numerous, but less widely distributed. As such they are not as important over basin scales, but can be critical to understanding wind–wave coupling in regions of particular applied interest such as cyclones, fronts and coastal areas. In this section, we will discuss recent advances in wind–wave coupling understanding in these non-homogeneous conditions.

2.3.3. *Swell waves*

Swell waves are those waves that are not under the direct influence of the wind and as such, one may be tempted to neglect them in a chapter on wind–wave interaction. That would be a significant oversight however, because the presence of swell can have a profound impact on the transfer of momentum from wind to shorter scale waves. In the case of very light winds, swell may even lead to upward momentum transfer. The effect of swell waves on air–sea coupling is strongly dependent upon the direction that the swell waves are traveling relative to the wind direction and also on the phase speed of the swell relative to the windsea as well as the wind.

For swell traveling in the same direction as the wind, there have been a number of laboratory studies that have looked at the momentum transfers from wind to waves by directly measuring the time average of the pressure-slope term. Makin *et al.* [2007] showed that paddle generated waves traveling in the same direction as a wind had a stress distribution that was wave steepness dependent. For higher wind conditions up to almost $30\,\mathrm{ms}^{-1}$, Savelyev *et al.* [2011] found that there was a strong dependence of the wind–wave momentum flux on the product of wind speed and wave slope. In lighter winds, where the swell may be traveling close to or faster than the wind speed, there is much less experimental data on the wind–wave momentum flux because it is a condition not conducive to laboratory work. Kamma *et al.* [2016] analyzed data collected in 1987 from a tower in Lake Ontario. They found direct evidence of an upward wave–wind flux over swell waves.

When the swell is traveling in the opposite direction to the wind, it has been shown to be attenuated by the wind. The attenuation rate reported from laboratory studies has varied considerably. Donelan [1999] measured attenuation rates that were about 1/2 the wave growth rate, while Pierson *et al.* [2003] observed rates over 3/2 times the growth rate. For swell traveling at an arbitrary angle to the wind–waves, there has been very limited experimental observation to establish the degree to which the wind–wave momentum flux is affected. This represents a significant gap because although the degree to which the two wave systems interact is likely limited, the situation is by far the most commonly observed situation on the open ocean, particularly in light winds or in tropical cyclones.

2.3.4. *Coastal regions*

There are two general conditions of interest for wind–wave coupling, that being offshore winds and onshore winds. For onshore winds, the usual assumption applied by those interested in representing nearshore wind fields is that offshore velocities and MOST stresses are appropriate. This has been demonstrated to be an erroneous assumption in some cases, which will be discussed later in this section. In the case of offshore winds, the simplest situation is wind blowing perpendicularly offshore to a long straight coastline with a smooth transition from water to land such that the wind flow remains attached at the shoreline.

Because of the aforementioned lack of confidence in the quantitative accuracy of theoretical wind–wave growth expressions, empirical studies to define the wave growth were necessary. The JONSWAP experiments [Hasselmann et al., 1973] led to widely used relationships for the growth of wind–waves with distance offshore (fetch). Later Donelan et al. [1992] used multiple offshore towers in Lake Ontario to provide empirical relationships for the wave growth rate as a function of non-dimensional parameters defining the non-dimensional fetch $\left(\tilde{X} = \frac{Xg}{U_{10}^2}\right)$ and energy $\left(\tilde{E} = \frac{Eg^2}{U_{10}^4}\right)$. The use of the non-dimensional terms enabled direct comparison of field and laboratory studies [e.g., Caulliez et al., 2008]. These empirically derived terms serve as a calibration for numerical model studies that contain input and dissipation terms, that have significant uncertainties.

The far more common situation when the wind is directed offshore is that it is blowing over a complex coastal topography, which could be due to natural topographic variability, structures built close to the shoreline or for laterally complex coastlines like bays, headlands and islands. In such situations, there have been limited studies. Exceptions to this are that wind–wave growth has been explored in situations such as gaps in mountain ranges [e.g., Romero and Melville, 2010] for which the wind field variability can be an important consideration for the wave development. It has also been shown by Ardhuin et al. [2007] that when the shoreline is not straight or the wind is not directly offshore then the so-called slanting fetch will alter the wave growth rate. It is clear that for many coastal applications, additional experimental efforts in this area are required.

When the wind is in the onshore directed then it may seem reasonable to assume that it can be treated as being unaffected by proximity to the shoreline. However, it has been observed that in shallow water, drag can be significantly increased due to wave shoaling which manifests itself as both a slowing of the waves and of increased wave steepness [Anctil and Donelan, 1996; Zachry et al.,

2013; Bi et al., 2015; Chen and Curcic, 2016]. Young et al. [2005] and Shabani et al. [2014] showed that as the waves slowed down due to depth limited effects, the relative wind–wave velocity increased and the corresponding momentum transfer rates (as expressed by the drag coefficient increased). Ortiz-Suslow et al. [2015] also found that the drag increased significantly as incoming waves interacted with an outgoing tidal flow. It is clear then that in coastal regions it will be necessary to consider the actual wave field when trying to compute wind–wave coupling.

Recent advances in fully-coupled atmosphere–wave–ocean modeling allowed for explicitly solving the wave energy balance as part of the coupled system, thus eliminating the dependency on a parametric C_D [Chen et al., 2013; Chen and Curcic, 2016; Breivik et al., 2015]. In these models, the air–sea momentum exchange is more finely parameterized through wind input and wave dissipation source functions. These coupled models have not been well-tested in high winds where the interface properties may be strongly impacted by intense wave breaking and sea-spray.

2.3.5. *Thermal fronts*

Frontal boundaries that separate water masses are often found in the ocean, particularly in coastal regions and they typically have gradients in sea surface temperature (SST) across them. While it is not common to directly consider SST in wave studies, the interaction between waves and the ASL above can be affected by such gradients. The fundamental MOST is based on the ratio of shear-forcing to buoyancy, as such it is dependent on the air–sea temperature difference. This is reflected in the different empirical expressions for the MOST parameter for stable and unstable conditions. Thus it is natural to assume that when the air-flow passes over water surface temperature gradients, the characteristics of the boundary layer could be changed and this could have an impact on wind–wave coupling.

Most previous studies have focused on the coupling between the marine atmospheric boundary layer (MABL) and SST variability across mesoscale fronts, O(50–500 km), that exist mainly near the major western boundary currents, e.g., Kuroshio current [Nonaka and Xie, 2002], Agulhas current [O'Neill et al., 2012] and Gulf stream [e.g., Song et al., 2006; Chelton and Xie, 2010; O'Neil, 2011; Plagge et al., 2016; O'Neil et al., 2017]. That there is a strong response of the MABL winds and surface fluxes to the temperature front is established from these previous observational and modeling studies. Wind speed and surface momentum, heat and moisture fluxes are usually enhanced over the warm region which has a thicker homogenized MABL. While these changes are evident over the scale of the MABL which is typically greater than 100 m, it is likely that they are not particularly relevant for surface waves, however, their quantitative effects if any on the momentum transfer from wind to waves have not been elucidated due to a lack of appropriate experimental guidance.

2.3.6. *Surface current effects*

The near-surface current velocity is generated by both local and non-local winds as well as surface waves through shear stresses and Stokes drift, respectively. Together they contribute to the wave breaking process that deposits extra momentum into the upper ocean that also produces currents.

Conversely, surface currents can affect the coupling between wind and waves in several different ways. The currents affect the wave field through two main mechanisms. The first occurs when the OBL currents are aligned with the wave propagation direction; in this case the relative velocity of the wind to the water surface is higher in opposing currents and lower in following currents which increases/decreases the wind stress on the surface. If there is a current gradient in the along-wave direction this can lead to current-induced shoaling of the waves which can lead to an increase or decrease in the local wave height and steepness depending on the wind and current direction and the direction of the gradient. This in turn can affect the wind–wave momentum coupling as observed by Ortiz-Suslow et al. [2015]. These types of situations are commonly observed in inlet and riverine conditions.

More interesting effects occur when there is an along-wave gradient in the cross-stream velocity. This occurs frequently in boundary currents as mentioned previously as well as in inlet outflows. In these cases, the waves refract over the current gradient, with the short-gravity waves that carry most of the wind stress [Laxague et al., 2015] responding quickly to the shear. Drennan and Shay [2006], with *in situ* measurements of wind, wind stress and surface currents, reported a steering of the stress away from the mean wind direction by almost 30°, which they attributed to strong off-wind currents. Haus [2007] demonstrated that off-wind current shear led to reduced wave growth rates, which could be explained by stress veering. This scale of waves also has a longer relaxation time than capillary waves and can thereby cause the direction of the surface stress to shift away from the wind direction [Zhang et al., 2010] which can have implications for wind–wave growth in these regions.

2.4. Summary

It was the intent of this chapter on wind–wave interaction to explore flow regimes where there are still significant knowledge gaps, while providing general context in areas that have been covered extensively elsewhere. Ongoing rapid advances in autonomous vehicles, miniaturized sensing technology, imaging, high fidelity direct numerical simulation and computer hardware make it likely that many of these previously under-observed situations will have a much richer experimental and computational database in the near future. This will enable moving from the "case-study" mode which is where the state of research for many of these specific subjects resides. The development of more generalizable approaches from which to extract comprehensive understanding and a robust modeling capability for non-stationary and non-homogeneous wind–wave regimes will follow.

References

Aijaz, S., Rogers, W.E. and Babanin, A.V. (2016). Wave spectral response to sudden changes in wind direction in finite-depth waters, *Ocean Model.*, **103**, 98–117.

Allender, J.H., Albrecht, J. and Hamilton, G. (1983). Observations of directional relaxation of wind sea spectra, *J. Phys. Oceangr.*, **13**, 1519–1525.

Anctil, F. and Donelan, M.A. (1996). Air–water momentum flux observations over shoaling waves, *J. Phys. Oceanogr.*, **26**(7), 1344–1353.

Ardhuin, F., Herbers, T.H.C., Van Vledder, G.Ph., Watts, K.P., Jensen, R. and Graber, H.C. (2007). Swell and slanting-fetch effects on wind wave growth, *J. Phys. Oceanogr.*, **37**, 908–930.

Belcher, S.E. and Hunt, J.C.R. (1993). Turbulent shear flow over slowing moving waves, *J. Fluid Mech.*, **251**, 109–148.

Bi, X., Gao, Z., Liu, Y., Liu, F., Song, Q., Huang, J., Huang, H., Mao, W. and Liu C. (2015). Observed drag coefficients in high winds in the near offshore of the South China Sea, *J. Geophys. Res.*, **120**, 6444–6459, doi:10.1002/2015JD023172.

Breivik, O., Morgensen, K., Bidlot, J.R., Balmaseda, M.A. and Janssen, P.A. (2015). Surface wave effects in NEMO ocean model: Forced and coupled experiments, *J. Geophys. Res.: Oceans*, **120**, 2973–2992, doi:10.1002/2014 JC010565.

Buckley, M.P. and Veron, F. (2016). Structure of the airflow above surface waves, *J. Phys. Oceanogr.*, **46**, 1377–1396.

Businger, J.A., Wyngaard, J.C., Izumi, Y. and Bradley, E.F. (1971). Flux-profile relationships in the atmospheric surface layer, *J. Atmos. Sci.*, **28**, 181–189.

Caulliez, G., Makin V. and Kudryavtsev, V. (2008). Drag of the water surface at very short fetches: Observations and modeling, *J. Phys. Oceanogr.*, **38**, 2038–2055.

Cavaleri, L., Alves, J.H.G.M., Ardhuin, F., Babanin, A., Banner, M., Belibassakis, K., Benoit, M., Donelan, M., Groeneweg, J., Herbers, T.H.C., Hwang, P., Janssen, P.A.E.M., Janssen, T., Lavrenov, I.V., Magne, R., Monbaliu, J., Onorato, M., Polnikov, V., Resio, D., Rogers, W.E., Sheremet, A., Smith, J.M., Tolman, H.L., van Vledder, G., Wolf, J. and Young, I. (2007). Wave modelling — The state of the art, *Progr. Oceanogr.*, **75**, 603–674.

Chalikov, D.V. (1978). The numerical simulation of wind–wave interaction, *J. Fluid Mech.*, **87**, 561–582, doi:10.1017/S0022112078001767.

Charnock, H. (1955). Wind stress on a water surface, *Q. J. Royal Meteorol. Soc.*, **81**, 639–640.

Chelton, D. and Xie, S.-P. (2010). Coupled ocean-atmosphere interaction at oceanic mesoscales. *Oceanography*, doi:10.5670/oceanog.2010.05.

Chen, S.S., Zhao, W., Donelan, M.A. and Tolman, H.L. (2013). Directional wind–wave coupling in fully coupled atmosphere-wave-ocean models: Results from CBLAST-Hurricane, *J. Atmos. Sci.*, **70**, 3198–3215.

Chen, S.S. and Curcic, M. (2016). Ocean surface waves in Hurricane Ike (2008) and Superstorm Sandy (2012): Coupled modeling and observations, *Ocean Model*, **103**, 161–176, doi:10.1016/j.ocemod.2015.08.005.

Clarke, A.J. and Van Gorder, S. (2018). The relationship of near-surface flow, Stokes drift and the wind stress, *J. Geophys. Res.: Oceans*, **123**, 4680–4692, doi: 10.1029/2018JC014102.

Cox, C. and Munk, W. (1954). Measurement of the roughness of the sea surface from photographs of the sun's glitter, *J. Opt. Soc. Am.*, **44**(11), 838–850.

Csanady, G.T. (1974). Equilibrium theory of the planetary boundary layer with an inversion lid, *Bound.-Layer Meteorol.*, **6**(1–2), 63, https://doi.org/10.1007/BF00232477.

Donelan, M.A. (1990). Air-sea Interaction, in *The Sea*, Ocean Engineering Science, Vol. 9, Wiley, New York, pp. 239–292.

Donelan, M.A., Skafel, M., Graber, H., Liu, P., Schwab, D. and Venkatesh, S. (1992). On the growth rate of wind-generated waves, *Atmos.-Ocean*, **30**(3), 457–478.

Donelan, M.A. (1999). Wind-induced growth and attenuation of laboratory waves, in *Wind-over-Wave Couplings: Perspectives and Prospects*, eds. S.G. Sajjadi, N.H. Thomas and J.C.R. Hunt, Oxford University Press, pp. 183–194.

Donelan, M.A., Haus, B.K., Reul, N., Plant, W.J., Stiassne, M., Graber, H.C., Brown, O.B. and Saltzman, E.S. (2004). On the limiting aerodynamic roughness in very strong winds, *Geophys. Res. Lett.*, **31**(18), L18306, doi:10.1029/2004GL019460.

Donelan, M.A., Babanin, A.V., Young, I.R. and Banner, M.L. (2006). Wave follower measurements of the wind-input spectral function. Part 2. Parameterization of the wind input, *J. Phys. Oceanogr.*, **36**, 1672–1688.

Donelan, M.A., Curcic, M., Chen, S.S. and Magnusson, A.K. (2012). Modeling waves and wind stress, *J. Geophys. Res.*, **117**, C00J23.

Drennan, W.M. and Shay, L.K. (2006). On the variability of the fluxes of momentum and sensible heat, *Bound.-Layer Meteorol.*, **119**, 81–107, doi:10.1007/s10546-005-9010-z.

Hara, T., Bock, E.J. and Lyzenda, D. (1994). *In situ* measurements of capillary-gravity wave spectra using a scanning laser slope gauge and microwave radars, *J. Geophys. Res.*, **99**(C6), 12593–12602, doi:10.1029/94JC00531.

Hasselmann, K. *et al.* (1973). Measurements of wind-wave growth and swell decay during the Joint North Sea Wave Project (JONSWAP), *Dtsch. Hydrogr. Z., Suppl. A*, **8**(l2), 1–95.

Haus, B.K. (2007). Surface current effects on the fetch limited growth of wave energy, *J. Geophys. Res.*, **112**, C03003, doi:10.1029/2006JC003924.

Högström, U., Sahlée, E., Smedman, A.-S., Rutgersson, A., Nilsson, E., Kahma, K.K. and Drennan, W.M. (2015). Surface stress over the ocean in swell-dominated conditions during moderate winds, *J. Atmos. Sci.*, **72**(12), 4777–4795.

Hristov, T., Miller, S. and Friehe, C. (2003). Dynamical coupling of wind and ocean waves through wave-induced air flow, *Nature*, **422**, 55–58, doi:10.1038/nature01382.

Hwang, P.A., Atakturk, S., Sletten, A. and Trizna, D.B. (1996). A study of the wave number spectra of short water waves in the ocean, *J. Phys. Oceanogr.*, **26**, 1266–1285, doi:10.1175/1520-0485(1996) <0261266:ASOTWS>2.0.CO;2.

Hwang, P.A. (2005). Wave number spectrum and mean square slope of intermediate-scale ocean surface waves, *J. Geophys. Res.: Oceans*, **110**, C10029, doi:10.1029/2005JC003002.

Janssen, P. (2004). *The Interaction of Ocean Waves and Wind*, Cambridge University Press.
Jeffreys, H. (1924). On the formation of waves by wind, *Proc. R. Soc. A*, **107**, 189–206.
Jeffreys, H. (1925). On the formation of waves by wind. II, *Proc. of the Roy. Soc. A*, **110**, 341–347.
Jimenez, P.A. and Dudhia, J. (2018). On the need to modify the sea surface roughness formulation over shallow waters, *J. Appl. Meteorol. Clim.*, **57**(5), 1101–1110, doi:10.1175/JAMC-D-17-0137.1.
Kahma, K.K., Donelan, M.A., Drennan, W.M. and Terray, E.A. (2016). Evidence of energy and momentum flux from swell to wind, *J. Phys. Oceanogr.*, **46**, 2143–2156.
Kenyon, K.E. (1971). Wave refraction in ocean currents, *Deep Sea Res.*, **638**(18), 1023–1034.
Large, W.G. and Pond, S. (1981). Open ocean momentum flux measurements in moderate to strong winds, *J. Phys. Oceanogr.*, **11**, 324–336.
Laxague, N.J.M., Haus, B.K., Bogucki, D. and Özgökmen, T. (2015). Spectral characterization of fine-scale wind waves using shipboard optical polarimetry, *J. Geophys. Res.: Oceans*, **120**, 3140–3156, doi:10.1002/2014JC010403.
Laxague, N.J.M., Özgökmen, T.M., Haus, B.K., Novelli, G., Shcherbina, A., Sutherland, P., Guigand, C.M., Lund, B., Mehta, S., Alday, M. and Molemaker, J. (2018). Observations of near-surface current shear help describe oceanic oil and plastic transport, *Geophys. Res. Lett.*, **45**(1), 245–249.
Makin, V. (2008). On the possible impact of a following-swell on the atmospheric boundary layer, *Bound.-Layer Meteorol.*, **129**, 469–478.
Miles, J.W. (1957). On the generation of surface waves by shear flows, *J. Fluid Mech.*, **3**, 185–204.
Miles, J.W. (1959). On the generation of surface waves by shear flows, Part 2. *J. Fluid Mech.*, **6**, 568–582.
Monin, A. and Obukhov, A. (1954). Basic laws of turbulent mixing in the surface layer of the atmosphere, *Tr. Geofiz. Inst. Akad. Nauk SSSR*, **24**, 163–187.
Nonaka, M. and Xie, S.-P. (2002). Covariations of sea surface temperature and wind over the Kuroshio and its extension: Evidence for ocean-to-atmosphere feedback, *J. Climate*, **16**(9), 1404–1413.
O'Neill, L.W. (2011). Wind speed and stability effects on coupling between surface wind stress and SST observed from buoys and satellite, *J. Climate*, **25**, 1544–1569, doi:10.1175/JCLI-D-11-00121.1.28.
O'Neill, L.W., Chelton, D.B. and Esbensen, S.K. (2012). Covariability of surface wind and stress responses to sea surface temperature fronts, *J. Climate*, **25**(17), 5916–5942, doi:10.1175/JCLI-D-11-00230.1.
O'Neill, L.W., Haack, T., Chelton, D.B. and Skyllingstad, E. (2017). The Gulf stream convergence zone in the time-mean winds, *J. Atmos. Sci.*, **74**(4), 2383–2412, doi:10.1175/JAS-D-16-0213.1.
Ortiz-Suslow, D.G., Haus, B.K., Williams, N.J., Laxague, N.J.M., Reniers, A.J.H.M. and Graber, H.C. (2015). The spatial-temporal variability of air-sea

momentum fluxes observed at a tidal inlet, *J. Geophys. Res.: Oceans*, **120**(2), 660–676, doi:10.1002/2014JC010412.

Phillips, O.M. (1957). On the generation of waves by a turbulent wind, *J. Fluid Mech.*, **4**, 426–434.

Phillips, O.M. (1985). Spectral and statistical properties of the equilibrium range in wind-generated gravity waves, *J. Fluid Mech.*, **156**, 505–531.

Pierson, W.L., Garcia, A.W. and Pells, S.E. (2003). Water wave attenuation due to opposing wind, *J. Fluid Mech.*, **487**, 345–365.

Plagge, A., Edson, J.B. and Vandemark, D. (2016). *In situ* and satellite evaluation of air–sea flux variation near ocean temperature gradients, *J. Climate*, **29**(4), 1583–1602, doi:10.1175/JCLI-D-15-0489.

Romero, L. and Melville, W.K. (2010). Airborne observations of fetch-limited waves in the Gulf of Tehuantepec, *J. Phys. Oceanogr.*, **40**, 441–465.

Savelyev, I.B., Haus, B.K. and Donelan, M.A. (2011). Experimental study on wind-wave momentum flux in strongly forced conditions, *J. Phys. Oceanogr.*, **41**, 1328–1344.

Shabani, B., Nielsen, P. and Baldock, T. (2014). Direct measurements of wind-stress over the surf zone, *J. Geophys. Res.: Oceans*, **119**(5), 2949–2973, doi:10.1002/2013JC009585.

Smith, S.D. (1980). Wind stress and heat flux over the ocean in gale force winds, *J. Phys. Oceanogr.*, **10**(5), 709–726.

Song, Q., Cornillon, P. and Hara, T. (2006). Surface wind response to oceanic fronts, *J. Geophys. Res.: Oceans*, **111**(12), 1–23, doi:10.1029/2006JC003680.

Stewart, R.H. and Joy, J.W. (1974). HF radio measurements of surface currents, *Deep Sea Res.*, **21**, 1039–1049.

Sullivan P.P., McWilliams, J.C. and Moeng, C.H. (2000). Simulation of turbulent flow over idealized water waves, *J. Fluid Mech.*, **404**, 47–85, doi:10.1017/S0022112099006965.

Sullivan, P.P., Edson, J.B., Hristov, T. and McWilliams, J.C. (2008). Large eddy simulations and observations of atmospheric marine boundary layers above nonequilibrium surface waves, *J. Atmos. Sci.*, **65**, 1225–1245, doi:10.1175/2007JAS2427.1.

Tamura, H., Miyazawa, Y. and Oey, L.-Y. (2012). The Stokes drift and wave-induced mass flux in the North Pacific, *J. Geophys. Res.: Oceans*, **117**, C08021.

Young, I.R., Hasselmann, S. and Hasselmann, K. (1987). Computations of the response of a wave spectrum to a sudden change in the wind direction, *J. Phys. Oceanogr.*, **17**, 1317–1338.

Young, I.R., Banner, M.L., Donelan, M.A., Babanin, A.V., Melville, W.K., Veron, F. and McCormick, C. (2005). An Integrated system for the study of wind wave source terms in finite depth water, *J. Atmos. Ocean. Technol.*, **22**(7), 814–828.

Webb, A. and Fox-Kemper, B. (2011). Wave spectral moments and Stokes drift estimation, *Ocean Model.*, **40**(3–4), 273–288, https://doi.org/10.1016/j.ocemod.2011.08.007.

Wu, J. (1983). Sea-surface drift currents induced by wind and waves, *J. Phys. Oceanogr.*, **13**, 1441–1451.

Zachry, B.C., Schroeder, J.L., Kennedy, A.B., Westerink, J.J., Letchford, C.W. and Hope, M.E. (2013). A case study of nearshore drag coefficient behavior during Hurricane Ike (2008), *J. Appl. Meteorol. Clim.*, **52**, 2139–2146.

Zhang, F., Drennan, W.M., Haus, B.K. and Graber, H.C. (2009). On wind wave-current interaction during the Shoaling Waves Experiment, *J. Geophys. Res.: Oceans*, **114**, C01018, doi:10.1029/2008JC004998.

Zhao, Z.-K., Liu, C.-X., Li, Q., Dai, G.-F., Song, Q.-T. and Lv, W.-H. (2015). Typhoon air-sea drag and coefficient in coastal regions, *J. Geophys. Res.: Oceans*, **120**(2), 716–727, doi:10.1002/2014JC010283.

Chapter 3

Wave Breaking and Ocean Mixing

Alex Babanin
University of Melbourne, Australia

Wave breaking is not a phenomenon specific only to the surface water waves. Electromagnetic waves, waves in plasmas and other media also break. This is how this phenomenon is described in Wikipedia (http://en.wikipedia.org/wiki/Wave_breaking): "In physics, a breaking wave is a wave whose amplitude reaches a critical level at which some process can suddenly start to occur that causes large amount of wave energy to be dissipated. At this point, simple physical models describing the dynamics of the wave will often become invalid, particularly those which assume linear behavior."

Beyond the general similarity, the dynamics of water-wave breaking is distinctly special. Water waves are propagating surface oscillations, and these oscillations spread at speed C, usually much faster than the actual physical motion of the fluid particles u along the wave orbits:

$$u/C \sim \varepsilon, \qquad (3.1)$$

where the typical steepness of ocean wind-generated waves is approximately

$$\varepsilon = ak \sim 0.1\text{--}0.12. \qquad (3.2)$$

If locally or instantaneously, the water overtakes the wave, the surface collapses and such a collapse by definition is a wave breaking.

The steeper a wave, as a result of nonlinear instabilities within wave trains or simply linear superposition of dispersive directional waves, the greater the magnitude of the fluid motion. Therefore, the fluid mechanics of the water below the surface imposes severe limits on possible magnitudes of the critical wave heights, by comparison with oscillation magnitudes in other physical media. In nonlinear optics, for example, light flashes can reach intensities of thousands of times above the mean power levels in an electromagnetic wave train [e.g., Solli et al., 2007]. For surface water waves, breaking in a one-dimensional wave train exists between the mean steepness of (3.2), when individual breaking events start to occur [e.g., Babanin et al., 2011a], and the ultimate steepness of

$$\varepsilon = ak \approx 0.44, \qquad (3.3)$$

where the surface inevitably turns over and any wave will break [Stokes, 1880; Michel, 1893]. On two-dimensional wave surfaces, the limiting steepness (3.3) is slightly increased [see, e.g., Babanin [2011]], but still the amplitude a can possibly grow only a few times above the average, before the wave breaks. Note that on the way to breaking onset, the wave number k also increases, and hence the possible variation of the wave height is even smaller than implied by the simple algebra of Eqs. (3.2) and (3.3).

The dissipation which occurs during wave breaking is also special. Usually, dissipation of mechanical energy means its conversion into heat. The wave-breaking energy, however, is not necessarily lost to the mechanical air–sea system. It is spent on the generation of turbulence, both above and below the surface, injection of bubbles into the water and production of spray in the air, suspension of sediment in shallow areas and other mechanical processes, but little of this energy is actually immediately transferred to warming of the environment. Nevertheless, regardless where the wave-breaking energy goes, it never goes back to regular wave orbital motion and is lost to waves. Therefore, in wave theories and models, wave breaking serves the role of dissipation and, together with the wind input and nonlinear interactions, is one of the three most important energy source/sink terms: as soon as the wind forcing brings waves to the steepness of (3.1), the breaking is initiated and starts balancing the input by taking energy out of the wave field.

Water waves also possess momentum, and reduction of wave height in the course of breaking signifies loss of momentum which is passed on to the mean currents or winds, to the bottom, the coast, or to vessels and engineering structures. Thus, ocean-wave breaking plays a critical role in air–sea exchanges of energy, momentum, as well as heat and mass (moisture above and gas below the interface), produces bubbles and aerosols. It is also of primary importance for coastal and maritime engineering, navigation, ocean remote sensing and other practical applications.

Wave-coupled processes in general, both in the ocean and atmosphere, are linked to large-scale phenomena which define ocean circulation, weather and climate: wave-induced mixing through the thermocline can cool the surface and thus the atmosphere (note that 2–3 m of the sea water has the same heat capacity as the entire dry atmosphere [e.g., Gill, 1982; Soloviev and Lukas, 2006]), breaking-produced aerosols can influence formation of clouds above the oceans and corrosion rates as far as the middle of continents, among many other influences.

Physically, the breaking of deep-water surface waves represents an interesting and one of the most challenging problems of fluid mechanics, oceanography, ocean engineering and marine weather forecast. It is a strongly nonlinear intermittent random process, very rapid compared to other processes in the system of surface waves in general and wind-generated waves in particular. The wave energy slowly accumulated under wind action over tens of hours and tens or even hundreds of kilometers of fetch, is suddenly released within a fraction of wave period, i.e., within seconds. The distribution of breaking on the water surface is not continuous, but its role in maintaining the energy balance within the continuous wind–wave field is critical.

Therefore, the significance of wave breaking in the context of surface ocean waves, and more broadly air–sea interactions is hard to overestimate. There is one aspect of the upper-ocean dynamics, however, where such significance is usually exaggerated or perhaps even misrepresented by scientists. This aspect is the mixing of the upper ocean where other wave-induced processes step forward and overpower wave breaking.

Waves play a primary role in the upper-ocean mixing which fast-tracks the thermodynamic balance between the atmosphere and the ocean in the spring-summer period when the vertical convection is blocked by stable stratification (warm air/colder ocean). Until recently, this role was often overlooked in large-scale models, and the wind was parameterized to directly drive the dynamics of the upper-ocean: i.e., wind stress provides energy/momentum flux to the ocean surface and this energy is then diffused down by means of parameterized turbulence. At moderate and strong winds, the dominant part of the wind stress, however, is supported by the flux of momentum from wind to waves [e.g., Kudryavtsev and Makin, 2002]. This means that, before the energy and momentum are received by the upper-ocean in the form of turbulence and mean currents, and thus enter the further cycle of air–sea interactions, they go through a stage of wave motion. In the course of events, the waves only accumulate a small fraction of the energy and momentum received [3–4%, Donelan, 1998], and the rest is dissipated or lost locally through wave breaking. Therefore, if the only role of the waves were to transfer the energy/momentum down under the ocean skin, the overall approach of disregarding the wave phase would be acceptable.

Such an approach, though, assumes the breaking is the only source of wave-induced turbulence across the interface. The breaking injects turbulence at the scale of the wave height [Terray et al., 1992; Babanin et al., 2005], this turbulence is then diffused down and the mixing of the upper layers is achieved. There are, nonetheless, two potential problems in such an approach.

Firstly, timescales of the turbulence lifetime and turbulence diffusion down to some ∼50 m (global mean depth of the mixed layer) should agree, and they do not. The lifetime of the small-scale turbulence (comparable with turnaround time of its vortex) is seconds, whereas the reaction time of the mixed layer to a storm above is of the order of hours [e.g., Smith, 1992; Toffoli et al., 2012]. Wave-breaking turbulence, however, is not an admixture injected at the ocean interface: admixtures will eventually be diffused regardless of the time required, but the turbulence will dissipate and hence disappear at timescales much less than those required to facilitate mixing through the thermocline.

Secondly, before the energy and momentum are received by the upper ocean in the form of turbulence and mean currents, they go through a stage of surface wave motion, and such motion can directly affect or influence the upper-ocean mixing and other processes. There are at least two processes in the upper ocean which can deliver turbulence straight to the depth of 50 m or so, instead of diffusing it from the top.

The first mechanism is due to turbulence induced by the wave orbital motion, which is unrelated to the breaking [Phillips, 1961; Benilov et al., 1993; McWilliams et al., 1997; Qiao et al., 2004; Babanin, 2006, 2017; Benilov, 2012]. This turbulence should persist at any depth where the wave motion is significant, and vertically this is a wavelength scale which is $\sim$100 m for ocean waves. Therefore, wave-orbital turbulence does not require any vertical diffusion in order to be present at the bottom of the mixed layer and facilitate mixing through the thermocline.

The second process is Langmuir circulation [Langmuir, 1938; Craik and Leibovich, 1976; Smith, 1992] which should not be confused with the concept of Langmuir turbulence introduced by McWilliams et al. [1997]. Because this circulation is dependent for its existence on the Stokes drift, its depth also scales with wavelength of the dominant surface waves. Langmuir cells can physically take the surface turbulence water and advect it down to the bottom of the mixed layer, but since the rate of rotation of these cells is necessarily slow (of the order of cm/s), the problem of dissipation of the small-scale wave-breaking turbulence remains.

Over the past years, our understanding of the breaking and dissipation of ocean waves has undergone a dramatic leap forward. From phenomena that were very poorly understood, there has emerged greater understanding of the physical processes and clarification to some extent of the magnitude. In this chapter, we will first address the topic of wave breaking, and then the issues of wave-induced mixing of the upper ocean, linked through the problem of wave-energy dissipation and wave-induced turbulence.

3.1. Wave Breaking

Historically, the importance of wave breaking in the context of wave evolution (because breaking controls the wave-energy dissipation) was always understood by ocean wave modelers. As early as the 1950s so-called second-generation wave forecast models (still used today for some applications) were based on a balance of the wind energy input and whitecap energy dissipation [Rogers *et al.*, 2014].

For many years (and until now), the whitecapping and wave breaking have been used as synonyms, but these are most certainly not the same features. Breaking waves are usually subdivided into three types: plunging, spilling and micro-breaking, and the latter does not involve air entrainment and hence does not produce whitecaps. Investigators of such micro-scale breaking, point out that this phenomenon is in fact much more widespread than the whitecaps [e.g., Jessup *et al.*, 1997; Sutherland and Melville, 2015]. According to Tulin and Landrini [2001], these are waves of

$$L \leq 25\,\text{cm}, \quad f \geq 2.5\,\text{Hz}, \tag{3.4}$$

i.e., in most cases, this is indeed a major part of the wave spectrum.

3.1.1. *Measurements of wave breaking*

Measurement and even detection of breaking is a challenging task due to their seemingly random occurrence and ferocity of the event in high seas. Breaking of large waves is most prominently exhibited through generation of whitecaps on the ocean surface, and as early as the 1940s, this breaking signature was employed to quantify the probability of occurrence of such events. Munk (1947) used observations of the white foam patches per unit of area in order to estimate the breaking probability. Nowadays, there is understanding that whitecap coverage area depends on many physical and chemical properties of the water and air, but it is still a sought after characteristic of the ocean surface. It is used in the remote sensing of the ocean, provides links to the gas and mass exchanges across the ocean interface, and of course to the basic characteristics of the breaking, both to the probability and severity [see e.g., Bortkovskii, 2003].

As wave-measurement instrumentation and techniques developed, methods of visual observations advanced, and for decades this remained the main means to investigate wave breaking. These methods can involve marking the breaking events in wave records by a human observer, who would trigger the mark when a breaker happened on a wave probe [e.g., Holthuijsen and Herbers, 1986], or using video records for repeated and more robust statistical analyses of whitecaps and even micro-breakers [e.g., Weissman et al., 1984]. In recent years, stereo photographing and video recording of the wavy surface have been used [Fedele et al., 2013].

Gradually, however, direct techniques to detect the breakers and measure their properties are employed. These methods use, for example, criteria related to impacts of the breaking at or below the water surface, such as a "sudden jump in surface elevation" [Longuet-Higgins and Smith, 1983], changes to electrical conductivity of the water [Lamarre and Melville, 1992] or void fraction below the surface [Gemmrich and Farmer, 1999].

Note that all the above methods deal with breaking in progress (see Section 3.1.2), i.e., consequences rather than causes of wave breaking. Laboratory experiments over the years enabled researchers to investigate breaking in its full complexity (with the exception, perhaps, of its three-dimensional features). Methods and techniques in this regard are numerous, from a simple measurement of surface elevations in front and behind a breaking [e.g., Galchenko et al., 2010], to sophisticated PIV mapping of turbulence velocities within a random breaking event [Alberello et al., 2018]. In this regard, we can broadly classify laboratory testing into two large groups, based on the way the laboratory breaking is created: those due to linear focusing of waves propagating with different speeds [e.g., Rapp and Melville, 1992; Brown and Jensen, 2001] and those due to modulational instability of nonlinear wave trains [e.g., Babanin et al., 2010].

In the field, where *in situ* methods of wave-breaking measurement are challenging, innovative remote-sensing techniques were proposed.

Examples include, but are not limited to, acoustic detection of bubble-formation events [e.g., Manasseh et al., 2006], "sea spikes" in the radar backscatter signals which are associated with the breakers [e.g., Smith et al., 1996], infrared signature of the breaking on the ocean's skin layer [Jessup et al., 1997]. Some such methods are useful in identifying breaking waves among those non-breaking, others can also be calibrated in terms of breaking severity (energy lost by a wave or wave group in the course of the breaking).

3.1.2. *Breaking onset and other phases of the breaking process*

Wave breaking is a fast process (a fraction of wave period in the deep water, i.e., Rapp and Melville [1990], Babanin et al. [2010]), but it is a process and not an instantaneous condition. Liu and Babanin [2004] classified this process into four stages: incipient breaking, developing breaking, subsiding breaking and residual breaking. Most breaking measurements deal with breaking in progress, i.e., what is happening with a wave and the water surface past the breaking onset — when the water surface is collapsing and whitecaps are visible. The incipient breakers, however, that is the lead-up to the breaking onset, will be detected, for example, by the wavelet method and will not be detected by such measurements, whereas the wavelets will fail to detect subsiding breakers. Therefore clarification, definitions and understanding of the relative duration and significance of the phases are needed.

The first phase is most essential in order to predict the breaking or its statistics, and most elusive. At this stage, no whitecapping is produced yet and in general, in the absence of criteria for the breaking onset, it is not clear whether a steep wave will break or recurrence to a less steep shape will take place, and the wave will propagate on without breaking.

Babanin et al. [2007a] suggested that the breaking onset is defined as such an instantaneous state of wave dynamics where a wave has already reached its limiting-stability state (for example, steepness (3.3) in the case of monochromatic one-dimensional wave trains), but has not yet started the irreversible breaking process characterized by rapid dissipation of wave energy. That is, the breaking onset is the ultimate point at which the wave dynamics caused by initial instabilities is still valid. This definition allows us to identify the onset and, once the location of the breaking onset can be predicted, to measure the physical properties of such waves. Note again that the breaking onset is instantaneous, unlike the stages of incipient breaking and breaking in progress.

Some analytical criteria for breaking onset have been offered long ago, but for a long time were disregarded as unrealistic. Criterion (3.3), for example, has a clear physical meaning: it signifies a wave whose crest propagates faster than the phase speed of the wave. Clearly, such a water surface will start collapsing and trigger the breaking, but such a steepness of individual waves was regarded as unrealistic, since the typical steepness of ocean waves is much smaller (3.2).

Lately, both experimental and theoretical evidence reinstated the steepness (3.3) as credible and in fact a robust criterion for wave breaking. Brown and Jensen [2001] for breaking achieved though linear focusing and Babanin et al. [2010] for breaking due to modulational instability both showed that in laboratory experiments the waves would break when $Hk/2 \approx 0.44$. The same limiting steepness was also revealed in numerical simulations by means of fully nonlinear wave models [Babanin et al., 2010].

The contradictions between these new results and the earlier measurements mentioned above seem to be rather straightforward to reconcile. Babanin et al. [2010] argued that the breaking onset is a very short-lived stage of wave evolution to the breaking, and thus both before and after the moment when the water surface starts to collapse the steepness of the wave-to-break and the breaker-in-progress is below that specified by (3.3). Therefore, in order to detect steepness (3.3), the very instant of the breaking onset has to be captured, rather than any other phase of wave breaking, which is apparently not the case in experiments other than those explicitly dedicated to studying the incipient-breaking point. For example, Brown and Jenkins [2001] who used "essentially the same experimental setup to generate and study breaking waves" as Rapp and Melville [1990] indeed found that "wave breaking occurs when the instantaneous local wave steepness $(ka)_{\max}$ exceeds ~ 0.44, whereas Rapp and Melville did not mention such large steepness in their study.

The field experiments, as far as observations of breaking onset are concerned, suffer from even more serious shortcomings. The majority of such measurements rely on whitecapping and associated features to detect the breaking, and the waves only produce whitecapping when the breaking is already well in progress. Therefore, most of the field observations deal with the developing and subsiding phases of the breaking process, and the steepness of breakers at these phases can be considerably below the limiting steepness (3.3) and even below the mean steepness of the wave field (3.2) [e.g., Holthuijsen and Herbers, 1986]. To address this issue, Toffoli et al. [2010] investigated the probability distribution of the steepness of individual waves, based on two different field datasets and two sets obtained in different directional wave tanks, and found a threshold for the ultimate steepness as

$$\varepsilon = \frac{H_{\text{front}} k}{2} = 0.55 \tag{3.5}$$

for the front steepness, and

$$\varepsilon = \frac{H_{\text{rear}} k}{2} = 0.45 \tag{3.6}$$

for the rear steepness. Here, since for realistic wave groups the front and rear troughs are not symmetric, height H is estimated as the vertical distance from the crest either to the front or to the rear troughs. Because the probability density function is terminated at the limiting-steepness value, i.e., does not extend to the higher values of steepness, this result should be interpreted as the ultimate steepness beyond which ocean waves will certainly break.

Thus, the limiting steepness beyond which breaking will definitely occur is $\varepsilon = ak \sim 0.45$ for progressive waves if defined for the rear face of waves. It should be mentioned that it is somewhat higher, $\varepsilon = ak \sim 0.65$, for standing waves [Babanin, 2011]. For progressive waves within a full spectrum, however, which is what the wind-forced waves in the ocean are, Chalikov and Babanin [2012] showed that (3.5) is the upper limit rather than the regular occurrence for the breaking onset. While the limiting steepness (3.5) can indeed be reached by dominant waves in the spectrum, on average they break at a steepness of

$$\varepsilon = ak \sim 0.2 \qquad (3.7)$$

due to the influence of short waves (spectrum tail). When a dominant wave is already steep enough, the short waves on its front face can produce a very steep water surface locally, which will make the surface collapse and trigger breaking: if the water crosses the vertical, it will not go back to the regular orbital motion.

Beyond the point of onset, the breaking process can be further subdivided into stages with different properties and different dynamics [Rapp and Melville, 1990; Liu and Babanin, 2004]. The breaking occurs very rapidly, but the wave may lose a significant proportion of its height or even completely disappear [Rapp and Melville, 1990; Babanin et al., 2010; Galchenko et al., 2010]. This is a highly nonlinear mechanism, conceptually very different from the processes leading to breaking onset and should be considered separately. As described by Babanin [2011], while the collapse is driven, to a greater extent, by gravity and inertia of the moving water mass and, to a lesser extent, by hydrodynamic forces, the breaking onset occurs mostly due to the dynamics of the wave motion in the water. Waves can break even in the total absence of wind forcing, provided hydrodynamic conditions are appropriate like in the flumes with wave maker [Rapp and Melville, 1990; Babanin et al., 2007a, 2010; Galchenko et al., 2010]. Therefore, processes leading to wave breaking, i.e., the stage of incipient breaking, can be simplified by studying only the water side of the surface behavior, whereas for breaking in progress, the air–sea interaction part, such as whitecapping [e.g., Guan et al., 2007], void fraction [e.g., Gemmrich and Farmer, 2004], work against buoyancy forces [e.g., Melville et al., 1992], wind–wave momentum/energy exchange [Babanin et al., 2007b; Iafrati et al., 2014] are of very essential importance.

The pre- and post-breaking physics are not entirely disconnected, and the outcome of the breaking collapse appears to "remember" the dynamic input which caused a wave to break. Furthermore, the breaking in progress can be further subdivided into distinctly different phases.

The developing stage is characterized by the lateral spread of breaking with a whitecapping appearance for the crest to pass over the measurement spot, so a developing breaker should be readily detected by the whitecap-oriented measurements. But the developing breaker also exhibits an increase of wave front steepness until it subsides. Rapp and Melville [1990] in their Section 3.4 defined the front steepness as the ratio of crest-to-front zero-crossing height to crest-to-front-zero-crossing length and showed that it is larger, compared to the incipient breaking front steepness, for both spilling and plunging breakers. As shown by Liu and Babanin [2004], even though the front steepness is not unambiguously linked to the maximal instantaneous downward acceleration, this is an indication that the over-limiting acceleration values may persist through the developing stage, and thus the developing breaker will be detected by the wavelet method as well.

The relaxing or subsiding stage of breaking has not received as much attention in the literature as the developing stage. Therefore, it is not clear, for example, what will happen to the breaking crest once its whitecap has reached its maximal lateral length according to Phillips *et al.* [2001] or when the front steepness of breakers, described by Rapp and Melville [1990], will start to decrease. But at some stage, it will start to decrease. Babanin [2011], by comparing the wave-breaking count obtained by means of wavelet analysis in Liu and Babanin [2004] (wave incipient breaking plus developing breaking) and through the concurrent whitecapping observations (developing breaking plus subsiding breaking), concluded that the second breaking stage lasts for approximately one-half compared to either incipient or subsiding phases. Thus, the developing phase is the shortest (fastest) stage and only accounts for some 20% of the wave breaking duration (and even less if the residual stage is included).

The last, residual stage is usually introduced formally following Rapp and Melville [1990], as the phase of the breaking process when the whitecap is already left behind, but the underwater turbulent front is still moving downstream. During this fourth stage of breaking, whitecaps decrease in size as entrained bubbles rise to the surface, but spatial evolution of mixing continues. Effectively, this is no longer breaking: the "wave dissipation" has already occurred and it is now dissipation of turbulent mechanical energy into heat happening. The broken wave has left the scene, and even if there is interaction of surface waves with this enhanced sub-surface turbulence, it is subsequent waves in the wave train which are now involved. Rapp and Melville [1990] found that the propagation of the turbulent front continues for many wave periods, although at a much slower speed by comparison with the wave celerity. It should be noted that the experiments of Rapp and Melville [1990] were conducted with wave breaking caused by superposition of linear waves. Post-breaking dynamic effects and outputs of such breaking can be very different to those due to other breaking mechanisms. For example, for breaking caused by modulational instability of wave trains, Iafrati *et al.* [2014] found that up to 80% of the energy lost by the waves is spent on air turbulence and hence most of the actual dissipation of mechanical energy takes place above the surface. Therefore, it is not clear whether the residual stage is a general feature of wave breaking and will persist in the case of breaking other than that caused by the linear superposition.

3.1.3. *Breaking in quasi-monochromatic one-dimensional wave trains*

In order to reveal the nature of breaking as such, numerical and laboratory tests were conducted in the simplest conditions: uniform monochromatic one-dimensional deep-water wave trains [Babanin *et al.*, 2007a, 2010, 2011a]. The breaking was not forced in any way: there was not any kind of focusing intended to facilitate appearance of a high (steep) wave, no winds, currents or bottom proximity able to enhance the wave steepness.

If background noise is present in numerical environments (or turbulence in the physical medium), such trains start interacting with the noise and modulate themselves, hence become quasi-monochromatic where the nonlinear group evolution can lead to a wave much higher (and steeper) than the original mean uniform waves. Such a wave will either subsequently break, or in the one-dimensional case the full recovery of the initial uniform wave train will occur. This is the so-called modulational instability [Zakharov, 1966, 1967; Benjamin and Feir, 1967; Yuen and Lake, 1982].

3.1.3.1. *Numerical modeling*

A numerical model employed to obtain the fully nonlinear solution of the Euler equation is the two-dimensional Chalikov–Sheinin Model (CSM) [Chalikov and Sheinin, 2005]. The CSM numerical approach is based on a non-stationary conformal mapping. This allows the principal equations of potential flow, with a free surface, to be written in a surface-following coordinate system. The Laplace equation retains its form, and the boundary of the flow domain (i.e., the free surface) is the coordinate surface in the new coordinate system. Accordingly, the velocity potential over the entire domain receives a standard representation based on its Fourier expansion on the free surface. As a result, the hydrodynamic system (without any simplifications) is represented by two evolutionary equations that can provide numerical solutions, stable for many hundreds of wave periods, with very high precision [Chalikov and Sheinin, 2005]. Most importantly for the wave-breaking study, is the model's ability to describe the evolution of very steep waves and to reproduce known strongly nonlinear features of real waves, such as wave asymmetry with respect to the vertical, which has been shown to be an inherent characteristic of wave breaking [Caulliez, 2002; Young and Babanin, 2006; see also Fig. 3.1 (left) where the waves move from right to left]. In the CSM, the wave model can be coupled with an atmospheric boundary layer model. Thus, it is possible to introduce wind forcing of the waves if necessary, which tends to accelerate the breaking process. Initially, the model was run for monochromatic waves without wind.

Two of the most commonly reported nonlinear features of a breaking wave are its asymmetry (i.e., the front face of the wave is steeper than the rear face) and its skewness (i.e., the crest elevation above

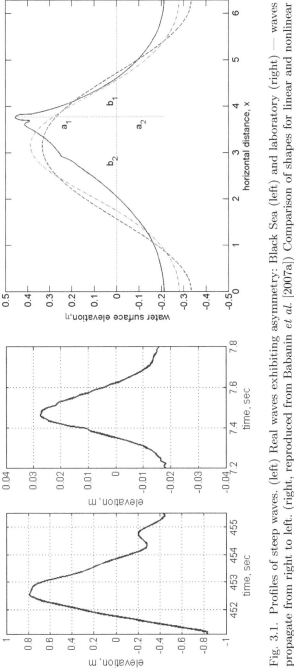

Fig. 3.1. Profiles of steep waves. (left) Real waves exhibiting asymmetry: Black Sea (left) and laboratory (right) — waves propagate from right to left. (right, reproduced from Babanin et al. [2007a]) Comparison of shapes for linear and nonlinear waves of the same steepness — waves propagate from left to right.

the mean water level is greater than the trough elevation below the mean water level). Geometric definitions of skewness, S_k, and asymmetry, A_s, are shown in Fig. 3.1 (right) where the waves move from left to right.

Following Fig. 3.1 (right), the skewness and asymmetry can be defined as follows:

$$S_k = a_1/a_2 - 1, \tag{3.8}$$

$$A_s = b_1/b_2 - 1. \tag{3.9}$$

Hence, positive skewness represents a wave with a crest elevation greater than the trough depth and negative asymmetry represents a wave tilted forward in the direction of propagation.

In Fig. 3.1, all waves have the same steepness for clarity, although if that is the same wave in the course of nonlinear evolution, its steepness will be changing (as further shown in Fig. 3.4). The incipient breaker in Fig. 3.1 (right, solid line, $S_k = 1.15$, $A_s = -0.51$) was determined from the CSM by commencing the simulation with a sinusoidal (linear) wave (dashed line, $S_k = 0$, $A_s = 0$) with initial monochromatic steepness $IMS = ak = 0.25$. Such a value of steepness is well above the limits of perturbation theory. The model is then allowed to evolve from this initial condition. It had been previously shown [Chalikov and Sheinin, 2005] that such a steep sinusoidal wave immediately transforms into a Stokes wave (i.e., dash-dotted wave of Fig. 3.1, $S_k = 0.39$, $A_s = 0$) whose further evolution is controlled by the modulational instability mechanism [Babanin et al., 2007a]. The instability leads to modulation of the initially monochromatic wave train, and as a result some waves can become very large (steep) and ultimately break.

The nonlinear properties of fully nonlinear waves have been shown to be connected and oscillate as the waves evolve [Babanin et al., 2007a]. Coherence and phase relationships of the steepness, skewness and asymmetry are analyzed in Fig. 3.2. Figure 3.2 (left) compares spectra of running instantaneous steepness $\varepsilon = ak$ (top subplot) and skewness S_k (second subplot), their coherence (third subplot) and phase (bottom subplot). Since the timescale of the simulations are dimensionless (i.e., presented in terms of wave periods), the frequency

Wave Breaking and Ocean Mixing 63

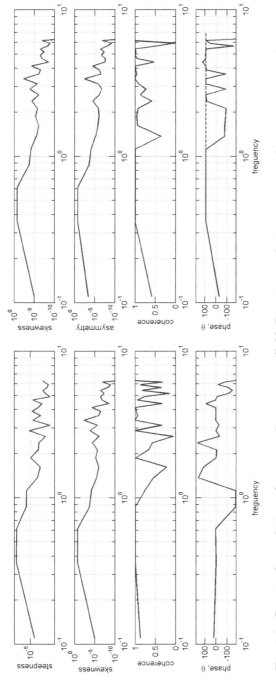

Fig. 3.2. Spectra for evolution of steep nonlinear waves. (left) From top to bottom, spectra of wave steepness, skewness, their coherence and phase. (right) Same for skewness and asymmetry.

scale is expressed in inverse wave periods. Therefore, the peak of the steepness/skewness modulation falls on the double wave period (0.5 of the inverse wave period). This frequency dominates the modulation in agreement with the theory [Longuet-Higgins and Cokelet, 1978], and the spectral density drops very rapidly away from the peak.

The peak is rather broad and covers a range of frequencies of 0.4–0.6 of inverse wave periods. Coherence of the steepness and skewness oscillations in this range is 100% as could have been expected for numerical simulations of the theory with very high precision. The phase shift between the dominant oscillations of steepness and skewness is zero. Thus, steepness and skewness are in phase, and the maximal steepness is achieved at the same instant as the maximal skewness.

Similarly, Fig. 3.2 (right) compares co-spectra of running instantaneous skewness S_k and asymmetry A_s. Again, the broad peak of asymmetry spectrum falls into the 0.4–0.6 range of inverse wave periods. Spectra of skewness and asymmetry are almost perfectly coherent, with a phase shift of approximately 90° (asymmetry is leading). The latter means that the asymmetry reaches its positive maximum (i.e., wave is tilted backward) when skewness is zero (wave crest and trough are of equal magnitude) and the local steepness is half-way trough raising from its minimal to the maximal value in an oscillation. From the positive maximum, the asymmetry begins to decrease and reaches zero, half a wave period later (quarter of the period of the oscillation) — i.e., the wave becomes symmetric with respect to the vertical. At this time, steepness and skewness are at their largest, and it is at this point that the wave may break. Whether the wave breaks or not, the asymmetry will keep decreasing into negative values (wave is tilting forward — compare with the developing breaking phase in Section 3.1.2), while the steepness/skewness start subsiding in quadrature with the asymmetry. It is interesting to look at this moment from a point of view of a sailor who encounters the breaker: they see a very tall crest which begins to break with the water mass falling down from the top and at the same time the front face of the wave steepens as the wave is learning forward — obviously a very dangerous position to be in.

As far as the wave breaking process is concerned, breaking onset characteristics are most elusive of all. In these numerical simulations, a wave is regarded as breaking if the water surface becomes vertical at any point. Since the model cannot reproduce the rotational phase of the breaking which will follow soon after, at such an instant of evolution it stops. Nonlinear characteristics of the wave at this stage apparently asymptote to properties of the incipient breaking in experiments (Section 3.1.3.2).

Statistics of these described features of incipient breaking are shown in Fig. 3.3. Estimates are obtained from a comprehensive set of numerical runs covering a range of initial monochromatic steepness $IMS = 0.10$–0.30 and wind forcing $U/c = 0$–11 (U is wind speed). Steepness is shown in terms of $kH = 2\varepsilon$.

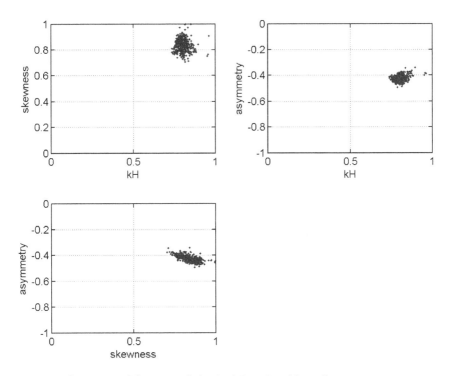

Fig. 3.3. Statistics of features of the incipient breaking: skewness versus steepness (top left), asymmetry versus steepness (right), asymmetry versus skewness (bottom).

Since the local steepness is expected to be the reason for wave collapse, it is most instructive to look at the limiting values of kH. The data points concentrate well (top subplots), although scatter is noticeable: $kH = 0.75$–0.85 with some outliers reaching over 0.9 values. This places mean value of the limiting steepness at around $\varepsilon \sim 0.4$ which is very high, particularly if we keep in mind that local steepness towards the crest is even higher than this mean value. This limiting steepness is much higher than the initial IMS, but given the fact that the original IMS values varied by a factor of 3, the convergence of steepness values at the breaking onset is remarkable and highlights the fact that, at such magnitudes of the water surface slope, the surface may simply collapse because of gravity, depending on the velocity field in the water (limiting steepness (3.3)). If so, the role of the modulational instability is not in generating the wave breaking as such, but rather in simply producing a very steep wave.

Skewness and asymmetry are not breaking criteria, but they apparently have some limiting values. In Fig. 3.3, the incipient-breaking skewness is scattered in the range of $S_k = 0.7$–1 (top left) and the asymmetry is $A_s = (-0.35)$–(-0.5) (bottom). Within the scatter, there is no dependence of one property on another except a tentative negative correlation between the skewness and asymmetry in the bottom panel. The latter signifies a dynamic feature apparent from Fig. 3.2 and is not necessarily an indicator of approaching breaking onset: larger negative asymmetry is likely to follow a greater skewness.

3.1.3.2. *Laboratory experiments*

The experimental investigations described here were conducted at the air–sea interaction tank at RSMAS, University of Miami [Babanin, 2007a, 2010]. Near-monochromatic deep-water two-dimensional wave trains were generated with a wave maker. With a tank length of 13.24 m, surface elevations were recorded at 4.55 m, 10.53 m, 11.59 m and 12.56 m from the paddle. For each record, the IMS was varied in such a way that the waves would consistently break just after one of the wave probes. In this way, the dimensional distance to breaking, wave train properties immediately prior to breaking and detailed properties of the incipient breaker could be determined. The fact that breaking could be predicted and controlled by manipulating only *IMS* is a good corroboration of the numerical model in its own right.

It should be pointed out that qualitative rather than an exact quantitative agreement between numerical model and experiment is expected. Although sophisticated, the model is still a simplification of nature. Importantly, the two-dimensional model predicts an immediate breaking onset at $IMS > 0.3$ whereas it has been previously shown that at $ak > 0.3$ waves can exist, but three-dimensional dynamics will stabilize the crest behavior [Melville, 1982].

Figure 3.4 shows a wave record with an initial monochromatic frequency, $IMF = 1.8\,\text{Hz}$ and an $IMS = 0.30$, with no wind forcing. It should be noted that there is a conceptual change to the numerical model results. In the case of the model, a single wave was followed as it approached the point of breaking. Here, observations are made at a single point as a succession of waves pass. One can approximately move from the fixed frame of reference in Fig. 3.4 to the moving frame by considering the waves shown propagating from right to left, as indicated by the arrow in the figure.

The top panel in Fig. 3.4 shows the measured water surface elevation (η) as a function of time (horizontal axis). Interpreting this as a wave moving from right to left demonstrates that, within each wave group, the maximum value of the water surface elevation gradually decreases and then suddenly increases until a point, where breaking occurs. This point of breaking was located immediately after the probe at a distance of $10.73 \pm 0.10\,\text{m}$ from the wave maker. Each successive wave passing the wave gauge was analyzed to determine its steepness, skewness, asymmetry and frequency, which are shown in the four panels of Fig. 3.4.

The major features seen in the numerical model are confirmed by the laboratory data. The incipient breaking waves are the steepest in the wave train, with the steepness oscillating in a periodic fashion. Skewness and asymmetry also oscillate, but behave in a less ordered fashion. However, at the point of breaking, skewness is positive (i.e., peaked up) and asymmetry is small (i.e., not tilted forward).

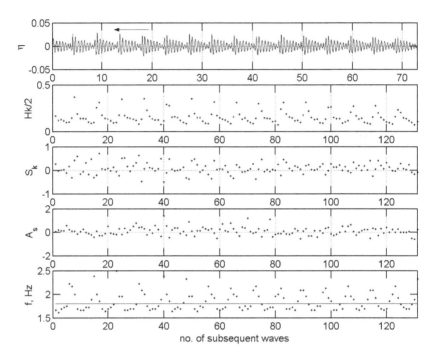

Fig. 3.4. Segment of a time series with $IMS = 1.8\,\text{Hz}$, $IMS = 0.30$ and $U/C = 0$. The top panel shows the water surface elevation, η as a function of time in seconds. The highest wave in each group is an incipient breaker at the measurement point, breaking immediately after the wave probe. The subsequent panels show properties determined for each of the waves: steepness (second panel), skewness (third panel), asymmetry (fourth panel) and frequency (bottom panel, solid line signifies $IFM = 1.8\,\text{Hz}$). Figure is reproduced from Babanin et al. [2007a].

A feature which could not be determined from the numerical model is that there is also a modulation in the frequency. At the point of breaking, the frequency increases rapidly, further increasing the steepness and hastening the onset of breaking.

These visual observations are summarized in Fig. 3.5, which shows data for the five steepest breakers. The analysis is limited to these steepest cases because wave quantities close to the breaking point change rapidly, as seen in Fig. 3.4. The steepest cases are considered to be on the point of breaking.

As in the numerical simulations, the steepness seems to be the single robust criteria for breaking. For the 20 steepest breakers (not

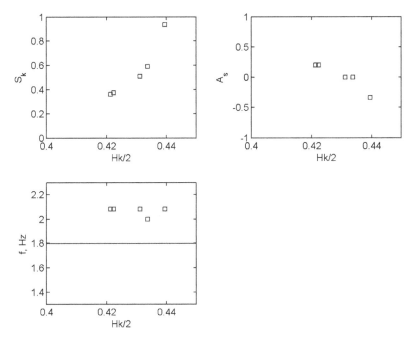

Fig. 3.5. Laboratory statistics for incipient breakers (five steepest waves). $IMF = 1.8\,\text{Hz}$, $IMS = 0.30$, $U/C = 0$. (top left) Skewness versus steepness. (right) Asymmetry versus steepness. (bottom) Frequency (inverse period) versus steepness. $IMF = 1.8\,\text{Hz}$ is shown with the solid line. Figure is reproduced from Babanin et al. [2007a].

all shown in Fig. 3.5), steepness was confined to the narrow range $H_k/2 = 0.37\text{--}0.44$, while skewness was scattered over the wide range $S_k = 0\text{--}1$ and asymmetry $A_s = 0.8$ to -0.4. Considering only those waves at the point of breaking, however, Fig. 3.5 shows a clearer trend. The steepness appears to approach an asymptotic limit of $H_k/2 \approx 0.44$. We should point out that this limit is remarkably close to the theoretical steady limiting steepness (3.3). Such an observation is very important because it signifies that the waves break once they achieve the well-established state beyond which the water surface cannot sustain its stability. It therefore can be postulated that, at least for one-dimensional wave trains, the other geometric, kinematic and dynamic criteria of breaking, explored in the literature, are indicative of a wave approaching this state, but are not a reason

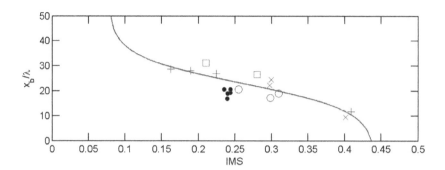

Fig. 3.6. Parameterization of the breaking probability, laboratory data. Number of wavelengths to breaking versus *IMS*. No wind forcing: o — $IMF = 1.6$ Hz; × — $IMF = 1.8$ Hz; + — $IMF = 2.0$ Hz. Filled circles represent — $IMF = 2.0$ Hz, with wind forcing applied. The parameterization (3.10) is shown with the solid line. Figure is reproduced from Babanin et al. [2007a], where $\lambda = L$.

or a cause for the breaking. As this limit is approached, the skewness increases very rapidly and immediately after the limit is reached the asymmetry becomes negative (i.e., the wave starts tilting forward at the point of breaking).

These laboratory results are summarized in Fig. 3.6, which shows the non-dimensional distance to breaking, $N = x_b/L$ as a function of *IMS*, where x_b is the distance to breaking. A range of values of *IMS* are shown, along with cases with and without wind forcing. As expected, the addition of wind forcing reduces the non-dimensional distance to breaking. However, this reduction is not so great as to make the data points deviate markedly from the functional relationship between N and *IMS*, i.e., the nonlinear effect dominates over the wind forcing.

In Babanin et al. (2007a), this result was approximated by the relationship

$$N = -11 \operatorname{atanh}[5.5(IMS - 0.26)] + 23 \quad \text{for } 0.08 \leq IMS \leq 0.04. \tag{3.10}$$

Consistent, with the model results, the formula imposes two threshold values of *IMS*. For $IMS > 0.44$, the wave breaks immediately (compared to $IMS \approx 0.3$ for the two-dimensional model) and if

$IMS < 0.08$, the wave, in the absence of wind forcing, will never break (compared to $IMS \approx 0.1$ for the model). In Fig. 3.6, two points (squares) are shown which were derived from Figs. 1 and 2 of Melville [1982] for comparison. The two measurements in Melville [1982] were conducted for initially uniform wave trains, their initial steepness and approximate dimensionless distance to breaking being known. Although recorded under different conditions, these points agree very well with the above parameterization and provide further support for (3.10).

Thus, we conclude that based on a combination of a nonlinear numerical model and laboratory data, the wave-breaking onset can be described and predicted: once waves reach the Stokes limiting steepness (3.3), they will break. The distance until breaking occurs is a function of the wave initial steepness, and both in the simulations and experiments this condition is achieved in the absence of wind forcing. Note again that there was no frequency or amplitude focusing of waves here, hence the hydrodynamic process, i.e., the modulational instability mechanism within the evolving nonlinear wave trains, leads to this limit, if the initial waves are sufficiently steep.

3.1.3.3. *When does nonlinear evolution of wave groups lead to the breaking, and when it does not*

The last sentence in the subsection above — if the initial waves are sufficiently steep — may hold the key to the nature of the breaking in the ocean as we will show below. Indeed, according to Fig. 3.6 and Eq. (3.10), if the mean steepness of the wave train is below $ak \approx 0.1$ nonlinear waves will modulate themselves like in Fig. 3.4, but will not reach the limiting steepness of Fig. 3.5 and will not break. This behavior was investigated and explained in Babanin *et al.* [2011a].

As already mentioned above, the modulational instability, which drove evolution of initially monochromatic wave trains in the numerical simulations and laboratory experiments above, happens even with infinitesimal waves [i.e., Benjamin and Feir, 1967] and it does lead to a wave which is much higher than background conditions, but if the limiting breaking onset steepness is not reached, recurrence occurs and this high wave subsides [see, e.g., Yuen and Lake, 1982, for a review]. If the mean steepness is higher, the nonlinearity is stronger and the maximal-height individual wave grows taller, in relative terms, with respect to the background wave train. That is, the ratio of the maximal individual-wave height in the nonlinear wave train H_m, to its significant wave height will keep increasing as a function of H_s.

This trend, however, should only be true until these highest individual waves start reaching the height H which signifies, for the wavelength of these individual waves, the limiting steepness (3.3). At this H_s, these individual waves will start breaking. Therefore, a further increase of H_s will not be accompanied by subsequent growth of H_m. H_s may be increasing now, but the individual waves which grow as part of the evolution of the nonlinear wave groups, will be breaking at the same steepness and at the same height H as before. As a result, the ratio H_m/H_s should start decreasing.

This is exactly the behavior observed in laboratory experiments by Hwung et al. [2005] with initially uniform wave trains of different mean steepness. In Fig. 3.7, reproduced from Hwung et al. [2005], the ratio of maximal wave-crest surface elevation η_m to initial amplitude a_0 of the initial uniform wave trains is plotted as a function of mean wave steepness $k_c a_c$. There is a clear growth of this ratio which peaks at $k_c a_c \approx 0.13$ and then starts decreasing.

Figure 3.7, if interpreted in terms of the wave height rather than wave-crest elevation, can lead to interesting conclusions as to the maximal possible wave height (in relative terms of H_s, which is how the rogue waves are defined: a wave is considered rogue if its height is at least $2H_s$). For nonlinear waves, the crest height a_{cr} is not twice the trough depth a_{tr}, i.e., wave height H_m is not equal to $2\eta_m$. Babanin et al. [2007a, 2010] demonstrated that at breaking onset due to modulational instability, skewness of the individual breaking wave asymptotes to $S_k \to 1$ where skewness is defined as $S_k = \frac{a_{cr}}{a_{tr}} - 1$ (see Eq. (3.8) above). We should stress that this is a dynamic asymptote, not a statistical asymptote.

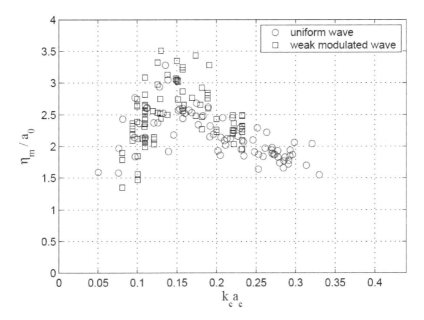

Fig. 3.7. Ratio of maximal crest height to the initial wave amplitude versus mean wave steepness, in the modulational-instability scenario of evolution of nonlinear wave groups. Figure is reproduced from Hwung et al. [2005].

Therefore, at the breaking point, the wave crest is twice the wave trough and the maximal wave height

$$H_m = a_{cr_m} + a_{tr_m} = \frac{3}{2}\eta_m. \qquad (3.11)$$

In Fig. 3.7, the maximal ratio is $\eta_m/a_0 \sim 3.5$ which converts into

$$\frac{H_m}{H_0} = \frac{(3/2)\eta_m}{2a_0} \approx 2.6. \qquad (3.12)$$

Expression (3.13) is for the highest wave in the initially uniform train with all the original waves of heights $H_0 = 2a_0$. In wave fields with a continuous spectrum, and for the definition of rogue waves, significant wave height is used as a mean wave characteristic, which is $H_s = 4\,std(\eta)$. For directional wave fields, the asymptotic skewness is less extreme, i.e., $S_k \to 0.7$ [Babanin et al., 2011b]. Then, assuming that the maximal ratio seen in Fig. 3.7 still holds,

$$\frac{H_m}{H_s} = \frac{\eta_m + \eta_m/1.7}{2\sqrt{2a}} \approx 2.0. \tag{3.13}$$

3.1.4. *Modulational instability or linear superposition?*

If the waves break because they reach some limiting steepness, then the key question concerning wave breaking is what can make water waves that steep? For the simple deep-water no-forcing condition, there are two possible processes: superposition of waves and instability of nonlinear wave trains.

It should be stressed that one mechanism does not prohibit or cancels the other: both are physically possible. In the laboratory they can be clearly separated, but in the field this is a matter of relative significance. No forcing in this case means low or moderate wind speeds (i.e., such winds which cannot directly make waves break — see, e.g., Babanin [2018]), and the absence of currents with strong horizontal or vertical gradients.

3.1.4.1. *Linear and nonlinear wave focusing*

Wave superposition is often regarded as a linear or quasi-linear process, as opposed to the nonlinear evolution of wave groups due to modulational instability (Section 3.1.3). Superposition, however, can occur due to three different reasons. First, surface waves in deep water are dispersive and hence waves of different scales propagate with different speeds. The ocean wave spectrum is continuous, therefore waves of all scales within the spectrum are present and propagate at the same time, and in a random wave field can potentially pile up into a very high surface elevation. This is the most commonly perceived cause for wave superposition, and this is the way breaking is most often simulated in laboratory [e.g., Rapp and Melville, 1990]. Depending on whether linear (sinusoidal) or nonlinearly-shaped (Stokes) waves are superposed, probability distributions for extreme wave heights, and particularly for wave crests and troughs are essentially different.

The second reason for wave dispersion and possible superposition is purely nonlinear. Waves of the same scale (same wavelength and period — as opposed to waves of different scales above) also can propagate with different speeds if they have different steepness. The maximum phase-speed difference due to such steepness corrections is small, about 5%, but even this makes waves of similar frequency more likely to overlap, and such potential overlap would be significant if it happened, since such waves have similar amplitudes (unlike waves of different frequencies). Indeed, in a wave field with a continuous spectrum, superposition happen often, but the problem is that waves with different speeds (frequencies) have very different amplitudes. As an approximate rule, if the Phillips spectrum of f^{-5} is used as a scale, then if the frequency f is doubled (i.e., phase speed is decreased two times), the amplitude a on average will drop by some $\sqrt{32} \approx 5.7$ times. With the typical mean steepness of ocean waves of (3.2), this would require superposition of a great number of individual waves in order to make a dominant wave reach the steepness defined by (3.3) or even (3.5), which then appears a low-probability event. For waves of the same scale, however, within steep wave fields, superposition of only two waves of the same amplitude maybe sufficient to reach value specified by (3.3). Therefore, this second reason should also be treated seriously.

In directional wave fields, even waves with the same frequency and same steepness, and therefore the same phase speed, can be focused and superposed if they come from different angles. Such situations are particularly likely in presence of swell, which is a typical condition in the ocean, but can also occur in storms and lead to enormously large waves like those with 32 m height in Typhoon Krosa [Babanin et al., 2011c]. This brings us to the third reason, directionality. In this scenario, again, waves of similar magnitudes can be added together.

It should be pointed out that while the focusing in the first and the last cases are apparently linear, the last stages of the focused-wave dynamics demonstrate essentially nonlinear behaviors if the steepness becomes large enough [Brown and Jensen, 2001]. Therefore, there is no clearly linear wave superposition, as there are no linear (non-Stokes) gravity surface waves.

3.1.4.2. *Instability of nonlinear wave trains*

Modulational instability of nonlinear wave trains has been discussed in detail in Section 3.1.3. Such trains are unstable with respect to some perturbations, even if very small, provided these fall within some parametric range in terms of wave steepness and spectral bandwidth [e.g., Tulin and Waseda, 1999; Waseda and Tulin, 1999]. In such cases, even perfectly uniform and initially monochromatic waves will start modulating themselves. In the ocean, which is usually turbulent, perturbations of all scales exist to initiate the instability, but then other features, such as directionality of wave fields, can prevent it.

If the wave fields are stable, then in terms of wave breaking, such conditions are described in the superposition context as in Section 3.1.4.1, with some correction for the sharper wave crests and flatter troughs. The probability distribution of wave crests should be approached through nonlinear approximations to the expected shape of Stokes waves, such as the second-order wave crest distribution. It should be noted that even at the second-order approximation the directionality of real wave fields complicates the conclusions and outcomes.

The third-order effects, however, apart from further (small) nonlinear corrections to the wave shape, bring about dynamic nonlinear exchanges, resonances and instabilities. The instabilities can then lead to rare transient events of high and steep waves, resulting in the breaking. These are, however, the events that can produce wave breaking unrelated to the superposition/focusing of linear/nonlinear waves. Therefore, it is important to understand first, which nonlinear wave trains and fields are stable and which are unstable.

3.1.4.3. *Instability of unidirectional wave trains*

For monochromatic one-dimensional wave trains, instability has been studied extensively over the years and its behavior is well-understood.

Diagrams of the instability range go back to the original paper by Benjamin and Feir [1967] and we can also refer the reader to a comprehensive review by Yuen and Lake [1982]. The original research was done for waves of small steepness, and later Tulin and Waseda [1999] also estimated the instability range for steeper waves.

Such diagrams show the growth rate of the instability β_x normalized by the initial steepness of uniform monochromatic waves $\varepsilon = a_0 k_0 = ak$; a_0, k_0 and $\omega = \omega_0$ are amplitude, wave number and radian frequency of the carrier wave, respectively, and $\delta\omega$ defines the frequencies of the perturbations, to which the original wave train is unstable, i.e., $\omega = \omega_0 \pm \delta\omega$. The combined parameter $\delta\omega/\varepsilon\omega_0$ determines whether the wave train is stable or unstable, subject to the perturbations. For waves of steepness $\varepsilon = 0.05$, for example, the instability range is from $\delta\omega/\varepsilon\omega_0 = 0$ to ~ 1.33, with the maximal growth occurring at $\delta\omega/\varepsilon\omega_0 \approx 1$, i.e., for $\delta\omega/\omega_0 \approx 0.05$. For steeper waves, the instability range shrinks. The respective index is usually called BFI (Benjamin–Feir Index), but since the Benjamin and Feir [1967] paper dealt with linear growth, for the general case we will call the inverse version of the same parameter the modulational index (M_I):

$$M_I = \frac{a_0 k_0}{\delta\omega/\omega_0}. \qquad (3.14)$$

3.1.4.4. *One-dimensional wave trains with full spectrum*

One-dimensional wave trains within a continuous wave spectrum have academic rather than practical interest. The wind-forced waves within a continuous spectrum are always directional. A discussion of such a one-dimensional full-spectrum academic case, however, is helpful as an intermediate step in the argument leading from the quasi-monochromatic one-dimensional wave trains to two-dimensional surfaces within the full frequency-directional wave spectrum.

It should be stressed that there is no strict analytical theory for instability of waves within a continuous spectrum. Credit has to be given to Onorato *et al.* [2001] who introduced an analogue of the index (3.14) for continuous wave spectra, where some characteristic mean steepness was used instead of ε and a relative width of the wave-spectrum peak instead of $\delta\omega/\omega_0$. Janssen [2003] conducted a detailed study of instability of such wave spectra by means of the Zakharov equation and concluded that $M_I \sim 1$ signals a transition from stable ($M_I \leq 1$) to unstable ($M_I \geq 1$) wave fields.

Ribal et al. [2013] investigated evolution of nonlinear wave trains within a full spectrum, both unidirectional and directional, by means of the Alber equation. Overall, their conclusions agree with the argument of Onorato et al. [2001] and Janssen [2003], but they further pointed out that, unlike monochromatic wave trains, steepness and bandwidth for spectral waves are not unambiguous and are not independent properties. For the most frequently used parameterization of wind–wave spectra, the JONSWAP spectrum, the same steepness can be achieved either by varying the equilibrium level of the spectrum α or the peak enhancement factor γ.

The two parameters, however, imply different physics with respect to the instability. If γ is fixed and the steepness is increased by raising the total spectrum energy level α, such a variation makes the spectrum broader and can slow down or even prevent the instability. If, on the contrary, α is fixed and γ is raised to increase the steepness, the effect is the opposite because the spectral peak becomes narrower which is what matters for instigating the instability. Thus, based on their numerical simulations, Ribal et al. [2013] introduced a new dimensionless property which describes transition to the conditions of instability in the case of JONSWAP spectrum:

$$\Pi_1 = \frac{\varepsilon}{\alpha\gamma} < 1. \qquad (3.15)$$

Gramstad [2017] verified and further refined this approach, and suggested more accurate values for this spectral criterion:

$$\Pi_1 < 1.3. \qquad (3.16)$$

Before applying this equation, note that there may be a factor of $\sqrt{2}$ difference between various definitions for the wave steepness mentioned throughout.

3.1.4.5. *Instability of directional wave fields*

Instability of directional wave fields is of key importance for the topic of breaking waves. Firstly, real ocean wind-forced waves are always directional. Second, it was recognized very early that the modulational instability is impaired or even suppressed in directional wave fields [McLean, 1982]. Therefore, it is still an open question, whether real ocean wave fields are too broad or are still narrow enough for the instability to be present in such wave fields.

Qualitatively, this question has been the subject of attention for some time. Quantitatively, Babanin *et al.* [2010] suggested a directional analogue of M_I, the Directional Modulational Index

$$M_{Id} = A \cdot ak, \tag{3.17}$$

which is effectively a ratio of steepness to normalized directional bandwidth A^{-1} [Babanin and Soloviev, 1998]:

$$A(f)^{-1} = \int_{-\pi}^{\pi} K(f, \varphi) d\varphi, \tag{3.18}$$

where φ is the wave direction, $K(f, \varphi)$ is the normalized directional spectrum

$$K(f, \varphi_{\max}) = 1. \tag{3.19}$$

Here, higher values of A correspond to narrower directional distributions. That is M_{Id} would signify whether the modulational instability takes place or not in the directional wave fields: if the directional spreading broadens, this can be compensated by increasing the characteristic steepness, and vice versa.

Babanin *et al.* [2011b] conducted a laboratory experiment in a directional wave tank in order to answer the quantitative question of whether the modulational instability is still active in directional wave trains with typical angular spreads and typical values of mean steepness comparable to those of the ocean waves. In the frequency/wave number space, the wave trains were quasi-monochromatic, but their directional spectrum varies. They found the modulational instability limit as

$$A^* ak \approx 0.18. \qquad (3.20)$$

That is, for $M_{Id} \geq 0.18$, directional wave trains are unstable. Such a limit is quite feasible [Babanin and Soloviev, 1998]. For dominant waves in realistic directional windseas, $A = 0.8$–1.8. Therefore, with $A \sim 1$ there should be $ak \sim 0.2$ which is possible, and for $A \sim 1.8$, there should be $ak \sim 0.11$ which is a typical steepness of ocean waves (3.2).

Ribal *et al.* [2013], as for the one-dimensional spectrum as described in Section 3.1.4.4, argued that for directional waves within a continuous spectrum the criterion for transition from stable to unstable seas has to account for the spectral shape rather than mean steepness alone. For a JONSWAP spectrum, their criterion is

$$\Pi_2 = \frac{\varepsilon}{\alpha \gamma} + \frac{\beta}{\varepsilon A} < 1.1, \qquad (3.21)$$

where $\beta = 0.00256$ is an empirical parameter.

Overall, the problem of directionality of ocean waves is much more complicated than outlined here. Apart from further issues related to the directional spreading of wave energy, it also connects to the three-dimensional structure of wave crests [Pinho and Babanin, 2014] and the presence of quasi-one-dimensional wave groups on the two-dimensional wave surfaces [Nieto-Borge *et al.*, 2013; Sanina *et al.*, 2016]. These issues are beyond the scope of this chapter.

3.1.5. *Threshold steepness and modulational instability*

The previous section describes possible physical reasons which can lead to a wave so steep that the water surface becomes unstable and the wave breaks. In the ocean, while superposition definitely occurs, there are indications that usually the breaking happens due to modulational instability. One such indications is the existence of a threshold mean steepness. If a wave train or wave field has mean steepness below such a threshold, the waves do not break. Typical examples of such waves are ocean swells.

If the breaking was due to superposition of waves, there should be no such threshold. As the mean wave steepness becomes lower, a superposition of a greater number of waves would be required to reach the limiting steepness (3.3) or (3.5), but the breaking probability would only become zero when there were no waves.

This is not so in the case of breaking due to modulational instability. As we have seen in Section 3.1.3.3, the breaking occurrence is zero for wave trains with steepness below some finite threshold. The instability is a nonlinear effect, but it takes its course even for infinitesimal waves. The initial wave trains become modulated, higher waves appear within the groups, but their steepness remain small by comparison with (3.3) and they do not break. As the mean steepness grows higher, the ratio of the maximal individual-wave height in the nonlinear wave train H_m, to its significant wave height will keep increasing as a function of significant wave height H_s, until the waves with H_m start breaking. A further increase of H_s will not be accompanied by subsequent growth of H_m, but rather by progressively higher rates of breaking occurrence.

This is indeed the behavior observed in the ocean [Banner et al., 2000; Babanin et al., 2001, 2007a]. Babanin et al. [2001], based on data for breaking of dominant waves obtained in the full range of possible wave conditions, from a small lake to the Southern Ocean and including finite-depth environments, proposed the following dependence for the breaking probability b_T:

$$b_T = 85.1 \cdot ((\varepsilon_p - 0.055) \cdot (1 + H_s/d))^{2.33}. \qquad (3.22)$$

Here, d is the water depth, and the last term approaches unity in the deep water. The peak steepness ε_p is a statistical property rather than the steepness of individual waves: this is the mean steepness of waves in the frequency range $\pm 0.3 f_p$ around the spectral peak f_p.

Thus, ocean waves do not break if $\varepsilon_p < 0.055$, and this fact is consistent with modulational-instability dynamics for such breaking and inconsistent with wave-superposition theory. Therefore, this observation alone is sufficient to infer that breaking of the dominant waves in the wave field is due to modulational instability. In turn, it also answers positively the decades-long question of whether the modulational instability is active in directional oceanic wave fields.

A number of other direct and indirect results in the dynamics and kinematics of ocean breakers also support such conclusion. The upshifting of the spectral energy prior to breaking (Fig. 3.5 (bottom)) is confirmed in field observations of wave breaking by means of wavelet analysis of the wave-breaking frequency [Liu and Babanin, 2004]. Oscillation of skewness and asymmetry, which is a feature of the nonlinear instability of wave trains (Section 3.1.3) is an observed behavior of ocean waves [Agnon et al., 2005].

A convincing argument in favor of instability-caused breaking in the field, also comes from the analysis of the dissipation process due to breaking. For example, Meza et al. [2000], who studied the dissipation of energy of laboratory two-dimensional waves with a narrow spectrum, found that the energy is lost almost entirely from the higher frequencies, whereas the spectral peak remained unchanged after breaking. In Meza et al. [2000], breaking was simulated by means of coalescing wave packets, i.e., superposition formed a high wave which broke. Babanin et al. [2010] analyzed the dissipation in nonlinear wave trains where the breaking was caused by modulational instability. The waves were quasi-monochromatic, and in the range of relevant frequencies up to 11 Hz, the average ratio of energy density before and after the breaking was 1.8, which translates into a loss of 45% of the energy. In both absolute and relative terms, most of the loss came from the peak which was reduced by a factor of 5. For frequencies above the spectral peak, the average ratio was approximately 1.7. Under field conditions, Young and Babanin [2006] observed that when the dominant waves break they lose some 30% of their energy and a similar amount of energy is also lost proportionally across the spectrum, the so-called cumulative effect. By obvious analogy, based on such dynamic consequences of waves breaking, this observation again suggests the nonlinear evolution mechanism, rather than wave superposition, as a likely cause of wave breaking in the field. We further refer the reader to Babanin et al. [2010], Babanin [2011] for more detailed discussions.

3.1.6. *Cumulative effect*

It should be emphasized that parameterization (3.23), and more generally the argument for wave breaking being either due to wave superposition or due to instability of nonlinear wave groups, is not directly applicable to waves at wavelengths much shorter than the spectral peak. While the argument is not incorrect, other processes may lead to breaking of such waves. Note that the notion of short waves here is relative, i.e., it should be short by comparison with the dominant wavelength in a wave field.

Short waves are affected by the longer waves in a number of different ways and this influence often leads them to break. For example, longer waves modulate trains of superposed shorter waves, the latter become steeper closer to the crests of long waves. If so, they will break more often in the presence of the long waves. Thus, breaking of waves with frequency $2f_p$, apart from the inherent superposition and instability reasons, will be influenced by waves with frequencies $f < 2f_p$, waves at $3f_p$ by a broader range of scales $f < 3f_p$ and so on. The further the wave scale from the peak, the higher the breaking rates that will be experienced as a result of the integral influence of all the progressively larger scales. In the end, breaking rates are determined by this cumulative integral rather than by the mean steepness of waves at this particular scale [Young and Babanin, 2006].

3.1.7. *Breaking severity*

Understanding breaking onset and the nature of the imminent breaking phase is essential for predicting and estimating breaking probability, whereas the breaking-in-progress phases, i.e., the developing breaking and the subsiding breaking phases allow us to estimate the breaking severity.

Combining the knowledge of breaking rates and breaking severity should lead to experimental descriptions of wave energy dissipation. This is a central task, for example, in wave forecasting, but with some exceptions this work is still to be done. Understanding breaking severity is also important for ocean engineering and coastal engineering applications.

Breaking severity, that is how much energy is lost in a breaking event, is defined by the process of the wave collapse. The dynamics of the collapse seems chaotic. In fact, the breaking waves appears to "remember" how it came to the point of collapse. For example, wave breaking due to linear focusing releases a certain fraction of the initial wave energy [Rapp and Melville, 1990], whereas wave breaking as a result of modulational instability can lose from 0% to 100% of wave energy, i.e., the severity can range from a mere toppling the tip of the crest all the way to the breaking wave disappearing completely. This also implies downshifting of wave energy as the nonlinear wave group which keeps pushing forward now consists of fewer carrier waves. While there have been attempts to parameterize the breaking severity in terms of properties of the modulated wave trains [e.g., Galchenko et al., 2010, 2012], the topic of breaking severity needs further attention of the research and ocean engineering communities.

3.1.8. *Forcing and wave breaking: Wind, currents, damping*

Ocean waves of practical interest are always subject to wind forcing. The wind makes the waves grow, hence increasing their steepness and breaking probability (which does not necessarily mean growth of the breaking severity and dissipation). Overall, however, unless the wind forcing is very strong, the effect of the wind forcing is indirect (through steepness and modulational instability) and direct correlation of the breaking probability with wind speed is marginal [Babanin et al., 2001, 2010].

While at certain conditions the wind can enhance the instability growth [Waseda and Tulin, 1999], generally the effect of the wind on waves appears to be that of a limiter. Strong forcing can blow the wave crests away and reduce the limiting wave-breaking steepness [e.g., Babanin et al., 2010]. Even weak forcing, however, can change the rate of instability growth, and the parameter range for the instability existence, such as the natural selection of the modulational frequency [e.g., Waseda and Tulin, 1999]. The shear currents and surface currents with velocity gradients can also impose forcing on the surface waves [e.g., Babanin et al., 2017].

Onorato and Proment [2012], by means of mapping of forced solutions (in both growing and damping conditions) of the Nonlinear Schrödinger Equation concluded that "an initially stable (unstable) wave packet could be destabilized (stabilized) by the wind (dissipation)". In experimental studies, Babanin et al. [2010] demonstrated that wind forcing affects both the frequency of occurrence of wave breaking and its severity, and slows down the instability growth. Galchenko et al. [2012] tested the effect of the wind forcing on the severity of wave breaking caused by the modulational instability, in laboratory experiments. Reduction of the breaking severity due to strong wind forcing was very significant.

At wind speeds in excess of $U_{10} \sim 33 \, \mathrm{ms}^{-1}$, changes of the physical regimes in all environments near the air–sea interface take place: in the atmospheric boundary layer, at the ocean interface and through the upper ocean. At the surface, the most apparent feature is the change of the physics of wave breaking. The wind forcing is so strong that the waves grow to the limiting steepness before the modulational instability or wave superposition can take their course. We refer the reader to Babanin [2018] for more detailed review of this issue.

3.2. Wave-Induced Mixing

As mentioned in the introduction to this chapter, wave orbital motion can produce turbulence unrelated to wave breaking. Since the orbital motion is distributed vertically at a scale comparable with the depth of the mixed layer, such turbulence then does not need vertical diffusion, convection or advection in order to facilitate the mixing through the thermo- or pycnocline well below the interface.

Note that in this section, we do not consider the Langmuir circulation (see the Introduction, and also discussion below), which can circulate the water at the vertical scale of the mixed layer, but does not necessarily produce vertically-distributed turbulence at this scale. This section is dedicated to the turbulence directly produced by the wave orbital motion (which only moves the water at the scale of wave orbits which is meters near the surface, and decays exponentially away from the surface). It should be mentioned that, once the surface waves are generated, such orbital motion is immediately (with the speed of sound) distributed through the water column at the wavelength scale and hence wave-induced turbulence is immediately present throughout (see Ghantous and Babanin [2014] for a recent review).

3.2.1. Shear-produced wave turbulence

There are three different physical mechanisms for wave-induced turbulence unrelated to wave breaking. The first is due to viscous shear [Phillips, 1961; Kinsman, 1965], which can generate turbulence if there is a wave-based Reynolds number greater than some critical value [Babanin, 2006].

Historically, works on this topic have persisted for almost a century [Jeffreys, 1920; Dobroklonskiy, 1947; Bowden, 1950; Longuet-Higgins, 1953; Cao, 1962; Ma, 1964, among others]. However, the success of potential flow theories in the 1960s rendered them forgotten. These new theories were very successful in discovering and describing a variety of new features and dynamics of nonlinear surface waves, for which they did not require viscosity and hence could not produce vorticity (turbulence). They did not disprove the viscous theories of water waves, but made them look unnecessary (see Babanin et al. [2012] for a discussion).

At the border between the two eras, however, Longuet-Higgins [1953] already demonstrated that the inclusion of viscosity in the equations of motion can for some problems radically alter the resulting calculations. It could not be taken as a given, therefore, that potential theory would always give useful results. Phillips [1961] enquired as to whether turbulence could appreciably alter the results of potential wave theory with regard to ocean swell. The abstract to his paper titled "A Note on Turbulence Generated by Gravity Waves" reads: "When gravity waves move across the surface of a liquid of small viscosity, an irregular vorticity (turbulence) field is generated which can attain a statistically steady state through the balance of vorticity generation by the straining associated with the waves and of viscous diffusion. The influence of this vorticity field on the wave motion is examined. It is found that in the equations describing the fluctuating properties of the wave motion, the influence is usually of the third order in the wave slope ..."

While clearly describing the theoretical mechanism for generation of turbulence by non-potential gravity waves, Phillips was disappointed with its weakness: such turbulence could not affect noticeably the dynamical predictions of potential theory, and what is important in the context of the present chapter — the turbulence produced in this way would be continuously damped by the very viscous forces which initiated its production. Kinsman in the 1965 textbook (Chapter 10, Perturbations of Irrotational Motion) offered an extended presentation of Phillips' ideas, explicitly showed derivations for infinitesimally small two-dimensional waves on the free surface of deep water with non-zero viscosity, and vorticity. He did not share Phillips' pessimism though, clearly articulating the limitations of the two-dimensional analytical approach and stated that in three dimensions the problem can be different. This foresight was very true (see the discussion of the Benilov [2012] theory of instability of three-dimensional vorticity and Section 3.2.2).

Here, we will also point out another important limitation of the early rotational theories of the linear waves. For infinitesimally small waves, the only length scale is wavelength, and a wave Reynolds number based on such scales is unrealistically large and, even more importantly, refers to the scale which does not encompass the actual motion of the water. Babanin [2006] suggested that the correct length scale for the wave-induced turbulence is wave orbit radius (wave amplitude a at the surface):

$$\text{Re} = \frac{av}{\nu_w} = \frac{a^2 \omega}{\nu_w}, \tag{3.23}$$

where $v = \omega a$ is orbital velocity, and ν_w is kinematic viscosity of the water. The orbits decay exponentially away from the surface and hence the depth-dependent wave Reynolds number (WRN) is

$$\text{Re} = \frac{\omega}{\nu_w} a^2 \exp(-2kz) = \frac{\omega}{\nu_w} a^2 \exp\left(-2\frac{\omega^2}{g}z\right). \tag{3.24}$$

If at the surface, this number is above the (unknown) critical value, the waves will produce turbulence, but at some depth below the surface, the number will drop below the critical value and this production will cease and the wave orbital motion should become laminar. For the spring-summer period, this should define the mixed layer depth (MLD) and vice versa (MLD can help to quantify the critical WRN). Based on measurements of the depth of the mixed layer in the Black Sea, Babanin [2006] estimated the critical WRN as $Re_{cr} \approx 3000$. This number was tested on mechanically-generated laboratory waves and was confirmed. Once the number was applied to ocean conditions, when mixing due to heating and cooling is less important than that due to the waves, quantitative and qualitative characteristics of the ocean's MLD were shown to be predicted with a good degree of agreement with observations.

A number of laboratory experiments followed, where the production of turbulence by surface non-breaking waves was simulated in undisturbed (non-turbulent) water. Dai et al. [2010] confirmed $Re_{cr} \sim 1000$, and Babanin and Haus [2009] for intermittent turbulence (i.e., below the critical Reynolds number) found the volumetric dissipation (production) rate ε_k for the wave-induced turbulence to be a cubic function of wave amplitude:

$$\varepsilon_k = 300 \cdot a^3. \tag{3.25}$$

The coefficient in (3.25) is dimensional because the measurements of Babanin and Haus were conducted for a single wave period (length) only, and the proper dimensional formulation for the turbulence dissipation/production was offered by Babanin [2011, 2012] later:

$$\varepsilon_k = b_1 k \omega^3 a^3 = b_1 k v^3 \tag{3.26}$$

with the empirical coefficient $b_1 \approx 0.0014$ [Young et al., 2013; Babanin, 2017]. Effectively, expression (3.26) confirms the dependence proposed earlier by Bowden [1950].

3.2.2. Interaction of waves with pre-existing turbulence

Thus, orbital motion of the surface gravity waves can and do produce turbulence, and this has been perceived and known for long time. But this says nothing about how a potential flow may respond to a pre-defined vorticity. As long as vorticity is imposed on a flow, regardless of viscosity, any shearing motion, even an otherwise irrotational one, can stretch the vortex lines. This gives a mechanism for generation of turbulence via the amplification of vorticity.

Such a mechanism was proposed by Benilov *et al.* [1993] and Benilov [2012]. It appears general enough as it does not involve any further assumptions if the waves (even linear waves) and three-dimensional vorticity are present. Since the ocean is usually turbulent, conditions for such dynamics are satisfied regardless of the process which produced the turbulence initially: the above-discussed shear wave turbulence, wind drift or another shear current, internal waves, convection, etc.

Benilov *et al.* [1993] considered a linear problem of stability of vortex disturbances in a velocity field of potential linear surface waves. Evolution of such initial disturbances is reduced to a coupled set of linear ordinary differential equations of the first order with periodic coefficients. The solution of this problem shows that initial vortex perturbations (even if small) of potential linear surface waves always grow.

We should note here that, while the wave orbits can be two-dimensional, the vorticity of the Benilov theory must be three-dimensional because the instability planes are perpendicular to the planes of the orbits. Although Phillips [1961] and Kinsman [1965] (see Section 3.2.1) could not have known this, because they were solving a two-dimensional coupled problem both for the waves and for the vorticity, Benilov's vortex instability answers their concern. Indeed, if the shear is able to generate turbulent vortexes, which outside of fine boundary layers are always three-dimensional, these vortexes will not be damped by viscosity like in their two-dimensional academic problem, but will grow in orthogonal directions on behalf of the energy of gravity waves.

Yet, the shear production of wave turbulence and the Benilov vortex instability are two distinctly different physical mechanisms: the former requires rotational waves and in the latter the wave motion can be treated as potential. Both utilize the wave energy for turbulence growth, and hence the important question is: can they be distinguished, for example, by the rate of this growth? For example, in the absence of water turbulence, surface waves would trigger the vorticity through the first mechanism and then amplify it through the second (presumably faster) mechanism. Perhaps, a reasonable analogy in this regard is the Phillips [1957] and Miles [1957] wave-growth theories: both rely on energy flux from the wind, but the latter cannot become active over an initially flat water surface until the former produces some waves, even if infinitesimally small. Growth rates of the two are also very different: Phillips' growth is linear and Miles' is exponential and therefore takes over and dominates once activated.

Figure 3.8 shows such a comparison: volumetric turbulence-dissipation rates vs. wave amplitude (for 1.5 Hz/0.7m-long waves) for the two wave-turbulence mechanisms. The shaded area corresponds to the range of the laboratory experiment by Babanin and Haus (2009) — dependence (3.25) with confidence limits. This experiment was done with no pre-existing turbulence, and hence it was created by the shear due to molecular viscosity (the first mechanism).

The dots are maximal values of ε_k, for a given amplitude a, obtained by means of the wave-turbulence coupled model of Babanin and Chalikov [2012]. In this model, nonlinear two-dimensional waves were coupled with the LES model of three-dimensional turbulence. The wave model was potential and could not produce the turbulence by itself, but once the turbulence was seeded (pre-existent) it grew. Note that the Babanin–Chalikov model does not rely on the Benilov theory to operate, it was solving basic fluid mechanics equations for variables which are a sum of the 2D potential orbital velocities and 3D non-potential disturbances, but effectively confirmed the second mechanism (interactions of waves with pre-existing turbulence).

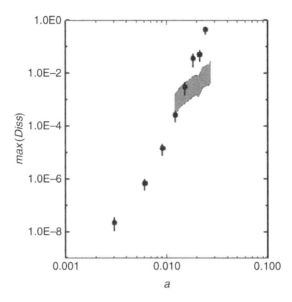

Fig. 3.8. Maximal volumetric dissipation rates max(Diss) versus wave amplitude a. The shaded area corresponds to the range and confidence limits of the laboratory experiment by Babanin and Haus [2009], obtained below the troughs. Dots are obtained by means of the wave-turbulence coupled model of Babanin and Chalikov [2012], vertical segments indicate their standard deviation. Figure is reproduced from Babanin [2011].

We should point out that the two estimates of ε_k are not exactly the same: PIV slices of Babanin and Haus [2009] had to be taken in the water and hence were below the troughs, whereas the maximal readings of Babanin and Chalikov [2012] are within the wave crests. Given this difference, the agreement between the turbulence production by the two mechanisms is very good.

It is therefore, fair to suggest that the two mechanisms work together: the viscous shear produces turbulence and the vortex instability amplifies it. The same symbiosis was proposed by Benilov [2012], Fujiwara and Yoshikawa [2018], and the quantitative consistency of the two processes in Fig. 3.8 supports the conjecture. The model points on the left bottom can be those where the shear production is stronger (like the linear versus exponential growth in case of wave generation by wind), or they may be just unrealistic academic cases where actual millimeter-high waves would be affected by surface tension (which was switched off in the modeling exercise). The model points on the right top, grow above the shear turbulence, as would be expected for an instability mechanism. Experimental and theoretical support for such behavior also exists: the steepness-dependent coefficient b_1 in Babanin [2011, 2017], Zieger *et al.* [2015].

3.2.3. *Other mechanisms for wave-induced turbulence*

There have also been other mechanisms proposed for wave-induced turbulence. One of them is the so-called Stokes shear, i.e., shear stresses brought about by the Stokes drift, mean current due to wave orbital motion. As such, the concept of Stokes shear is not unphysical, but should be applied and treated with caution and understanding.

The Stokes drift is a second-order solution of the potential theory, and hence does not involve viscosity and as such cannot create shear stresses. If the turbulence is pre-existing, however, and the water particles randomly travel between vertical levels of the Stokes-current profile, they will exchange horizontal momentum between these levels which is the classical concept of turbulent shear stresses. Thus, like the Benilov mechanism above, Stokes drift will not produce turbulence, but will amplify it, if it is present. We refer to Ardhuin and Jenkins [2006] for a description of this phenomenon, who also concluded that its intensity is weak for practical applications. We should also mention that Stokes drift is a small phase-average residual of the large (order of magnitude larger) wave orbital velocities. Therefore, if the orbital motion is able to produce turbulence directly (e.g., the Benilov mechanism), it should be expected to be more efficient since both its velocities and corresponding velocity gradients are much greater.

Another mechanism, frequently referred to in the context of wave-induced turbulence is the Langmuir turbulence [McWilliams *et al.*, 1997; Teixeira and Belcher, 2002; among others]. We note that Langmuir turbulence is not the same phenomenon as the Langmuir circulation [Langmuir, 1938; Craik and Leibovich, 1976], and this difference has created some confusion. The empirical and theoretical basis of the Langmuir turbulence follows from the analogy with Langmuir circulation which masks the explicit physics behind this concept. It may be that Langmuir turbulence is a phase-average representation of the Benilov instability, particularly as both of them predict the growth of vorticity filaments in the direction of wave propagation. Here is what Benilov [2012] wrote about his mechanism in the conclusions to his paper: "The analysis shows that the initial small disturbances of the vorticity field grow in time under the influences of the wave motions. The vortex disturbances are analogous to Langmuir circulation, and they have the growing in time the mean component in the direction of wave propagation." This interesting issue, the potential link between the explicit Benilov vortex-instability physical mechanism for turbulence amplification, and the phase average concept of Langmuir turbulence, remains to be clarified.

3.3. Summary

This chapter discussed two main topics: wave breaking and ocean mixing (due to wave-induced turbulence because of the wave orbital motion on the water side of the interface). These are different phenomena, and what brings them together in this book about ocean waves, is that from the point of view of the waves, both of them are dissipative mechanism in the wave-energy balance.

The wave-breaking Section 3.1 is a review of the present understanding of the wave-breaking phenomenon whose importance extends far beyond the topics of wave dynamics and Metocean Engineering: into air–sea interactions at all scales from turbulence to climate. We discuss the concept and causes of the breaking onset, dynamics of wave breaking — starting from academic cases of one-dimensional quasi-monochromatic waves trains, through wave trains with continuous wave spectra, and all the way to realistic two-dimensional wavy surfaces. It is argued that in typical open-ocean deep-water conditions there are two mechanisms which can lead to the breaking onset — linear (or nonlinear) superposition and modulational instability of the wave trains/fields, and the latter appears the main reason behind observed ocean wave breaking. Wind forcing affects, but does not change these dynamics. Important features of breaking, including its severity (which then defines the wave-energy dissipation) and the cumulative effect of small-scale breaking in the wave spectrum, are also discussed.

The ocean mixing due to wave-induced turbulence has captured the attention of the wave- and ocean-modeling communities only relatively recently, over the last 10 years or so, but historically this topic is not young and has been discussed for about a century in the literature. There are three different mechanisms proposed for the generation of such turbulence, and as it is often the case in Metocean application, none of them cancels the other two — all are feasible, so it is effectively a matter of their relative significance.

Historically, the first one was introduced by Phillips [1961]: viscous solutions of wave equations produce vorticity which, being stretched by random waves, becomes turbulence. The second mechanism was originally proposed by Benilov et al. [1993] and published in a mainstream journal by Benilov [2012]. This is turbulence generated by potential (non-viscous) waves, hence the turbulence must be pre-existing. The Benilov theory is an instability theory for three-dimensional turbulence subject to forcing by two-dimensional wave orbital motion. In this chapter, it is argued that the popular Langmuir turbulence [McWilliams et al., 1997], which requires Stokes drift and does not deal directly with turbulence vorticity and its instability, can be a phase-average representation of the Benilov mechanism. The third mechanism is the "Stokes shear" — i.e., the process when the Stokes drift shear plays a role of mean flow shear [Ardhuin and Jenkins, 2006].

All the theories predict enhancement of turbulence once the wind/waves increase, albeit the third mechanisms appears to be weak. The Benilov mechanism was verified in laboratory, phase-resolving modeling of nonlinear waves coupled with three-dimensional turbulence, and is quantitatively consistent with remote-sensing observations of swell. Thus, this phenomenon is sufficient to describe the wave-induced mixing in the ocean. As such, it requires turbulence to be pre-existing, but if combined with the first theory is self-sufficient, the

Phillips mechanism can continuously produce some background levels of turbulence, which is then amplified by the Benilov instability.

Finally, it should be mentioned that wave breaking provides extensive levels of turbulence, but it is injected down at the scale of wave height, whereas the turbulence produced by the wave orbital motions is distributed vertically at the scale of the wavelength (order of magnitude larger). Hence, wave-breaking turbulence would not contribute to the ocean mixing under regular deep-water circumstances. Thus, the wave-produced turbulence both links and separates the two phenomena described in this chapter.

References

Agnon, Y., Babanin, A.V., Young, I.R. and Chalikov, D. (2005). Fine scale inhomogeneity of wind-wave energy input, skewness and asymmetry, *Geophys. Res. Lett.*, **32**(12), L12603, doi:10.1029/2005GL022701.

Alberello, A., Chabchoub, A., Monty, J.P., Nelli, F., Lee, J.H., Elsnab, J. and Toffoli, A. (2018). An experimental comparison of velocities underneath focused breaking waves, *Ocean Eng.*, **155**, 201–210.

Ardhuin, F. and Jenkins, A.D. (2006). On the interaction of surface waves and upper ocean turbulence, *J. Phys. Oceanogr.*, **36**, 551–557.

Babanin, A.V. (2006). On a wave-induced turbulence and a wave-mixed upper ocean layer, *Geophys. Res. Lett.*, **33**, L20605, doi:10.1029/2006GL027308.

Babanin, A.V. (2011). *Breaking and Dissipation of Ocean Surface Waves*, Cambridge University Press.

Babanin, A.V. (2012). Swell attenuation due to wave-induced turbulence, in *Proc. ASME 2012 31st Int. Conf. Ocean, Offshore and Arctic Engineering OMAE2012*, July 1–6, 2012, Rio de Janeiro, Brazil.

Babanin, A.V. (2017). Similarity theory for turbulence, induced by orbital motion of surface water waves, *Procedia IUTAM*, **20**, 99–102.

Babanin, A.V. (2018). Change of regime of air-sea dynamics in extreme Metocean conditions, in *Proc. ASME 2018 37th Int. Conf. Ocean, Offshore and Arctic Engineering OMAE2018*, June 17–22, 2018, Madrid, Spain.

Babanin, A.V. and Soloviev, Yu.P. (1998). Variability of directional spectra of wind-generated waves, studied by means of wave staff arrays, *Mar. Freshwater Res.*, **49**(2), 89–101.

Babanin, A.V. and Haus, B.K. (2009). On the existence of water turbulence induced by non-breaking surface waves, *J. Phys. Oceanogr.*, **39**, 2675–2679.

Babanin, A.V. and Chalikov, D. (2012). Numerical investigation of turbulence generation in non-breaking potential waves, *J. Geophys. Res.: Oceans*, **117**, C00J17, doi:10.1029/2012JC007929.

Babanin, A.V., Young, I.R. and Banner, M.L. (2001). Breaking probabilities for dominant surface waves on water of finite constant depth, *J. Geophys. Res.: Oceans*, **106**(C6), 11659–11676.

Babanin, A.V., Young, I.R. and Mirfenderesk, H. (2005). Field and laboratory measurements of wave–bottom interaction, in *Proc. 17th Australasian Coastal and Ocean Engineering Conf. and 10th Australasian Port and Harbour Conf.*, 20–23 September 2005, Adelaide, South Australia, eds. M. Townsend, D. Walker, pp. 293–298.

Babanin, A.V., Chalikov, D., Young, I.R. and Savelyev, I. (2007a). Predicting the breaking onset of surface water waves, *Geophys. Res. Lett.*, **34**, L07605, doi:10.1029/2006GL029135.

Babanin, A.V., Banner, M.L., Young, I.R. and Donelan, M.A. (2007b). Wave-follower measurements of the wind-input spectral function. Part III. Parameterization of the wind-input enhancement due to wave breaking, *J. Phys. Oceanogr.*, **37**(11), 2764–2775, doi:10.1175/JPO3147.1.

Babanin, A.V., Chalikov, D., Young, I.R. and Savelyev, I. (2010). Numerical and laboratory investigation of breaking of steep two-dimensional waves in deep water, *J. Fluid Mech.*, **644**, 433–463.

Babanin, A.V., Waseda, T., Shugan, I. and Hwung, H.-H. (2011a). Modulational instability in directional wave fields and extreme wave events, in *Proc. ASME 2011 30th Int. Conf. Ocean, Offshore and Arctic Engineering OMAE2011*, July 19–24, 2011, Rotterdam, The Netherlands.

Babanin, A.V., Waseda, T., Kinoshita, T. and Toffoli, A. (2011b). Wave breaking in directional fields, *J. Phys. Oceanogr.*, **41**, 145–156.

Babanin, A.V., Hsu, T.-W., Roland, A., Ou, S.-H., Doong, D.-J. and Kao, C.C. (2011c). Spectral modelling of Typhoon Krosa, *Nat. Hazards Earth Syst. Sci.*, **11**, 501–511.

Babanin, A.V., Onorato, M. and Qiao, F. (2012). Surface waves and wave-coupled effects in lower atmosphere and upper ocean, *J. Geophys. Res.: Oceans*, **117**(C11), C00J01, doi:10.1029/2012JC007932.

Banner, M.L., Babanin, A.V. and Young, I.R. (2000). Breaking probability for dominant waves on the sea surface, *J. Phys. Oceanogr.*, **30**(12), 3145–3160.

Benjamin, T.B. and Feir, J.E. (1967). The disintegration of wave trains in deep water. Part 1. Theory, *J. Fluid Mech.*, **27**, 417–430.

Benilov, A.T. (2012). On the turbulence generated by the potential surface waves, *J. Geophys. Res.: Oceans*, **117**, C00J30, doi:10.1029/2012JC007948.

Benilov, A., McKee, T.G. and Safray, A.S. (1993). On the vortex instability of the linear surface wave, in *Numerical Methods in Laminar and Turbulent Flow*, ed. C. Taylor, Pineridge, pp. 1323–1334.

Bortkovskii, R.S. (2003). Gas transfer across the ocean surface under a strong wind and its contribution to the mean gas exchange, *Izvestiya Akad. Nauk. Fizika Atmosferi I Okeana*, **39**, 809–816.

Bowden, K.F. (1950). The effect of eddy viscosity on ocean waves, *Lond. Edinb. Dubl. Phil. Mag. J. Sci*, **41**(320), 907–917.

Brown, M.G. and Jensen, A. (2001). Experiments in focusing unidirectional water waves, *J. Geophys. Res: Oceans*, **106**(C8), 16917–16928.

Cao, Z. (1962). Theory of wave-induced turbulence, *Trans. Tianjin University*, 5.

Caulliez, G. (2002). Self-similarity of near-breaking short gravity wind waves, *Phys. Fluids*, **14**, 2917–2920.

Chalikov, D. and Sheinin, D. (2005). Modeling extreme waves based on equations of potential flow with a free surface, *J. Comput. Phys.*, **210**, 247–273.

Chalikov, D. and Babanin, A.V. (2012). Simulation of breaking in spectral environment, *J. Phys. Oceanogr.*, **42**, 1745–1761.

Craik, A.D.D. and Leibovich, S. (1976). A rational model for Langmuir circulations, *J. Fluid Mech.*, **73**, 401–426.

Dai, D., Qiao, F., Sulisz, W., Han, L. and Babanin, A.V. (2010). An experiment on the non-breaking surface-wave-induced vertical mixing, *J. Phys. Oceanogr.*, **40**, 2180–2188.

Dobroklonskiy, S.V. (1947). Eddy viscosity in the surface layer of the ocean and waves, *Dokl. Akad. Nauk SSSR*, **58**(7), 1345–1348.

Donelan, M.A. (1998). Air–water exchange processes, in *Physical Processes in Lakes and Oceans*, Vol. 54, Coastal and Estuarine Studies, ed. J. Imberger, AGU, pp. 19–36.

Fedele, F., Benetazzo, A., Gallego, G., Shih, P.-C., Yezzi, A., Barbariol, F. and Ardhuin, F. (2013). Space–time measurements of oceanic sea states, *Ocean Model.*, **70**, 103–115, doi:10.1016/j.ocemod.2013.01.001.

Galchenko, A., Babanin, A.V., Chalikov, D., Young, I.R. and Hsu, T.-W. (2010). Modulational instabilities and breaking strength for deep-water wave groups, *J. Phys. Oceanogr.*, **40**, 2313–2324.

Gemmrich, J.R. and Farmer, D.M. (1999). Observations of the scale and occurrence of breaking surface waves, *J. Phys. Oceanogr.*, **29**, 2595–2606.

Gemmrich, J.R. and Farmer, D.M. (2004). Near-surface turbulence in the presence of breaking waves, *J. Phys. Oceanogr.*, **34**, 1067–1086.

Ghantous, M. and Babanin, A.V. (2014). Ocean mixing by wave orbital motion, *Acta Phys. Slovaca*, **64**, 1–56, http://www.physics.sk/aps/pub.php?y=2014\&pub=aps-14-01.

Gill, A.E. (1982). *Atmosphere–Ocean Dynamics*, Elsevier.

Gramstad, O. (2017). Modulational instability in JONSWAP sea states using the Alber equation, in *Proc. ASME 2017 36th Int. Conf. Ocean, Offshore and Arctic Engineering OMAE2017*, June 25–30, 2017, Trondheim, Norway.

Guan, C., Hu, W., Sun, J. and Li, R. (2007). The whitecap coverage model from breaking dissipation parameterizations of wind waves, *J. Geophys. Res.: Oceans*, **112**(C5), C05031, doi:10.1029/2006JC003714.

Holthuijsen, L.H. and Herbers, T.H.C. (1986). Statistics of breaking waves observed as whitecaps in the open sea, *J. Phys. Oceanogr.*, **16**, 290–297.

Hwung, H.-H., Huang, L.-H., Lee, J.-F. and Tsai, C.-P. (2005). The characteristics of nonlinear wave transformation on sloping bottoms, *Overall Performance Report*, 91-E-FA09-7-3, National Cheng Kung University.

Iafrati, A., Babanin, A.V. and Onorato, M. (2014). Modelling of ocean-atmosphere interaction phenomena during the breaking of modulated wave trains, *J. Comput. Phys.*, **271**, 151–171.

Janssen, P.A.E.M. (2003). Nonlinear four-wave interaction and freak waves, *J. Phys. Oceanogr.*, **33**, 863 884.

Jeffreys, H. (1920). On turbulence in the ocean, *Lond. Edinb. Dubl. Phil. Mag. J. Sci.*, **39**(233), 578–586.

Jessup, A.T., Zappa, C.J. and Yeh, H. (1997). Defining and quantifying microscale wave breaking with infrared imagery, *J. Geophys. Res.: Oceans*, **102**(C10), 23145–23153.

Kinsman, B. (1965). *Wind Waves: Their Generation and Propagation on the Ocean Surface*. Prentice-Hall, Englewood Cliffs, NJ.

Kudryavtsev, V.N. and Makin, V.K. (2002). Coupled dynamics of short waves and the airflow over long surface waves, *J. Geophys. Res.: Oceans*, **107**(C12), 3209, doi:10.1029/2001JC001251.

Lamarre, E. and Melville, W.K. (1991). Void-fraction measurements and sound-speed fields in bubble flumes generated by breaking waves, *J. Acoust. Soc. Am.*, **95**, 1317–1328.

Langmuir, I. (1938). Surface motion of water induced by the wind, *Science*, **87**, 119–123.

Liu, P.C. and Babanin, A.V. (2004). Using wavelet spectrum analysis to resolve breaking events in the wind wave time series, *Ann. Geophys.*, **22**, 3335–3345.

Longuet-Higgins, M.S. (1953). Mass transport in water waves, *Philos. Trans. Royal Soc. Lond. Ser. A: Math. Phys. Sci.*, **245**(903), 535–581.

Longuet-Higgins, M.S. and Smith, N.D. (1983). Measurement of breaking waves by a surface jump meter, *J. Geophys. Res.: Oceans*, **88**(C14), 9823–9831.

Ma, D. (1965). Turbulence beneath waves and its parameterization, *Trans. Tianjin University*, **21**, 5.

Manasseh, R., Babanin, A.V., Forbes, C., Rickards, K., Bobevski, I. and Ooi, A. (2006). Passive acoustic determination of wave-breaking events and their severity across the spectrum, *J. Atmos. Ocean. Technol.*, **23**(4), 599–618.

McLean, J. (1982). Instabilities of finite-amplitude gravity waves, *J. Fluid Mech.*, **114**, 315–330.

McWilliams, J.C., Sullivan, P.P. and Moeng, C.-H. (1997). Langmuir turbulence in the ocean, *J. Fluid Mech.*, **334**, 1–30, https://doi.org/10.1017/S0022112096004375.

Melville, W.K. (1982). Instability and breaking of deep-water waves, *J. Fluid Mech.*, **115**, 165–185.

Melville, W.K., Loewen, M.R. and Lamarre, E. (1992). Sound production and air entrainment by breaking waves: A review of recent laboratory experiments, in *Breaking waves. IUTAM Symposium*, Sydney, Australia, 1991, Springer-Verlag, Berlin, pp. 139–146.

Meza, E., Zhang, J. and Seymour, R.J. (2000). Free-wave energy dissipation in experimental breaking waves, *J. Phys. Oceanogr.*, **30**, 2404–2418.

Michell, J.H. (1893). On the highest waves in water, *Lond. Edinb. Dubl. Phil. Mag. J. Sci.*, **36**(222), 430–437.

Miles, J.W. (1957). On the generation of surface waves by shear flows, *J. Fluid Mech.*, **3**, 185–204.

Munk, W.H. (1947). A critical wind speed for air–sea boundary processes, *J. Mar. Res.*, **6**, 203–218.

Nieto-Borge, J.C., Reichert, K. and Hessner, K. (2013). Detection of spatio-temporal wave grouping properties by using temporal sequences of X-band radar images of the sea surface, *Ocean Model.*, **61**, 21–37, doi:10.1016/j.ocemod.2012.10.004.

Onorato, M., Osborne, A.R., Serio, M. and Bertone, S. (2001). Freak wave in random oceanic sea states, *Phys. Rev. Lett.*, **86**, 5831–5834.

Phillips, O.M. (1957). On the generation of waves by turbulent wind, *J. Fluid Mech.*, **2**, 417–445.

Phillips, O.M. (1961). A note on the turbulence generated by gravity waves, *J. Geophys. Res.*, **66**, 2889–2893, https://doi.org/10.1029/JZ066i009p02889.

Phillips, O.M., Posner, F.L. and Hansen, J.P. (2001). High resolution radar measurements of the speed distribution of breaking events in wind-generated ocean waves: surface impulse and wave energy dissipation rates, *J. Phys. Oceanogr.*, **31**, 450–460.

Pinho, U.F. and Babanin, A.V. (2015). Emergence of short-crestedness in originally unidirectional nonlinear waves, *Geophysi. Res. Lett.*, **42**(10), 4110–4115, doi:10.1002/2015GL063973.

Qiao, F., Yeli, Y., Yang, Y., Zheng, Q., Xia, C. and Ma, J. (2004). Wave-induced mixing in the upper ocean: Distribution and application to a global ocean circulation model, *Geophys. Res. Lett.*, **31**, L11303, doi:10.1029/2004GL019824.

Rapp, R.J. and Melville, W.K. (1990). Laboratory measurements of deep-water breaking waves, *Philos. Trans. Royal Soc. Lond. Ser. A: Math. Phys. Sci.*, **311**(1622), 735–800.

Ribal, A., Babanin, A.V., Young, I.R., Toffoli, A. and Stiassnie, M. (2013). Recurrent solutions of the Alber equation initialized by JONSWAP spectra, *J. Fluid Mech.*, **719**, 314–344.

Rogers, W.E., Dykes, J.D. and Wittmann, P.A. (2014). US Navy global and regional wave modeling, *Oceanography*, **27**, 56–67, doi:10.5670/oceanog.2014.68.

Sanina, E.V., Suslov, S.A., Chalikov, D. and Babanin, A.V. (2016). Detection and analysis of coherent groups in three-dimensional fully-nonlinear potential wave fields, *Ocean Model.*, **103**, 73–86.

Smith, J. (1992). Observed growth of Langmuir circulation, *J. Geophys. Res.: Oceans*, **97**(C4), 5651–5664.

Smith, M.J., Poulter, E.M. and McGregor, J.A. (1996). Doppler radar measurements of wave groups and breaking waves, *J. Geophys. Res.: Oceans*, **101**(C6), 14269–14282.

Solli, D.R., Ropers, C., Koonath, P. and Jalali, B. (2007). Optical rogue waves, *Nature*, **450**, 1054–1057.
Soloviev, A. and Lukas, R. (2006). *The Near-Surface Layer of the Ocean: Structure Dynamics and Applications*, Springer.
Stokes, G.G. (1880). Considerations relative to the greatest height of oscillatory irrotational waves which can be propagated without change of form, in *On the Theory of Oscillatory Waves*, Cambridge University Press, pp. 225–229.
Sutherland, P. and Melville, W.K. (2015). Field measurements of surface and near-surface turbulence in the presence of breaking waves, *J. Phys. Oceanogr.*, **45**, 943–965.
Teixeira, M.A.C. and Belcher, S.E. (2002). On the distortion of turbulence by a progressive surface wave, *J. Fluid Mech.*, **458**, 229–267.
Terray, E.A., Donelan, M.A., Agrawal, Y.C., Drennan, W.M., Kahma, K.K., Williams III, A.J., Hwang, P.A. and Kitaigorodskii, S.A. (1996). Estimates of kinetic energy dissipation under breaking waves, *J. Phys. Oceanogr.*, **26**, 792–807.
Toffoli, A., Babanin, A.V., Waseda, T. and Onorato, M. (2010). Maximum steepness of oceanic waves, *Geophys. Res. Lett.*, **37**, L05603, doi:10.1029/2009GL 041771.
Toffoli, A., McConochie, J., Ghantous, M., Loffredo, L. and Babanin, A.V. (2012). The effect of turbulence induced by non-breaking waves on the ocean mixed layer: Field observations on the Australian North-West Shelf, *J. Geophys. Res.*, **117**, C00J24, doi:10.1029/2011JC007780.
Tulin, M.P. and Waseda, T. (1999). Laboratory observations of wave group evolution, including breaking effects, *J. Fluid Mech.*, **378**, 197–232,
Tulin, M.P. and Landrini, M. (2001). Breaking waves in the ocean and around ships, in *Proc. of the Twenty-Third Symposium of Naval Hydrodynamics*, pp. 713–745.
Waseda, T. and Tulin, M.P. (1999). Experimental study of the stability of deepwater wave trains including breaking effects, *J. Fluid Mech.*, **401**, 55–84,
Weissman, M.A., Atakturk, S.S. and Katsaros, K.B. (1984). Detection of breaking events in a wind-generated wave field, *J. Phys. Oceanogr.*, **14**, 1608–1619.
Fujiwara, Y., Yoshikawa, Y. and Matsumura, Y. (2018). A wave-resolving simulation of Langmuir circulations with a Nonhydrostatic free-surface model: Comparison with Craik–Leibovich theory and an alternative Eulerian view of the driving mechanism, *J. Phys. Oceanogr.*, **48**(8), 1691–1708.
Young, I.R. and Babanin, A.V. (2006). Spectral distribution of energy dissipation of wind-generated waves due to dominant wave breaking, *J. Phys. Oceanogr.*, **36**(3), 376–394.
Young, I.R., Babanin, A.V. and Zieger, S. (2013). The decay rate of ocean swell observed by altimeter, *J. Phys. Oceanogr.*, **43**, 2322–2333.
Yuen, H.C. and Lake, M. (1982). Nonlinear dynamics of deep-water gravity waves, *Adv. Appl. Mech.*, **22**, 67–229.

Zakharov, V.E. (1966). The instability of waves in nonlinear dispersive media, *Zh. Eksp. Teor. Fiz*, **51**, 1107–1114, (in Russian).

Zakharov, V.E. (1967). The instability of waves in nonlinear dispersive media, *Sov. Phys. JETP (Engl. Transl.)*, **24**, 740–744.

Zieger, S., Babanin, A.V., Rogers, W.E. and Young, I.R. (2015). Observation-based source terms in the third-generation wave model WAVEWATCH, *Ocean Model.*, **96**(1), 2–25.

Chapter 4
Nonlinear Processes

Takuji Waseda
University of Tokyo, Japan

4.1. Introduction

Ocean waves generated by wind, grow as they propagate down wind, and concurrently the wavelengths elongate. The empirical law that governs the evolution of the windsea is called the fetch law which was first derived by Sverdrup and Munk [1946]. The fetch law describes the evolution of the non-dimensional significant wave height and the significant wave period in both space and time, and serves as a constraint on the model that describes the evolution of the windsea. The evolution following the fetch law is equivalent to the downshifting of the spectral peak. Ocean wave spectra are known to be self-equivalent and are characterized by the existence of the equilibrium spectral tail [Phillips, 1957; Zakharov and Filonenko, 1966; Toba, 1972]. While retaining the high-frequency spectral tail, the spectral peak keeps downshifting as the energy is pumped into the wave due to wind.

The key to understand the fetch law and the equilibrium spectrum is the weak nonlinearity of the water waves. Phillips first suggested a possibility of the nonlinear interaction of a combination of four propagating surface gravity waves [Phillips, 1960]. Independently, the interaction among all the possible combinations of wave modes in a spectrum was derived by Hasselmann [1962]. The spectral downshifting is considered a result of the four-wave resonance among the spectral modes. The four-wave resonance is also responsible for spreading the energy into direction away from the main wind direction. As a result, the directional spectrum of windsea spreads widely in direction, which was first documented by Mitsuyasu et al. [1975].

Ocean waves are well represented by a random Gaussian process, and weak nonlinearity of waves introduces only a small deviation [Cartwright and Longuet-Higgins, 1956]. Nonetheless, the statistical descriptions of ocean waves have been revisited extensively in the last two decades to understand freak or rogue waves in the ocean. Numerous observational occurrences of freak waves have been reported, and theories based on second- and third-order nonlinearity were developed to explain the slight enhancement of the tail of the distribution. The notion of the significance of detuned resonance further developed into the modification of the Hasselman's theory. But after all these efforts, there seem to be no solid conclusion on the generation mechanism of freak waves.

The goal of this chapter is to provide a bird's-eye view of nonlinear processes related to ocean waves, and to highlight the outstanding questions that are yet to be answered. Important equations are introduced without mathematical derivations. Rather, the chapter is designed to give an overview of how they are related to each other and what we have learned about ocean waves from them. Appropriate references will be included for interested readers.

4.2. Resonant Interaction

4.2.1. *Fundamentals*

The aim of this subsection is to provide a physical interpretation of resonant interaction based on elementary mathematics rather than going through the details of the sophisticated mathematical derivations. Readers who are interested in learning about it, should refer to the textbooks by Phillips [1977], Kinsman [1984], Janssen [2004], Onorato et al. [2016] and other literature referenced therein. The linear free surface gravity wave satisfies the following wave equation, when extended to include a spatially and temporary changing atmospheric pressure field $P_\text{atm}(x,t)$,

$$-c_w^2 \zeta_{xx} + \zeta_{tt} = -\frac{P_\text{atm}}{\rho_w/k}, \qquad (4.1)$$

where infinite depth is assumed. The linear dispersion relation with $P_\text{atm} = 0$ defines $c_w^2 \equiv g/k$. The modified dispersion relation reads,

$$c^2 = c_w^2 + \frac{P_\text{atm}}{\rho_w k \zeta}. \qquad (4.2)$$

Known dispersion relations can be readily derived from Eq. (4.2). For example, $P_\text{atm} = \tau k^2 \zeta$ results in the dispersion relation of the capillary-gravity waves, i.e., $c^2 = \frac{1}{k}(g + \frac{\tau k^2}{\rho_w})$, where τ is the surface tension. The atmospheric pressure or the normal pressure acting on the free surface modifies the gravitational restoring force. Likewise, the dispersion relation of waves under a thin elastic ice sheet can be readily derived from $P_\text{ice} = Dk^4 \zeta$, where D is the bending stiffness of the ice sheet: $c^2 = \frac{1}{k}(g + \frac{Dk^4}{\rho_w})$. By considering a complex form of $P_\text{atm} = (\alpha + i\beta)\rho U^2 ak$, where U is the wind speed and a is the amplitude of the wave, the wave growth due to wind can be expressed in different forms.

When the nonlinear terms are retained in the derivation of a free surface propagating wave, the wave equation can be written in the same form,

$$-c_w^2 \zeta_{xx} + \zeta_{tt} = -\frac{P_\text{nl}}{\rho_w/k}, \qquad (4.3)$$

where all the nonlinear terms are contained in the pseudo-pressure term, P_nl. Now, unlike the pressure terms due to surface tension or ice-sheet that are linearly dependent on the surface undulation of the propagating wave itself, the pseudo-pressure term[1] P_nl emanates because of the interaction of the free waves that are independent of the propagating wave under consideration. Phillips [1960], in his

---

[1] To the third order, the P_nl satisfies the following relation (e.g., Phillips [1977]):
$\frac{1}{\rho_w}\frac{\partial}{\partial t}(P_\text{nl}) = -\eta \frac{\partial}{\partial z}\left(\frac{\partial^2 \phi}{\partial t^2} + g\frac{\partial \phi}{\partial z}\right) - \frac{1}{2}\eta^2 \frac{\partial^2}{\partial z^2}\left(\frac{\partial^2 \phi}{\partial t^2} + g\frac{\partial \phi}{\partial z}\right) - \frac{\partial \vec{u}^2}{\partial t} - \eta\frac{\partial^2 \vec{u}^2}{\partial z \partial t} - \vec{u}\cdot\nabla(\frac{1}{2}\vec{u}^2)$.

celebrated paper, showed that for deep water free gravity waves, satisfying the dispersion relation,

$$\omega^2 = gk, \tag{4.4}$$

the combined phases of the three waves can match exactly the phase of the propagating wave when the four-wave resonance condition is satisfied:

$$\begin{cases} \omega_1 + \omega_2 = \omega_3 + \omega_4, \\ \vec{k}_1 + \vec{k}_2 = \vec{k}_3 + \vec{k}_4. \end{cases} \tag{4.5}$$

When the propagation of the wave ζ_1 is considered, the pseudo-pressure term P_{nl} contains terms that are multiplications of the other three waves, ζ_2, ζ_3, and ζ_4. In other words, when the propagation speed of the pseudo-pressure field constructed by the three free waves coincides with that of the first wave, i.e., $(-\omega_2 + \omega_3 + \omega_4)/(|-\vec{k}_2 + \vec{k}_3 + \vec{k}_4|) = \omega_1/|\vec{k}_1|$, the fourth wave will grow at the expense of the energy of the other three waves. This is the physical interpretation of the resonant quartet interaction of free surface gravity waves.

4.2.2. *Experimental evidence*

Despite its paramount importance in understanding the evolution of ocean waves, experimental verification of the four-wave interaction of the gravity waves is rather limited. Two historical experiments were conducted by the group of Longuet-Higgins and Phillips independently in the mid-1960s [Longuet-Higgins and Smith, 1966; McGoldrick *et al.*, 1966]. Counting one of the quadruplets twice, one can construct a resonant triad in the framework of four-wave interaction:

$$\begin{cases} 2\omega_1 = \omega_2 + \omega_3, \\ 2\vec{k}_1 = \vec{k}_2 + \vec{k}_3, \end{cases} \tag{4.6}$$

in which the two waves propagating at normal angles satisfy $\omega_1/\omega_3 = 1.7356$, $k_1/k_3 = 3.0113$ and the third wave propagates at an angle $\theta = \tan^{-1}(k_3/2k_1)$ to wave k_1. The ingenious idea was to construct a wave tank whose two sides are equipped with a wave maker. Therefore, the

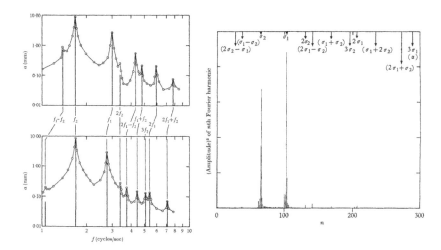

Fig. 4.1. Four-wave resonant interaction experiments by McGoldrick *et al.* [1966] (left plate) and Longuet-Higgins and Smith [1966] (right plate).

two waves propagating normal to each other are generated, and as a result of their interaction, an oblique wave will emerge. The result is most notable when the wave spectrum of the resonant and the non-resonant cases are compared. The figures are reproduced here from Longuet-Higgins and Smith [1966] and McGoldrick *et al.* [1966] in Fig. 4.1.

When the exact resonant condition is satisfied, the spectral peak at frequency $2f_1 - f_2$ appears but when the resonant condition is not satisfied, the spectral peak does not appear. The appearance of these spectral peaks is evidence to the existence of the resonant interaction of surface gravity waves. However, these experiments were limited in their experimental conditions. The representative wavelengths were around a few tens of centimeters, a marginal condition to neglect the effect of surface tension. These experiments were repeated in a much larger tank in Japan by Tomita [1989]. Tomita has repeated the experiment at a basin 80 m × 80 m with waves in the gravity wave regime (1.016 Hz, 0.566 Hz and 1.466 Hz). A direct comparison of the experimental result with the evolution equation based on Zakahrov's equation was made, and confirmed its validity. Waseda *et al.* [2015] conducted experiments in a narrow channel equipped with a directional wave maker in one end. With the effect of sidewall reflection taken into consideration, the growth of the initially non-existent wave was confirmed for various conditions including the renowned combination of three waves used in the Discrete Interaction Approximation of Hasselmann and Hasselmann [1985].

4.2.3. Deterministic evolution

The evolution equations of four waves, resonantly interacting with each other, was first derived by Benney [1962]. The rate of change of the complex amplitudes of the four waves are expressed as follows [Phillips 1977]:

$$\dot{a}_1 = ia_1(g_{11}a_1a_1^* + g_{12}a_2a_2^* + g_{13}a_3a_3^* + g_{14}a_4a_4^*) + ih\omega_1 a_2^* a_3 a_4,$$
$$\dot{a}_2 = ia_2(g_{21}a_1a_1^* + g_{22}a_2a_2^* + g_{23}a_3a_3^* + g_{24}a_4a_4^*) + ih\omega_2 a_1^* a_3 a_4,$$
$$\dot{a}_3 = ia_3(g_{31}a_1a_1^* + g_{32}a_2a_2^* + g_{33}a_3a_3^* + g_{34}a_4a_4^*) + ih\omega_3 a_1^* a_2 a_4,$$
$$\dot{a}_4 = ia_4(g_{41}a_1a_1^* + g_{42}a_2a_2^* + g_{43}a_3a_3^* + g_{44}a_4a_4^*) + ih\omega_4 a_1^* a_2 a_3,$$
(4.7)

where surface elevation is given as $\eta = \sum_{n=1}^{4}[a_n \exp(i(k_n x - \omega_n t)) +$ c.c.]. The coefficients g_{ij} and h are formally given in Benney [1962], and here, the compact notation of Phillips [1977] is adopted. The coefficients will eventually re-derived, for example, by Krasistkii [1994] in the framework of Zakharov's derivation (see Section 4.2.4). Although Eq. (4.7) looks rather complicated, their physical meaning can be easily understood by grouping the terms:

$$\dot{a}_i = \underbrace{i(\text{real coefficient})a_i}_{\text{amplitude dispersion}} + \underbrace{ih\omega_i(a_j^* a_k a_l)}_{\text{resonant forcing}}. \quad (4.8)$$

The coefficient of the first term is a real value and will only change the phase of the wave; it is called the amplitude dispersion. Within the frame work of Benney's derivation, this term is equivalent to the Stokes correction. Since the magnitude of the second term depends on the amplitudes of the other three independent waves, it represents the resonant forcing term. For waves satisfying the exact resonant condition (4.5), the phase evolution of this forcing term matches exactly that of the first wave, and therefore, forces the first wave to grow linearly if $a_i \ll a_j a_k a_l$ where $i,j,k,l = 1,2,3,4$. The evolution equation (4.7) can be solved for the four finite amplitude waves, in which case, the forcing term is no longer "independent" of the evolution of the waves in consideration. Thereby, the growth of the first wave gradually deviates from the linear one.

The evolution equation equivalent to the Benney's equation can be derived from the Zakharov's equation [Zakharov, 1968; Mei, Stiassnie and Yue, 1989]. Here, an evolution equation for the four waves based on the extension of the Zakharov's equation will be introduced using the notation by Krasitskii [1994]:

$$i\frac{\partial B_1}{\partial t} = (T_{1111}B_1^*B_1 + T_{1212}B_2^*B_2 + T_{1313}B_3^*B_3 + T_{1414}B_4^*B_4)B_1$$
$$+ T_{1234}B_2^*B_3B_4 \exp(-i\Delta_{1234}t),$$
$$i\frac{\partial B_2}{\partial t} = (T_{2121}B_1^*B_1 + T_{2222}B_2^*B_2 + T_{2323}B_3^*B_3 + T_{2424}B_4^*B_4)B_2$$
$$+ T_{2134}B_1^*B_3B_4 \exp(-i\Delta_{1234}t),$$
$$i\frac{\partial B_3}{\partial t} = (T_{3131}B_1^*B_1 + T_{3232}B_2^*B_2 + T_{3333}B_3^*B_3 + T_{3434}B_4^*B_4)B_3$$
$$+ T_{3412}B_4^*B_1B_2 \exp(i\Delta_{1234}t),$$
$$i\frac{\partial B_4}{\partial t} = (T_{4141}B_1^*B_1 + T_{4242}B_2^*B_2 + T_{4343}B_3^*B_3 + T_{4444}B_4^*B_4)B_4$$
$$+ T_{4312}B_3^*B_1B_2 \exp(i\Delta_{1234}t). \qquad (4.9)$$

The equation resembles the Benney's equation (4.7) in the form presented in Phillips [1977] except for the last term that includes the resonance detuning, $\exp(\Delta_{1234}t)$, where $\Delta_{1234} = \omega_1 + \omega_2 - \omega_3 - \omega_4$. When $\Delta_{1234} = 0$, the Benney's equation (4.7) is retrieved. On the other hand, when $\Delta_{1234} \neq 0$, the resonance condition is not exactly satisfied,

$$\begin{cases} \omega_1 + \omega_2 = \omega_3 + \omega_4 + \Delta_{1234}, \\ \vec{k}_1 + \vec{k}_2 = \vec{k}_3 + \vec{k}_4, \end{cases} \qquad (4.10)$$

and the interaction is called the "detuned resonance". In the literature, it is also referred to as "non-resonance", "near-resonance", or "quasi-resonance". While Tulin and Waseda [1997] use detuned resonance, Tanaka [2001], Janssen [2003] and Annenkov and Shrira [2018] use non-resonance, Gramstad and Stiassnie [2009] uses near-resonance, and Onorato et al. [2004] uses quasi-resonance. In other literature, quasi-resonance seems most commonly used in the context of freak waves. Kartashova [2010] distinguishes non-resonance and quasi-resonance and considers those as a part of detuned resonance. The term quasi-resonance is used specifically for a discrete wave system whose wave numbers are arranged on a lattice in the wave number space. In the rest of this chapter, we will use detuned-resonance to avoid ambiguity. The remarkable consequence of the detuned-resonance is the so-called modulational instability of a unidirectional wave train whose relevance extends beyond the field of water waves and is now flourishing in the field of nonlinear optics (see also Chapter 3).

4.2.4. *Spectral evolution*

In this section, we provide an overview of how the spectral evolution equation of Hasselmann [1962] is derived from the deterministic Zakaharov's equation (by Krasitskii). The interaction coefficients will be defined and are to be referred in the rest of the text. The reduced gravity equation reads

$$i\frac{\partial b_0}{\partial t} = \omega_0 b_0 + \int \tilde{V}^{(2)}_{0,1,2,3} b_1^* b_2 b_3 \delta_{0+1-2-3} dk_{123},$$

$$\delta_{0+1-2-3} = \delta(\boldsymbol{k}_0 + \boldsymbol{k}_1 - \boldsymbol{k}_2 - \boldsymbol{k}_3),$$

(4.11)

where definitions of the variables follow Krasitskii [1994]. The b_i are canonical variables related to the surface elevation and velocity potential at the free surface via the canonical transformation of the complex amplitude function $a(\boldsymbol{k},t) = \sqrt{\omega(\boldsymbol{k})/2k}\hat{\eta}(\boldsymbol{k}) + i\sqrt{k/2\omega(\boldsymbol{k})}\hat{\psi}(\boldsymbol{k})$; i.e., $b(\boldsymbol{k},t) = a(\boldsymbol{k},t) + \text{H.O.T.}$ [see, e.g., Krasitskii, 1994]. The $\hat{\eta}(\boldsymbol{k})$ and $\hat{\psi}(\boldsymbol{k})$ are the Fourier coefficients of the surface elevation and the velocity potential at the free surface.

With a quasi-Gaussian closure, $\langle b(\boldsymbol{k})b^*(\boldsymbol{k}')\rangle = n(\boldsymbol{k})\delta(\boldsymbol{k}' - \boldsymbol{k}')$, the Hasselmann's integral is retrieved:

$$\frac{\partial n_0}{\partial t} = 4\pi \int [\tilde{V}_{0,1,2,3}^{(2)}]^2 n_0 n_1 n_2 n_3 \left[\frac{1}{n_0} + \frac{1}{n_1} - \frac{1}{n_2} - \frac{1}{n_3}\right]$$
$$\times \delta(\omega_0 + \omega_1 - \omega_2 - \omega_3)\delta_{0+1-2-3} dk_{123}. \qquad (4.12)$$

Derivation and the expressions for the coefficients of Zakharov's equation (4.11) and the Hasselmann's integral (4.12) are concisely provided in the textbook of Janssen [2004]. The third-generation wave models (see Chapter 5) numerically solve (4.12) and the right-hand side of (4.12) is called the nonlinear source term, S_{nl}. Up to this point, S_{nl} can only be solved as an approximation. Recent implementation of the Web–Resio–Tracy (WRT) scheme [Web et al., 1978; Tracy and Resio, 1982] by Van Vledder [2006] allowed users of NOAA WAVE-WATCH III [Tolman, 2009] to implement WRT schemes in realistic wave hindcasting simulations [Romero, 2008; Liu et al., 2017; Ponce de León et al., 2018]. However, most other studies still use the Discrete Interaction Approximation (DIA) for computational efficiency. Some other studies use the extended versions of the DIA for realistic wave simulations: SRIAM [Tamura et al., 2008, 2010] and mDIA [Tolman, 2004].

The Zakharov's equation (4.11) can be rewritten by change of variable, $b(\boldsymbol{k},t) = B(\boldsymbol{k},t)\exp[-i\omega(\boldsymbol{k})t]$, as follows:

$$i\frac{\partial B_0}{\partial t} = \int \tilde{V}_{0,1,2,3}^{(2)} B_1^* B_2 B_3 \exp[i\Delta_{0123}t]\delta_{0+1-2-3} dk_{123}. \qquad (4.13)$$

The variable $B(\boldsymbol{k},t)$ was used earlier in Eq. (4.9), evaluated at discrete wave numbers, $B_i = B(\boldsymbol{k}_i,t)$. For a discrete spectrum, the right-hand side of (4.13) takes a form of a summation [e.g., Mei, Stiassnie and Yue, 1989]:

$$i\frac{\partial B_0}{\partial t} = \sum_{0+1-2-3} T_{0123} B_1^* B_2 B_3 \exp[i\Delta_{0123}t]. \qquad (4.14)$$

The equivalent of Benney's equation (4.9) derives from (4.14) evaluated for a quartet of resonant waves.

4.3. Modulational Instability

A sensational consequence of the detuned resonance is that the Stokes wave train is unstable to a pair of sideband waves. Stokes waves will eventually disintegrate into groups of waves whose envelope evolves in time. The evolution can be separated into stages of initial instability, modulation, and recurrence. The condition for the modulational instability relates to the balance of nonlinearity and dispersion, and therefore, is generally considered a characteristic of a spectrally narrow wave system.

4.3.1. Benjamin–Feir instability: Disintegration of Stokes wave

The disintegration of Stokes wave train was first discovered in a towing tank by Feir [Benjamin, 1967]. As the mechanically generated monochromatic waves propagated down the tank, the amplitude gradually modulated, and eventually the wave train disintegrated into groups. Benjamin and Feir [1967] showed theoretically that this phenomenon is related to the stability of the Stokes wave. They showed that Stokes waves are unstable to a pair of sideband waves whose frequencies are separated from the carrier wave frequency ω_0 by $\pm\delta\omega$. The initially infinitesimal sidebands grow exponentially when the stability condition $0 \leq \delta \leq \sqrt{2}ak$ is satisfied. Here, $\delta = \delta\omega/\omega_0 = O(ak)$ is assumed.

The most physically comprehensible account of the Benjamin–Feir instability is given in Phillips [1977] and here, a solution is derived from (4.9). The Benjamin–Feir instability is a degenerate case of the four-wave resonant interaction:

$$B_4 = iB_3; \quad A_3 \equiv B_3 + B_4; \quad B_1, B_2 \ll A_3, \tag{4.15}$$

where A_3 is the carrier wave amplitude and B_1, B_2 are the infinitesimal sideband wave amplitudes. Here, the carrier wave at ω_3 is counted twice by adding B_3 and B_4 whose phases are shifted by $\pi/2$. By neglecting quadratic terms in B_1, B_2, we obtain the following evolution equations:

$$i\frac{\partial B_1}{\partial t} = T_{1313}|A_3|^2 B_1 + \frac{1}{2}T_{1233} B_2^* A_3^2 \exp(-i\Delta_{1233} t),$$

$$i\frac{\partial B_2}{\partial t} = T_{2323}|A_3|^2 B_2 + \frac{1}{2}T_{1233} B_1^* A_3^2 \exp(-i\Delta_{1233} t), \quad (4.16)$$

$$i\frac{\partial A_3}{\partial t} = T_{3333}|A_3|^2 A_3,$$

where

$$\begin{cases} \omega_1 + \omega_2 = 2\omega_3 + \Delta_{1233}, \\ k_1 + k_2 = 2k_3. \end{cases} \quad (4.17)$$

The sideband waves grow exponentially at the expense of the carrier wave energy, and when the amplitude dispersion completely balances the resonance detuning,

$$\underbrace{(2T_{3333}|A_3|^2 - T_{1313}|A_3|^2 - T_{2323}|A_3|^2)}_{\text{amplitude dispersion}} + \underbrace{\Delta_{1233}}_{\text{detuning}} = 0, \quad (4.18)$$

the growth rate of the sideband waves attains the largest value:

$$\beta = \frac{1}{2}T_{1233}|A_3|^2. \quad (4.19)$$

The Benjamin–Feir instability is a special case of the detuned resonant interaction of waves satisfying (4.15). The carrier wave interacts with the two free waves at frequencies deviating in the order of the wave steepness. An alternative interpretation of (4.15) may be possible as interactions of the sideband waves with the second harmonic of the carrier wave. However, as will be shown later, such a consideration leads to an incorrect estimation of the growth rate.

The growth rate of the sidebands was experimentally verified by Waseda and Tulin [1999]. In the limit of infinitesimal steepness, the solution based on resonant interaction theory asymptotes to the solution by Benjamin and Feir. For finite steepness, the stability range is smaller than $0 \leq \delta \leq \sqrt{2}ak$, and the maximum growth condition

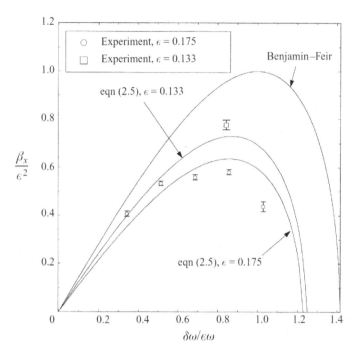

Fig. 4.2. Reproduced from Waseda and Tulin [1999]. Growth rate of the sidebands (seeded run). For waves 1.0 Hz, $\varepsilon = 0.175$, $\hat{\delta} = 0.343$, and for 0.514, 0.686, 0.857, 1.03, and 1.0 Hz, $\varepsilon = 0.133$, $\hat{\delta} = 0.842$. The curves are theoretical predictions from Benjamin and Feir [1967], and from (4.19) based on Krasitskii [1994].

is not when $\delta = ak$ as in the case of Benjamin and Feir [1967]. Here, we define a parameter $\hat{\delta} \equiv \delta/(ak)$ which is the ratio of the spectral bandwidth and the steepness, or the relative magnitude of dispersion and nonlinearity. Figure 4.2 plots the sideband growth rate as a function of $\hat{\delta}$. As the steepness changes, the range of unstable sidebands changes and for a typical steepness of around 0.1–0.2, the maximum growth condition is achieved at $\hat{\delta} \sim 0.8$.

4.3.2. *Long-term evolution of modulated wave train*

Immediately after the discovery of the Benjamin–Feir instability, numerous works followed studying the long-term evolution of the unstable Stokes wave train. The remarkable feature of the modulated wave train is the Fermi–Pasta–Ulam recurrence, in which the amplitude spectrum returns to its original state. Lake et al. [1977] observed the recurrence in a laboratory wave tank. However, their experiment, limited in scale ($0.9 \times 0.9 \times 12$ m), could not reproduce a complete recurrence as a spectral downshifting had occurred. Tulin and Waseda [1999] conducted an experiment in a larger wave tank ($4.2 \times 2.1 \times 50$ m) seeding the sideband waves at the wave maker and observed a complete recurrence. Figure 4.3 shows the wave spectra and the surface elevation time series from the two experiments. In the case of the Lake et al. [1977] experiment, while the amplitude modulation recurs, the spectral peak shifts to the lower sideband. While in the Tulin and Waseda [1999] experiment, downshifting does not occur and a return of energy to the carrier wave is observed.

The long-term evolution of the modulated wave train can be obtained as a solution to the Nonlinear Schrödinger Equation (NLS):

$$i\frac{\partial A}{\partial T} - \frac{\omega_0}{8k_0^2}\frac{\partial^2 A}{\partial x^2} - \frac{\omega_0 k_0^2}{2}|A|^2 A = 0, \qquad (4.20)$$

where $A(x,T)$ is a slowly varying wave amplitude or the envelope of the carrier wave: ω_0 and k_0. Note, Eq. (4.20) is given in a moving frame of reference: $T = t - C_g x$, where C_g is the group velocity. Yuen and Ferguson [1978] numerically solved the NLS and studied the long-term evolution of the modulated wave train, including not only the case of a single sideband pair, but also cases of a set of sideband pairs. Akhmediev and Korneev [1986] derived analytical solutions to the NLS, of which the simplest case of a single pair of sidebands is called the Akhmediev Breather. The wave train undergoes a cycle of modulation and demodulation, and the largest possible amplification is three times the initial amplitude. Stiassnie and Kroszynski [1982] derived an analytical expression for the recurrence period, which depends on three parameters: initial phase, initial amplitude of the sideband waves, and the parameter $\gamma = (\delta k/k)/(ak)$, which is identical to $\hat{\delta}$ that appeared in the stability theory of Benjamin and Feir [1967]. Once again, the relative ratio of the dispersion and nonlinearity was the key parameter determining the period of the recurrence cycle.

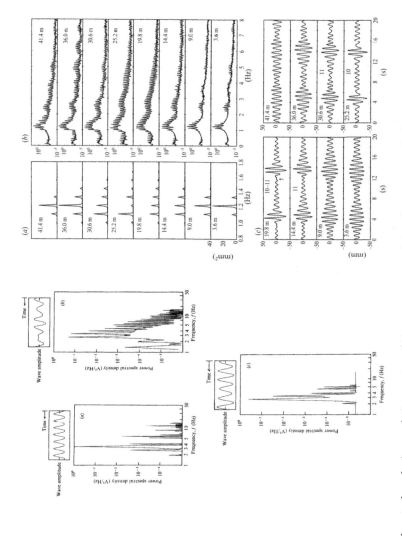

Fig. 4.3. Spectral evolution and time series of the recurrent modulated wave train. Left: Reproduced from Lake *et al.* [1977]. Right: Reproduced from Tulin and Waseda [1999].

4.3.3. *Spectral broadening*

A remarkable feature of the long-term evolution of the unstable Stokes wave is the appearance of multiple energy peaks at a regular interval (Melville, 1982, and Fig. 4.3). Earlier studies showed that the phase speeds of these waves were bound to the carrier wave [Lake and Yuen, 1978]. However, these waves are actually free waves, but near the peak modulation, the dispersion relation alters such that these free waves are propagating at the same speed as the carrier wave [Houtani et al., 2018]. The possible cascading of energy through sequences of detuned resonance was presented in Tulin and Waseda [1999]:

$$\omega_0 \mp (\omega_0 - \delta\omega) \pm (\omega_0 + \delta\omega) = \omega_0 \pm 2\delta\omega$$

$$\underbrace{k_0}_{\text{carrier}} \mp \underbrace{(k_0 - \delta k)}_{\text{lower sideband}} \pm \underbrace{(k_0 + \delta k)}_{\text{upper sideband}} = \underbrace{k \pm 2\delta k}_{\text{cascade}} - \Delta k. \quad (4.21)$$

The process continues transferring energy to $(\omega_0 \pm n\delta\omega, k_0 \pm n\delta k)$. This process is now recognized as a Dynamic Cascade where a set of resonant quartets can induce a different set of resonance conditions and the process continues spreading energy far beyond the original spectral energy peaks [Kartashova and Shugan, 2011; Shugan et al., 2018]. The generation of spectral peaks at high frequencies is therefore a consequence of quasi-resonance in which waves on a discrete wave number lattice interact.

As a result of the dynamic cascade, the initially monochromatic Stokes wave train spreads energy into a broader spectrum. When the spectrum broadens, the condition that the spectral bandwidth and the steepness are of the same order, i.e., $\delta\omega/\omega_0 = O(ak)$, is violated implying that the NLS will fail to express the evolution.

The operator acting on the complex amplitude function A of the NLS:

$$i\frac{\partial}{\partial t} + i\frac{\omega_0}{2k_0}\frac{\partial}{\partial x} - \frac{\omega_0}{8k_0^2}\frac{\partial^2}{\partial x^2} - \frac{\omega_0}{2}k_0^2|A|^2 \quad (4.22)$$

can be derived formally from the dispersion relation of the Stokes wave $\omega = \sqrt{gk}(1 + \frac{1}{2}(ak)^2)$ replacing $\omega = \omega_0 + \omega'$ and $k = k_0 + k'$ and retaining terms up to the second-order, and substituting $\omega' \to i\partial/\partial t$ and $k' \to -i\partial/\partial x$. The wave train is assumed to have a dominant wave period and a wave number (ω_0, k_0) and a narrow energy spread in spectrum represented by the perturbation wave period and wave number (ω', k'). The perturbation is considered to be the same order as the wave steepness: $O(\omega'/\omega_0, k'/k_0) = O(ak)$.

As the modulation of the Stokes wave progresses, the spectrum broadens, and the assumption $O(\omega'/\omega_0, k'/k_0) = O(ak)$ no longer holds. Thereby, a higher-order terms in (ω', k') should be retained. Dysthe's equation [Dysthe, 1979] is derived retaining the high-order terms in spectral bandwidth $\delta = O(\omega'/\omega_0, k'/k_0)$ up to the fourth-order and the nonlinear terms up to the third-order in steepness $\varepsilon = O(ak)$.

$$\underbrace{i\frac{\partial A}{\partial T} - \frac{\omega_0}{8k_0^2}\frac{\partial^2 A}{\partial x^2} - \frac{\omega_0 k_0^2}{2}|A|^2}_{\text{Nonlinear Schrödinger Equation}} \underbrace{-\frac{\omega_0}{16k_0^3}\frac{\partial^3 A}{\partial x^3}}_{\text{Fourth-order dispersive term}}$$

$$\underbrace{-\frac{\omega_0 k_0}{4}A^2\frac{\partial A^*}{\partial x} + \frac{3}{2}\omega_0 k_0|A|^2\frac{\partial A^*}{\partial x} + ik_0 A\frac{\partial \phi}{\partial x}\bigg|_{z=0} = 0.}_{\text{Fourth-order nonlinear terms}}$$

(4.23)

Here, the velocity potential of the mean current satisfies:

$$\nabla^2 \phi = 0 \, (-h < z < 0); \quad \frac{\partial \phi}{\partial z} = \frac{\omega}{2}\frac{\partial |A|^2}{\partial x} \, (z=0); \quad \frac{\partial \phi}{\partial z} = 0 \, (z=-h).$$

(4.24)

Stiassnie [1984] applied the narrow-banded approximation to Zakharov's equation and re-derived the Dysthe's equation. The fourth-order nonlinear terms originate from the resonant interaction terms. Alternative forms of the modified Nonlinear Schrödinger Equations were derived. Trulsen and Dysthe [1996] considered the balance $\delta\omega/\omega_0 = O((ak)^{1/2})$ allowing an even broader spectral bandwidth.

The reader should be reminded here that the evolution of the modulated wave train can be derived as either a solution to the NLS (e.g., Akhmediev Breather) or as a solution to the discrete Zakharov's equation. In the latter case, the Benjamin–Feir instability was explained as an interaction of three free waves: the carrier and the two sideband waves. However, as the wave train evolves, spectral broadening occurs and consideration of multiple free waves at higher and lower frequencies become relevant. In fact, one can show that the solution to the NLS is equivalent to the evolution of the three interacting free waves in which the amplitudes of the sidebands remain symmetric. The impact of spectral broadening can be demonstrated numerically by gradually changing the number of free modes that are in action. Figure 4.4 shows such an example derived from the Dysthe's equation reducing the active modes from 32 to 4. The same result can be obtained as a solution to the discrete Zakharov equation.

As the spectrum broadens, the asymmetry appears not only in the spectral evolution but in the wave shape as well. Lo and Mei [1987] numerically solved the two-dimensional Dysthe equation and showed that the initially symmetric solitary wave pulse deforms into a crescent shape. Fore-and-aft trough depth asymmetry of the modulated wave train was obtained from Dysthe's equation as well [Clamond et al., 2006]. As shown by Stiassnie [1984], the asymmetric terms originate from the narrow banded approximation of the nonlinear interaction term of the Zakharov equation. Therefore, the asymmetric features of the wave shape are the consequences of the resonant interaction. This conjecture becomes important in the study of freak waves in the ocean (see Chapter 3).

4.3.4. *Temporally and spatially periodic modulated wave trains*

Temporal evolution of modulated wave trains has been studied extensively using the Zakharov equation, Nonlinear Schrödinger Equation, and the Dysthe equation. On the other hand, the observed modulated

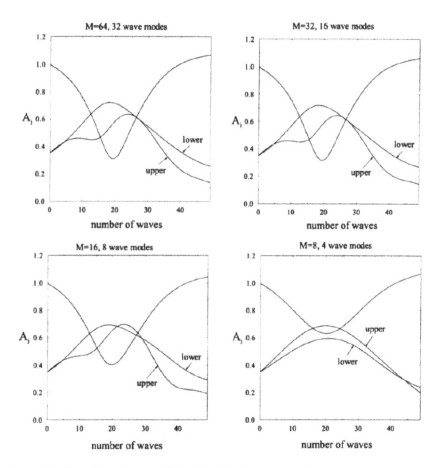

Fig. 4.4. The effect of spectral bandwidth demonstrated by Dysthe's equation. From upper left to lower right figures, the number of wave modes are 32, 16, 8, and 4. $ak = 0.1$, $df/f = 0.9$, $a/a = 0.3535$, and initial phases are $-pi/4$. Figure adopted from Waseda [1997].

wave train in physical wave tanks evolve in space. For a unidirectional wave train, the space and time coordinate can be exchanged by $x = C_g t$. The spatial NLS can be readily derived:

$$i\frac{\partial A}{\partial x} - \frac{k_0}{\omega_0^2}\frac{\partial^2 A}{\partial T^2} - k_0^3 |A|^2 A = 0. \quad (4.25)$$

This form of the NLS is used extensively in the field of nonlinear optics, apparently with different coefficients, where a transmission of a laser in a kilometer-long optical fiber cable is studied [e.g., Dudley *et al.*, 2009]. This form of NLS is called the spatial-NLS while the original form (4.20) is called the temporal-NLS. Spatial-Dysthe equations as well as spatial-Zakharov's equation for a narrow spectrum were derived and used to study evolution of waves in a wave flume [Lo and Mei, 1985; Shemer *et al.*, 2001; Shemer *et al.*, 2002].

Since the form of the equations are formally the same, solutions to these are interchangeable by switching space and time. Wave profiles and time series from the temporal and spatial Akhmediev Breathers are shown in Fig. 4.5 [Houtani *et al.*, 2018]. The envelope of the wave profile of the temporal NLS, and the envelope of the time series of the spatial NLS are identical. Likewise, envelopes of the time series and wave profile, of the temporal and spatial NLS, are identical. Furthermore, Houtani *et al.* [2018] have shown that the time series and the wave profiles are the same in the vicinity of the peak of the modulation (Fig. 4.6). The coincidence of the wave shapes that extends for a group length originates from the detuned resonant interaction of free

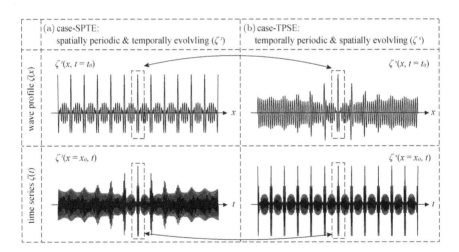

Fig. 4.5. Classification of a temporally evolving [(a) case-SPTE] and a spatially evolving [(b) case-TPSE] unstable Stokes wave train. The wave profile and time series are given for each case. Double-headed arrows indicate the comparison in after Houtani *et al.* [2018].

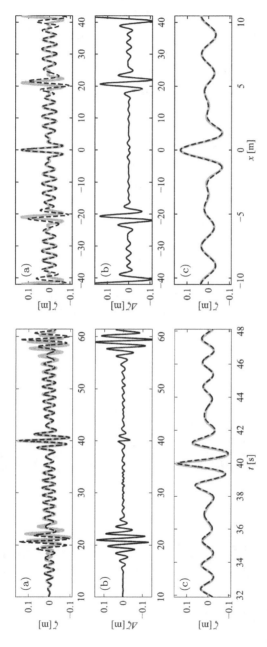

Fig. 4.6. The time series (a), the difference between SPTE and TPSE (b) and enlarged view of the surface elevations, from the tank experiment (left), and from the Akhmediev solution (right). Solid gray line: SPTE; dashed black line: TPSE. Adapted from Houtani et al. [2018].

waves. As a result, the instantaneous dispersion relation of discrete spectral peaks of the modulated wave train alters and satisfies the following relationships: $k_n = k_c \pm n\Delta k$ and $\omega_n = \omega_c \pm n\Delta\omega$. This was also the case for random spectral waves. In this case, the dispersion relation was a linear line tangent to the quadratic dispersion curve at the peak wave number. The detuned resonance is at play in the random wave field as well.

4.3.5. *Generation of modulated wave train in the wave tank*

The serendipitous discovery of modulational instability by Benjamin and Feir suggests nature will select the most unstable perturbation from the background random disturbance as the Stokes wave propagate down the tank. Inevitably, the tank has to be long enough for the perturbation to grow. Such "unseeded" generation of a modulated wave train was used in the early studies in wave tanks [e.g., Melville, 1982; Bliven, Huang and Long, 1986]. However, Tulin and Waseda [1999] have noticed that such a perturbation is associated with the repeated reflections of the leading wave front of the wave train between the beach and the wave maker, and as a result, the sideband waves that developed did not coincide with the waves of maximum growth rate.

Lake and Yuen [1977], in a relatively short tank, observed the appearance of a small perturbation in the end of the tank and "seeded" the disturbance to the wave generator signal in the next experiment. By repeating this procedure, they were eventually able to generate largely modulated waves. The systematic "seeded" MI study is reported in Waseda and Tulin [1999] for the initial stability problem and in Tulin and Waseda [1999] for long-term evolution leading to a breaking wave. The initial steepness, and sideband frequencies and amplitude were systematically adjusted to generate a wide variety of modulated wave trains near their peak modulated state. Breaking waves of different strength were generated and were used for studying radar observation of breaking waves [Fuchs *et al.*, 1999;

Lamont-Smith et al., 2003], and the impact of the modulated wave train on ocean structures [Levi et al., 1998; Welch et al., 1999]. Note that the initial sideband wave amplitude should be less than about 30% of the carrier wave to assure that the nonlinear change of phase is minimal.

The shortcoming of seeding the sideband waves is the neglect of the numerous free waves that appear due to dynamical cascade. There are two methods to overcome this problem. First is to control the wave maker motion by the amplitude function given as a solution to the envelope equation. The second is to use a solution of a fully nonlinear numerical simulation result.

The use of the envelope function derived from the NLS was extensively used in the study of "Super rogue waves". In addition to the Akhmediev Breather, a collection of solutions superposing multiple breathers was discovered [Akhmediev et al., 2009]; the maximum amplification is 3 for a single breather, 5 for the case of two colliding breathers, and 7 for the case of three colliding breathers. Extensive studies on experimental realizations are reported in a series of papers by Chabchoub et al. [2011, 2012a,b] which are concisely summarized in Chabchoub et al. [2016].

A scheme that reproduces nonlinear directional wave fields based on a phase-resolved numerical simulation was presented recently. The HOSM-WG makes use of the High-Order Spectral Method (HOSM) simulation results to generate a Wave Generator (WG) signal [Houtani et al., 2018]. A spatially periodic wave field mimicking the HOSM result was reproduced in a wave flume and in a directional wave tank. The unidirectional or long-crested wave train can be generated in a typical seakeeping tank with a single wave paddle and an example of such is shown in Fig. 4.6. A more complicated wave field of fully directional seas was also demonstrated in a rectangular basin with 382 wave makers surrounding the periphery of the wave tank. The method allows the selection of a desired wave field nonlinearly, which evolves in time and reproduces the wave field in a wave basin. Therefore, a long distance from the wave maker is not required to allow nonlinear evolution.

4.3.6. *Observed evidence of modulated wave trains*

One of the earliest notions of the existence of modulated wave trains in the ocean dates back to the 1990s when an extensive study on understanding the radar signature of breaking waves was conducted. Tulin hypothesized that the breaking wave occurs in a group, and analyzed the radar image obtained by a low grazing angle radar [Tulin, 1996]. His depiction of the breaking events forming a pattern in the low-grazing angle radar observations is reproduced here (Fig. 4.7). While the sea spike itself propagates at the phase speed, the group line that connects breaking events propagates at the group speed. The depiction is consistent with the earlier notion of wave breaking in a group by Donelan *et al.* [1972]. Compilation of low-grazing angle radar observations at a loch is summarized by Werle [1995] in a study motivated by Tulin's depiction. At about the same time, low-grazing angle radar images of Smith *et al.* [1996] revealed a systematic pattern of sea-spikes associated with breaking events, that is consistent with Tulin's depiction.

To confirm, a controlled tank experiments with a C-band low-grazing angle radar was conducted at the Ocean Engineering Laboratory Tank of the University of California at Santa Barbara using modulated wave trains [Fuchs *et al.*, 1999; Tulin and Waseda 1999]. Likewise, a signature of group formation in a wind-generated wave

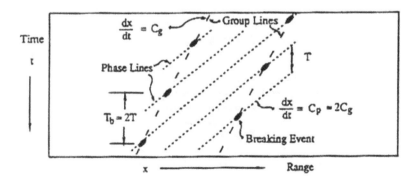

Fig. 4.7. A range-time schematic allowing a quantitative analysis of the range-time radar return image from a low-grazing angle radar. This is a schematic depiction presented by Tulin [1996].

detected by the low-grazing angle radar is documented by Lamont-Smith et al. [2003] for both the tank and ocean waves. These radar images, including the one from the open ocean, bare the same characteristics as depicted by Tulin and thereby suggest that modulational instabilities do exist in the random directional seas. The low-grazing angle radar images from the tank with controlled modulated wave trains, tank measurement of wind-generated waves, and the open waters are reproduced here for comparison (Fig. 4.8).

An interesting work by Hatori [1984] depicts the appearance of spectral peaks in the wind–wave spectrum in a laboratory wind–wave tank. Seemingly, equally spaced spectral peaks appear in the detailed structure of the spectrum, and a continuous downshifting of the spectral peaks among these peaks are observed (Fig. 4.9). The separation of the peaks is relatively narrow and is much smaller than the order of the wave steepness. Therefore, these peaks do not satisfy the maximum growth condition of the Benjamin–Feir instability. Waseda [1997] observed a similar downshifting process in a different wind–wave facility.

The appearance of these spectral peaks imply the formation of coherent wave groups in the physical domain. The discovery of the existence of wave groups dates back to the 1970s in coastal waters [Nolte and Hsu, 1972; Rye, 1975; Goda, 1976] and in deep waters [Donelan et al., 1972]. The persistence and coherence of a wave group is a result of nonlinearity and high-order dispersion, and can be attributed to a narrow-banded process expressed by the NLS. Once a wave group forms, then, a breather-like evolution is expected, as depicted by the evolution of a wave packet by Su [1982]. Sedivy and Thompson [1978] plotted the largest wave height in a group relative to the number of waves in a group. This interesting study of observed waves indicated that the individual wave height reaches the largest value when the number of waves per group is around 8. This corresponds to a spectral bandwidth of the order 0.12 or so, which is a reasonable number for waves in a modulated wave train. Further study is warranted, and the use of satellites or ship-borne radar images may be useful as well [Nieto-Borge et al., 2004, 2013].

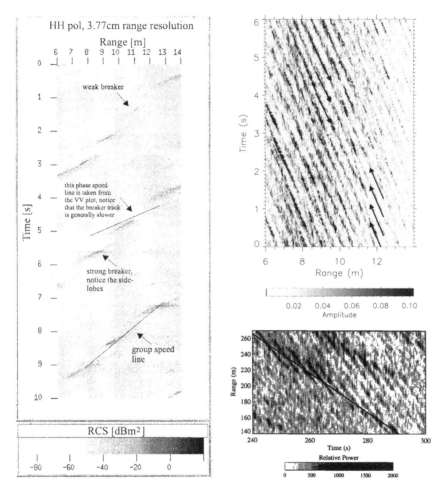

Fig. 4.8. (Left) Range-time-intensity diagram of a 2.3-m wave group with 0.165 initial steepness, 12 m/s wind. Tracks of individual breaking waves can be identified in HH polarization. [After Fuchs *et al.*, 1999] (Right-top) Range-time amplitude image of the range-time radar data collected in a wind–wave tank. [After Lamont-Smith *et al.*, 2003] (Right-bottom) Image of the backscattered power for resolved waves. The set of short solid lines outlines the propagation of 3.1-s waves toward the radar. The single line shows an additional modulation propagating at the group velocity. [After Smith *et al.*, 1996]

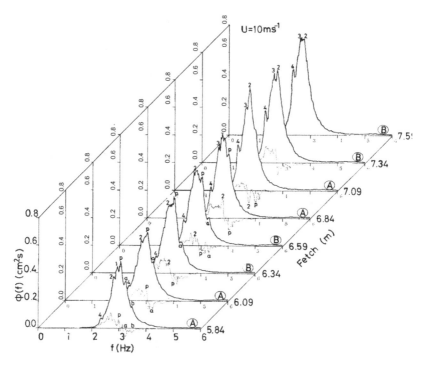

Fig. 4.9. Sideband peaks and continuous downshifting. Reproduced from Hatori [1984].

4.4. Freak Waves in the Ocean

The objective of this section is to report the most recent developments in the use of the third-generation wave models to assess the possible mechanism of freak wave generation in realistic sea conditions. The connection between the spectral model (see Chapter 5) and the phase-resolving model (see Chapter 6) will be discussed. Under rapidly changing weather condition, the spectral geometry changes and thereby the significance of the weakly nonlinear and narrow-banded process changes.

4.4.1. *Modulational instability in a random directional wave field: pdf and kurtosis*

Ocean waves are considered to follow a random Gaussian process, and the probability density function of the crest height and wave height follows the Rayleigh distribution. A slight deviation from the Gaussian process can be expressed by the Gram–Charlier expansion where skewness μ_3 and excess kurtosis $\mu_{40} \equiv \mu_4 - 3.0$ are non-zero [Cartwright and Longuet-Higgins, 1956]. Apparently, the nonlinear interaction among spectral components leads to non-zero values of the skewness and kurtosis, the former due to second-order nonlinearity or three wave interaction and the latter due to the third-order nonlinearity or four-wave interaction. Hatori [1984a] studied laboratory wind waves and showed that the excess kurotosis is negative, and that was confirmed by other experiments as well [e.g., Waseda, 1997]. A statistical description of the random wave field based on the Zakharov's equation is presented in Yuen and Lake [1982] in a quasi-Gaussian closure form, and will be the basis of the nonlinear modification of the distribution.

To the second-order, and thereby considering the three-wave interaction, the surface elevation distribution becomes skewed, and the corresponding crest-height distribution deviates from the Rayleigh distribution, while the wave height remains Rayleigh [Tayfun, 1980]. When the third-order nonlinearity is considered, kurtosis becomes important. Mori et al. [2002] investigated observed wave records in the interest of understanding "freak waves" and noted that the excess kurtosis was positive. "Freak wave" was defined as an individual wave that exceeds twice the significant wave height, $H_{\max} > 2.0\,H_s$, representing a clear exceedance of the maximum wave height based on the Rayleigh distribution [proposed by Kjeldsen and Sand, in Torum and Gudmestad, 1990]. In the engineering community, a waves exceeding twice or 2.2 times the significant wave height is called a freak wave. The importance, however, is the enhanced probability at the tail of the distribution, and Janssen [2003] theoretically showed extending the work of Yuen and Lake [1982], that the change of kurtosis can be attributed to the detuned resonance of the spectrally narrow wave field. Almost simultaneously and independently, Onorato et al. [2004] experimentally showed the systematic dependence of the tail enhancement on the spectral narrowness.

Both Janssen [2003] and Onorato et al. [2004] investigated an initial value problem where nonlinear evolution of the random wave field was investigated. The wave field started to deviate from the Gaussian distribution and notably the kurtosis increased in time. The corresponding expression of the probability density function was derived by Mori and Janssen [2006]:

$$p(H) = p_{\text{Rayleigh}}(H) \times [1 + f(H)\mu_{40}]. \quad (4.26)$$

The initial spectral geometry was represented by the nondimensional parameter representing the relative balance of dispersion and nonlinearity. Onorato et al. [2001] called it the Ursell number, while Janssen [2003] called it the Benjamin–Feir Index (BFI) for the obvious reason that the waves modulated due to detuned resonance. Janssen [2003] noted that the excess kurtosis is proportional to the BFI:

$$\mu_{40} \propto \text{BFI}^2. \quad (4.27)$$

The reader should be reminded that these parameters (Table 4.1) are identical to the parameters introduced by Benjamin and Feir [1967], and Stiassnie and Kroszynski [1982].

The shortcoming of these earlier works with the lack of directionality was immediately noted. Socquet-Juglard et al. [2005] showed that the short-crested wave field did not evolve as the long-crested waves and the Tayfun distribution was adequate to represent the distribution. The result was substantiated by Gramstad and Trulsen

Table 4.1. Parameters appearing in literature representing the relative significance of nonlinearity and dispersion (unidirectional).

	Benjamin and Feir [1967]	Stiassnie and Kroszynski [1982]	Onorato et al. [2012]	Onorato et al. [2001]	Janssen [2003], Janssen and Bidlot [2003]
Index	$\hat{\delta} \equiv \frac{\delta\omega/\omega_0}{ak}$	$\gamma = \frac{\delta k/k}{ak}$	$\rho = \frac{N}{ak}$	$Ur = \left(\frac{ak}{\delta\omega/\omega_0}\right)^2$	$\text{BFI} = \frac{\sqrt{2}ak}{\sigma_\omega/\omega_0}$ $\text{BFI} = k_0 m_0^{1/2} Q_p \sqrt{2\pi}$
Use	Sideband growth rate	Recurrence period	Maximum wave height	Freak wave index	Freak wave index

[2007] indicating a sharp change in the statistics between the short-crested wave field and long-crested wave field of approximately 10 crest wavelengths and longer. However, a systematic experiment by Waseda [2006] at the University of Tokyo Ocean Engineering Basin showed that the excess kurtosis gradually increased as the directional spread narrowed, indicating a possibility of parameterizing the effects of directional spread. Simultaneously, a similar experiment was conducted at Marintek and the result compared well with each other [Waseda *et al.*, 2009; Onorato *et al.*, 2009a,b]. An expression for the probability density function including the kurtosis was given by Mori *et al.* [2011] and independently by Tayfun and Fedele [2007]. Ribal *et al.* [2013] provided a solution to Alber's equation which is a stochastic counterpart of the Fermi–Pasta–Ulam recurrence and the associated freak wave occurrence probability was derived. Extensions for the BFI were provided in all these works to include directional effects and are summarized in Table 4.2. Waseda *et al.* [2009] made use of the transformation which maps the perturbation wave number vector to a coordinate that takes into account the effect of the bounding locus at an angle $35.26°$ by Albers [1978]. Janssen and Bidlot [2009] derived a simple form when the spectrum can be represented by a Gaussian shape. The numerical solutions to Albers equation led to the derivation by Ribal *et al.* [2013] specifically for the JONSWAP and cosine power directional spectrum.

4.4.2. *Modulational instability in realistic sea states — Spectral evolution*

The classical views of ocean waves were re-examined in the last few decades, in studies of freak wave research, and modification to the probability density function was suggested. With the advanced wave forecast models and observations, the directional spectrum was discovered to change in time and space. The spectral evolutions were documented in detail mostly based on hindcast simulation by the third-generation wave models for known freak wave events and marine accident cases [e.g., Kharif *et al.*, 2009].

Table 4.2. Parameters appearing in literature extending the BFI to include directionality.

	Waseda et al. [2009]	Janssen and Bidlot [2009], Fedele and Tayfun [2009], Mori et al. [2011]	Xiao et al. [2013]	Ribal et al. [2013]
BFI	$\dfrac{\epsilon_{\text{eff}}}{\sqrt{(\delta k/k)^2 - 2(\delta l/k)^2}}$	$\dfrac{\sqrt{2}ak}{(\sigma_\omega/\omega_0)\sqrt{1+\alpha_2 R}}$	$\left[\dfrac{1+\tan^2\Theta/2}{1+4\tan^2\Theta/2}\right]\dfrac{ak}{\delta k/k}$	$\dfrac{ak}{\alpha\gamma} + \dfrac{\beta}{akA_d}$
	$\epsilon_{\text{eff}} \sim ak$	$R = \dfrac{\delta_\theta^2}{2\delta_\omega^2}$ (JB and M)	$\Theta = 2\tan^2(K_x/K_y)$	$A_d = \dfrac{\Gamma(1+\frac{n}{2})}{\sqrt{\pi}\Gamma(1/2+\frac{n}{2})}$
	$\times\sqrt{1 - \delta\kappa/4 - 5/8\delta\kappa^2}$	or $R = \dfrac{\sigma_\theta^2}{2\nu^2}$ (FT)		α, γ: JONSWAP
	$\delta\kappa = \sqrt{\delta k^2 + \delta l^2}$			$\cos^n\theta$ distribution

The earliest report of an observed "freak wave" is by Sand et al. [1990] on the time series data from the Gorm Field at the Danish sector of the North Sea in the 1980s. Kinematic properties of "freak waves" were investigated and the exceedance of the crest height from theoretical estimates were pointed out. The renowned "Draupner Wave" event was reported by Haver [2004] in association with the presence of a major low and a smaller low migrating southwards in the North Sea. Simulations of sea states during the Draunper event based on reanalysis winds were conducted by numerous authors, Adcock et al. [2011], Janssen [2015], Cavaleri et al. [2016, 2017] and references therein, revealing that the event was associated with an anomalous migration and strengthening of the low.

The rapid narrowing of the directional spectrum in realistic sea state occurs under a variety of weather conditions. During marine accidents, Tamura et al. [2009] showed that the spectral narrowing occurred when a low pressure system migrated along the seasonal stationary front near Japan, while Waseda et al. [2014] showed spectral narrowing in a running fetch condition with a moving gale system. A systematic narrowing of the directional spectrum occurred due to a meridional shift of the storm track in the North Sea [Waseda et al., 2012]. The spectral narrowing is represented by the increase of the quality factor or Goda's Q_p, and the decrease of the directional spreading σ_θ. In a majority of the marine accident cases near Japan, spectral narrowing occurred and the rapid change was depicted in a trajectory of the spectral geometry in the $Q_p - \sigma_\theta$ diagram [Waseda et al., 2012]. Albeit there is no direct connection between the generation of freak waves and the marine accident, the coincidence of the timing of the incident and the narrowing of the directional spectrum indicates that such parameters might serve as a warning system (Fig. 4.10). Indeed, in most cases, the directional spectrum narrowed when a bimodal mixed sea state shifted to a unimodal windsea. For the case of the Suwa-Maru incident studied by Tamura et al. [2009], the swell system grew at the expense of the windsea energy, around 30° off the direction and around 80% of the wavelength. This peculiar spectral evolution was accounted for by the nonlinear interaction between Swell and windsea [Masson et al., 1993].

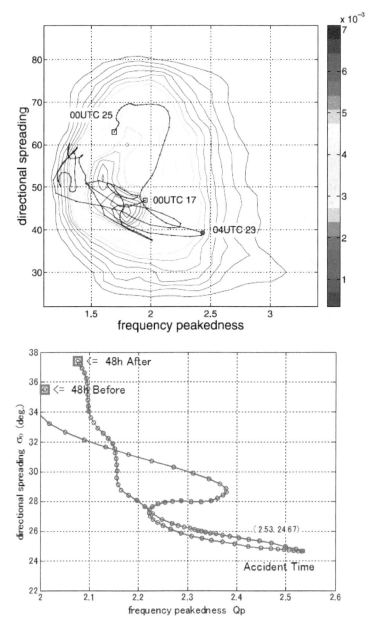

Fig. 4.10. The $Q_p - \sigma_\theta$ diagram during two marine incidents. The timing of the Suwa-Maru accident in June 29, 2008 and Onomichi-Maru accident on December 30, 1980 are indicated by red dots. (Adapted from Tamura *et al.*, 2009; Waseda *et al.*, 2014)

The geometry of the directional spectrum varies within the typhoon footprint due to enhanced wind speed in the right side of the typhoon [Young, 1999; Mori, 2012] and within the extra-tropical cyclone footprints due to the presence of fronts [Kita *et al.*, 2018]. The typhoon translation speed is relatively slow, and therefore, the wave severity coincides with the typhoon strength, whereas in the extra-tropical cyclone case, because of its swift translation, the wave energy enhances with a delay to the deepening of the mean sea level pressure of the cyclone.

Numerous other studies were made looking into the wave spectrum during marine accidents and observed freak wave cases, and the majority show a unimodal windsea system. The Andreas wave [Magnusson and Donelan, 2013] is associated with a unimodal windsea [Bitner-Gregersen *et al.*, 2014; Dias *et al.*, 2015], and the Killard event as well [Fedele *et al.*, 2016]. The Prestige incident occurs during a crossing sea state [Trulsen *et al.*, 2015]. In an extensive analysis of marine accidents, Toffoli *et al.* [2005] found that the sea states vary from following and opposing unimodal wave system to crossing sea states. The spectral geometries of the marine accident cases studied in Waseda *et al.* [2012] and the two observed freak wave cases [JKEO-Narrow and JKEO-Broad; Waseda *et al.*, 2014] are summarized in Fig. 4.11. Qualitatively, a larger percentage of the marine accidents occur when the directional spectra are narrow. The next step is to quantify the spectral narrowness.

4.4.3. *Monte Carlo simulations by phase-resolved wave models of a known freak wave events*

To evaluate the statistical quantities of a given directional spectrum, a Monte Carlo simulation with a phase-resolved model is becoming a standard technique with advanced computational resources. Tanaka [2001a,b] has conducted simulations using the so-called High-Order Spectral Method [HOSM: West *et al.*, 1987; Dommermuth and Yue, 1987; Choi, 1995] to investigate deterministic evolutions of directional wave fields. Toffoli *et al.* [2008a,b] and Bitner-Gregersen [2012] made use of HOSM to study the statistics of unidirectional

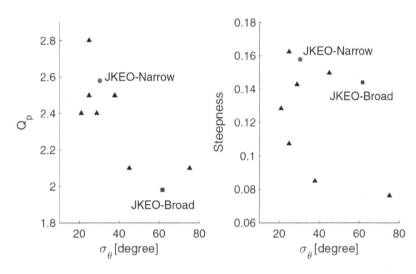

Fig. 4.11. The spectral parameters during marine accident cases. Adapted from Fujimoto et al. [2018].

as well as directional wave fields. An extensive parameter sweep numerical experiment was conducted by Xiao et al. [2013] revealing that the modulational instability is suppressed as the directional spectrum broadens, consistent with the earlier conjecture based on tank experiments limited in parameter range [Waseda et al., 2006, 2009; Onorato et al., 2009a,b]. In all these studies using HOSM simulation, parameterized wave spectra such as JONSWAP and cosine to the power directional spreads were used.

Now, the question is raised regarding realistic spectra. HOSM simulations based on the directional spectra from third-generation wave models were conducted by Bitner-Gregersen et al. [2014] and Dias et al. [2015] for the Andrea wave case, Fedele et al. [2016] for the Draupner, Andrea and Killard cases, and Trulsen et al. [2015] for the Prestige incident. The values of total kurtosis were small, and were comparable to Janssen's bound kurtosis; $\mu_{40}^b = (9/2)(ak)^2$ [Janssen, 2009]. The wave steepness was from 0.08 to 0.1, the directional spread was around 20° for the Draupner, Andrea and Killard cases (Table 4.3). What is notable is that all these cases are observed in relatively shallow waters where the $k_p d$ is close to or less than the

Table 4.3. The steepness ak, frequency bandwidth Q_p, directional spreading σ_θ, the Benjamin–Feir Index (unidirectional) BFI, non-dimensional depth $k_p d$, and kurtosis κ_4, for BG 2014 [Bitner-Gregersen et al., 2014], D 2015 [Dias et al., 2015], F 2016 [Fedele et al., 2016], T 2015 [Trulsen et al., 2015] and F 2018 [Fujimoto et al., 2018].

	Cases	spectrum	ak	Q_p	BFI	σ_θ	k_{pd}	κ_4
BG 2014								3.0
D 2015	Andrea	unimodal	0.099	N/A	0.24	21.2	1.47	3.03
								3.04
F 2016	Killard	unimodal	0.078	N/A	0.18	19.5	0.90	2.99
	Draupner	unimodal	0.095	N/A	0.23	22.3	1.30	3.03
T 2015	Prestige	bimodal	0.08	N/A	~0.4	N/A	Deep	3.01
F 2018	JKEO-N	unimodal	0.115	2.58	0.24	30.4	Deep	3.05
	JKEO-B	bimodal	0.083	1.98	0.16	61.7	Deep	3.03

minimum value of modulational instability to occur: $k_p d = 1.36$. The Prestige incident occurred in water deeper than 1000 m, but in a crossing sea state. Therefore, the kurtosis remained small and the distribution was well-explained by the second-order theory [Tayfun and Fedele, 2007]. The study by Fujimoto et al. [2018] analyzed two freak wave cases observed at the deep mooring buoy (5400 m depth) during a passage of a typhoon [Waseda et al., 2011]. The directional spectrum, however, changed significantly in a day: the JKEO-Narrow case with $\sigma_\theta = 30.4$ and JKEO-Broad case with $\sigma_\theta = 61.7$. These two events served as ideal cases to compare the nonlinear evolutions of distinct directional spectra.

In October 2009, within a day, two freak waves were observed at the JAMSTEC Kuroshio Extension Observatory (JKEO) east of Japan (Fig. 4.12). The freak waves were 12.3 m (Hs = 5.8 m) and 13.3 m (Hs = 6.6 m), respectively. As the typhoon Lupit moved north-eastward passing over the JKEO site, the wave field changed drastically from unimodal to bimodal directional spectra. The steepness was lower for the JKEO-Broad case as well. Therefore, the relative significance of the nonlinearity to the dispersion is much more pronounced for the JKEO-Narrow case than the JKEO-Broad case. HOSM ensemble simulations were conducted and the estimated

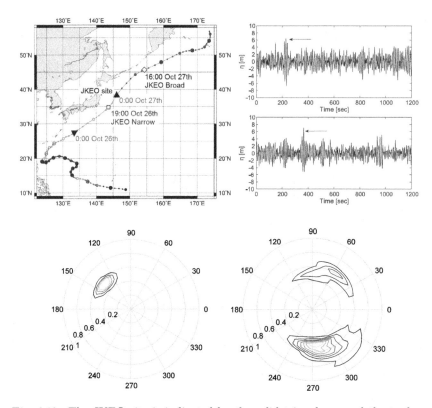

Fig. 4.12. The JKEO site is indicated by the solid triangle ▲, and the typhoon center locations when the two freak waves were observed (□ and ◇). The JKEO-Narrow and JKEO-Broad freak wave time series are shown in the right diagrams. Directional spectra are shown below.

evolutions of the excess kurtosis were distinct (Fig. 4.13). While the kurtosis of the JKEO-Broad case remains unchanged around 0.03 over 50 wave periods, the JKEO-Narrow case shows a gradual increase from 0.03 to 0.05 and slight decrease to 0.04. A notable difference is the separation between the second-order simulation ($M = 2$) and the third-order simulation ($M = 3$) of the HOSM. For the broader case, the difference remains unchanged whereas for the narrower case, the difference changes over time and occasionally disappears (compare solid and dotted lines of Fig. 4.13). The importance of this in separating the total kurtosis to the bound and dynamic parts will be discussed in Section 4.4.4.

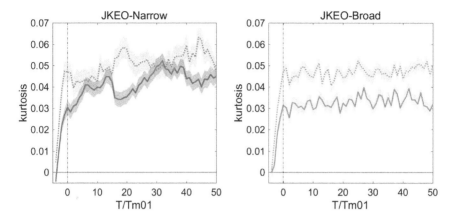

Fig. 4.13. Kurtosis time-series of simulated JKEO-Narrow (red, left) and JKEO-Broad (blue, right) wave fields for HOSM $M = 3$ (solid lines) and $M = 2$ (dotted lines). Shaded area is the standard deviation of ensemble mean kurtosis. Observed on October 26 19:00 and 27 16:00, 2009.

4.4.4. *Dynamic and canonical kurtosis: Connecting the spectral model and phase-resolved model*

The ease of controlling the degree of nonlinearity is the attraction of the High-Order Spectral Method. The order $M = 3$ is typically used, and, at that order, the HOSM formulation is equivalent to the Zakharov's equation at the third-order [Tanaka, 2001b]. There are studies that employed the higher-order HOSM to assure full-nonlinearity, e.g., $M = 6$ [Slunyaev et al., 2014; Clamond et al., 2016]. A comprehensive account of the HOSM formulation is given in Tanaka [2001b] and here the formulation is introduced to define the nonlinear order M. The motion of a horizontally unbounded body of irrotational and inviscid fluid with a flat bottom can be expressed by a velocity potential satisfying the Laplace's equation and the free surface boundary conditions [Zakharov, 1967]:

$$\frac{\partial \eta}{\partial t} + \nabla_H \Phi \cdot \nabla_H \eta = (1 + (\nabla_H \eta)^2)W,$$
$$\frac{\partial \Phi}{\partial t} + \frac{1}{2}(\nabla_H \Phi)^2 + g\eta = \frac{1}{2}(1 + (\nabla_H \eta)^2)W^2, \quad (4.28)$$

where $\Phi(x,y,t) = \phi(x,y,\eta,t)$ is the velocity potential evaluated at the free surface η, and W is the vertical velocity at the free surface. The velocity potential is expanded in a series $\phi(x,y,z,t) = \sum_{m=1}^{M} \phi^m(x,y,z,t)$ where $O(\phi^m) = O((ak)^m)$, and $\phi^m(x,y,\eta,t)$ is Taylor expanded around $z=0$:

$$\phi^m(x,y,\eta,t) = -\sum_{k=1}^{m-1} \frac{\eta^k \partial^k}{k! \partial z^k} \phi^{m-k}(x,y,0,t). \tag{4.29}$$

The W is expanded in a series in ascending order of wave steepness:

$$W(x,y,t) = \sum_{m=1}^{M} W^{(m)}, \quad W^{(m)} = \sum_{k=0}^{m-1} \frac{\eta^k}{k!} \frac{\partial^{k+1}}{\partial z^{k+1}} \phi^{m-k}(x,y,0,t). \tag{4.30}$$

Typically, $M = 3$ is used to include four-wave resonance. Numerous authors [Tanaka, 2001; Onorato et al., 2007; Fujimoto et al., 2018] have noted that HOSM $M = 3$ is equivalent to the Zakharov's equation for the complex amplitude function $a(\boldsymbol{k},t) = \sqrt{\omega(\boldsymbol{k})/2k}\hat{\eta}(\boldsymbol{k}) + i\sqrt{k/2\omega(\boldsymbol{k})}\hat{\psi}(\boldsymbol{k})$. With the tedious canonical transformation $b(\boldsymbol{k},t) = a(\boldsymbol{k},t) + \text{H.O.T.}$, the reduced gravity equation (4.11) is derived. While the reduced gravity equation describes the interaction of "free waves" that satisfies the dispersion relationship of propagating waves $\omega(\boldsymbol{k})^2 = gk$, the Zakharov equation for the $a(\boldsymbol{k},t)$ involves the interaction of "free waves" and "bound waves". That is equivalent to the fact that the HOSM formulation does not distinguish free and bound waves in $\phi^m(x,y,z,t)$. Bound waves result from the second-order nonlinearity and account for the crest and trough asymmetry.

Alternatively, as a result of the canonical transformation, the interaction coefficient of the reduced gravity equation includes contribution of bound components [Janssen, 2004]:

$$T_{0123} = W^{(2)}_{0123} + f(V^{(-)}, V^{(+)}). \tag{4.31}$$

The $W^{(2)}_{0123}$ is the direct interaction and $f(V^{(-)}, V^{(+)})$ is the virtual-state interaction that involves the second harmonic of the free waves.

Consequently, the virtual-state interaction appears at the $M = 2$ HOSM simulation. For that reason, the Benjamin–Feir instability occurs at the $M = 2$ HOSM simulation, but the growth rate is half of the true growth rate obtained at $M = 2$ [Fujimoto and Waseda, 2016]. This is why the interpretation of the Benjamin–Feir instability, (4.17), as the interaction of two sideband waves and second harmonic of the carrier wave is incorrect.

The HOSM $M = 2$, therefore, would include the effects of both the canonical transformation and a part of the four-wave resonance via the second harmonics. The four-wave resonance of the four free waves appear only at HOSM $M = 3$. Accordingly, the difference of $M = 2$ and $M = 3$ HOSM simulations will indicate the contribution of the four-wave resonance, that is the $W_{0123}^{(2)}$. It is incorrect to state that the kurtosis estimate of the $M = 2$ HOSM as the canonical or bound kurtosis μ_{40}^b and the $M = 3$ as the dynamic kurtosis μ_{40}^d (personal communication W. Choi, May 24, 2018). It should be noted that a part of the dynamic kurtosis originating in $f(V^{(-)}, V^{(+)})$ is included in the kurtosis estimate of $M = 2$ HOSM. However, it is correct to interpret the difference as a dynamic contribution of the four-wave resonance. In the case of the JKEO freak waves, this contribution was negative and varied in time for the narrow spectrum but remained unchanged for the broad spectrum. The result can be interpreted as the enhanced contribution of the detuned resonance as the spectrum narrows. A possible consequence of this result for the wave group formation will be discussed next.

4.4.5. *Beyond the occurrence probability: Lifetime of freak wave groups and freak wave shape*

Over the last two decades, freak wave research has been centered around the understanding of the occurrence probability. Agreement of numerous numerical simulations as well as tank experiments of unidirectional wave trains with weak nonlinear theory convincingly demonstrated the significant role of detuned resonance. However, the effect diminishes for short-crested wave fields, and did not seem to explain the observed freak waves [Christou and Ewans, 2014].

The conjecture, however, seems to contradict the fact that the observed breaking wave patterns in the ocean bear the characteristics of modulated wave trains (see Section 4.3.6 and Chapter 3). Indeed, recent work employing nonlinear Fourier analysis [Osborne, 2010] suggests that ocean waves consist of nonlinear coherent structures (Stokes waves and breather), Osborne et al. [2017].

The HOSM simulation results of the JKEO-Narrow and JKEO-Broad were analyzed to identify freak wave groups [Fujimoto et al., 2018]. Examples of the identified freak wave groups are shown in Fig. 4.14. In contrast to the scattered spots in the JKEO-Broad case (right figure), elongated freak wave groups are identified in the JKEO-Narrow case (left figure). These freak wave groups have a much longer lifetime than the ones found in the JKEO-Broad case. Noteworthy is the coexistence of the freak wave groups of longer and shorter lifetimes in the case of JKEO-Narrow. Probability density functions of the lifetimes of the identified freak wave groups are shown in Fig. 4.15. Unlike the probability density function of the wave height, the pdf of the lifetime seems to show a clear distinction between narrow and broad spectral cases. A part of this difference can be explained by linear superposition (as shown by the comparison of $M = 1$ simulations), but the enhancement due to

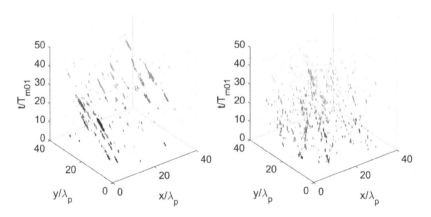

Fig. 4.14. Freak wave groups simulated by HOSM $M = 3$. Left and right panels show JKEO-Narrow and JKEO-Broad cases, respectively. Color indicates indices assigned to each freak wave group. After Fujimoto et al. [2018].

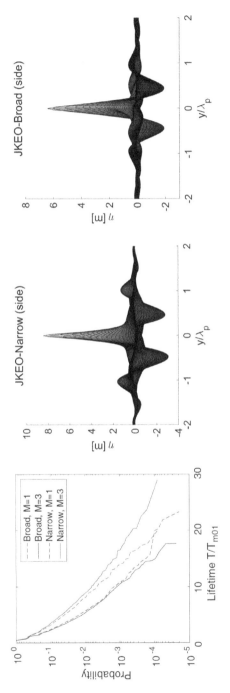

Fig. 4.15. The pdf of lifetime (left). Side views of average freak wave shapes for (middle) JKEO-Narrow and (right) JKEO-Broad of HOSM $M = 3$ calculation. After Fujimoto et al. [2018].

nonlinearity is pronounced for the narrow spectral case ($M = 3$ simulations). The long lifetime freak wave group corresponds to the breather, and the short lifetime freak wave group corresponds to the random superposition of the Stokes waves. Hence, the depiction is consistent with the breather turbulence concept presented by Osborne *et al.* [2017].

Moreover, the averaged shape of the freak waves are distinct (Fig. 4.15). For the JKEO-Narrow case, the front trough is shallower than the rear trough, whereas the JKEO-Broad case remains symmetric. The asymmetry of the wave shape is related to the weakly nonlinear narrow-banded process (Section 4.3.3). However, even the JKEO-Narrow case may not be narrow enough to be represented by a single Nonlinear Schrödinger Equation. To represent the ocean waves with wave groups propagating at different angles is an outstanding question.

4.5. Outstanding Questions

More than 50 years have passed since the pioneers have provided descriptions of ocean waves such as the fetch law, Gaussian probability density function, equilibrium spectrum, directional spread, and four-wave resonance. Owing partly to the research on freak waves, these classical views of ocean waves have been revisited, and questions are raised as to their validity. Is the Fourier spectral representation sufficient or not? Can we represent ocean waves as a superposition of wave groups instead? In this section, some mathematical foundation in existing literature will be reviewed together with a unique representation of ocean waves without nonlinear wave–wave interaction.

4.5.1. *Representing wave field by wave groups*

Observational evidence as well as the analysis of HOSM simulation results suggested the existence of coherent wave groups. Wave group appears as a solution to the Nonlinear Schrödinger Equation [e.g., Su, 1982; Lake *et al.*, 1977], but recent studies reveal that a realistic wave spectrum may be too broad to be represented by the

NLS [Janssen, 2019]. To represent ocean waves as a superposition of coherent wave groups, the appropriate mathematical representation will be the coupled NLS. Rasmussen and Stiassnie [1998] derived the coupled NLS by discretizing the Zakharov equation. Considering a lattice of countable and regularly distributed wave numbers, the integral was replaced by a superposition of sub-integrals within the wave number grids. By projecting the perturbation wave number space into spatial domain, the amplitudes of wave components can be expressed by a function slowly varying in space. As a result, a coupled NLS is derived (nomenclature following Rasmussen and Stiassnie, 1998):

$$i\frac{\partial B_{M,N}}{\partial t} + i\boldsymbol{c_g} \cdot \nabla B_{M,N}$$
$$- \frac{g}{8k_{M,N}\omega_{M,N}} \left(\frac{M^2 - 2N^2}{M^2 + N^2} \frac{\partial^2 B_{M,N}}{\partial x^2} + \frac{N^2 - 2M^2}{M^2 + N^2} \frac{\partial^2 B_{M,N}}{\partial y^2} \right.$$
$$\left. + \frac{6MN}{M^2 + N^2} \frac{\partial^2 B_{M,N}}{\partial x \partial y} \right)$$
$$= \Delta^4 \sum_{m_1,n_1} \sum_{m_2,n_2} \sum_{m_3,n_3} T(\boldsymbol{k}_{M,N}, \boldsymbol{k}_{m_1,n_1}, \boldsymbol{k}_{m_2,n_2}, \boldsymbol{k}_{m_3,n_3})$$
$$\times B_{m_1,n_1} B_{m_2,n_2} B_{m_3,n_3} \delta_K(\boldsymbol{k}_{M,N} + \boldsymbol{k}_{m_1,n_1} - \boldsymbol{k}_{m_2,n_2} - \boldsymbol{k}_{m_3,n_3})$$
$$\times e^{i(\omega_{M,N} + \omega_{m_1,n_1} - \omega_{m_2,n_2} - \omega_{m_3,n_3})t}. \quad (4.32)$$

The M and N are the grid number in wave number space: $\boldsymbol{k}_{M,N} = (M\Delta, N\Delta)$ and Δ is the grid spacing of the wave number lattice. The third term on the left-hand side is the dispersion of each mode and the right-hand side is the nonlinear interaction among the modes. The multiple NLSs are coupled through the nonlinear interaction term.

Similar equations for a unidirectional three wave system of the carrier and two unstable sideband modes were derived and the coupled NLS were solved numerically by Waseda [1997]. The Benjamin–Feir instability and the subsequent recurrent evolution occur almost without the influence of finite spectral bandwidth of each mode. However, at the second cycle of recurrence, the solution starts to deviate

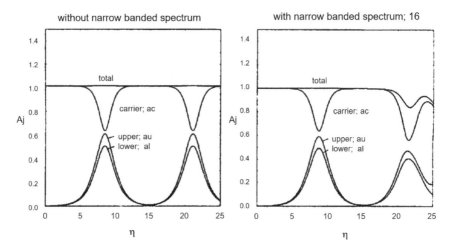

Fig. 4.16. Effect of narrow bandedness to the coupled NLS simulation of an unstable wave train of the carrier, and the two sidebands. After Waseda [1997].

from the purely three-wave system due to spreading of the energy (Fig. 4.16). Whether the coupled NLSs can model a spectrally broad directional spectrum and represent it as a superposition of multiple wave groups is an outstanding research topic.

4.5.2. *Spectral downshifting*

If the wave field is represented by superposition of wave groups, can the spectral downshifting be reproduced? Conventionally, the spectral downshifting was described as a consequence of nonlinear interactions of Fourier modes of the wave spectrum. In the literature, the word "downshifting" represents either the shift of the spectral peak or the shift of the moment period which is an integrated quantity.

Trulsen and Dysthe [1997] noted that by use of the modified NLS, the spectral peak permanently downshifted due to spreading of the energy in direction. The energy was conserved and therefore energy dissipation or the wave breaking did not play a significant role. However, the moment period, or the wave momentum was conserved. The total energy and the momentum of the wave system can be represented by the spectral density:

$$E_T = \iint F(\omega,\theta)d\omega d\theta,$$
$$M_T = \frac{1}{g}\iint \omega F(\omega,\theta)d\omega d\theta. \qquad (4.33)$$

To understand the fetch law, the evolution of both the total energy and the momentum needs to be considered [Toba, 1973; Tulin, 1994; Fontaine, 2013]. When there is an imbalance of energy and momentum increase or decrease, the frequency as defined as $\omega_T \equiv gM/E$ will change. Studies showed that the loss of energy due to energetic breakers work to reduce the ω_T [Tulin and Waseda, 1999; Waseda et al., 2009]. The modulated wave train permanently shifted energy to the lower sideband when the wave train undergoes energetic breaking. However, the permanent downshift was observed in the early study of Lake et al. [1978] where no energetic breaking was reported. A possible explanation for this is the loss of energy due to wall friction, as Tulin and Waseda [1999] demonstrated recurrence in a much wider tank without breaking. The permanent downshifting due to energy loss due to breaking was indicated in Melville [1982], and confirmed by Tulin and Waseda [1999] for a longer fetch (Fig. 4.17). In addition, permanent downshift was further demonstrated experimentally by Hwung et al. [2011] and theoretically by Shugan et al. [2018].

Overall, the downshifting mechanism works as follows. Due to weakly nonlinear process, the energy is spread to low and high frequencies. As depicted by Trulsen and Dysthe [1997], the spectral peak appears to downshift while the total energy and momentum are conserved. Now, when energetic breaking occurs, the energy is lost not as a wave satisfying $\Delta E = \Delta M(g/\omega)$ and therefore the remaining wave system downshifts. For a simple modulated wave train, wave breaking leaves the lower sideband permanently being the dominant wave. Thus, the energy appears to shift from the carrier wave to the lower sideband wave.

Donelan et al. [2012] implemented a wave model that disregards the Hasselman nonlinear source term and replace it with a downshifting mechanism based on breaking energy loss. For each wave mode,

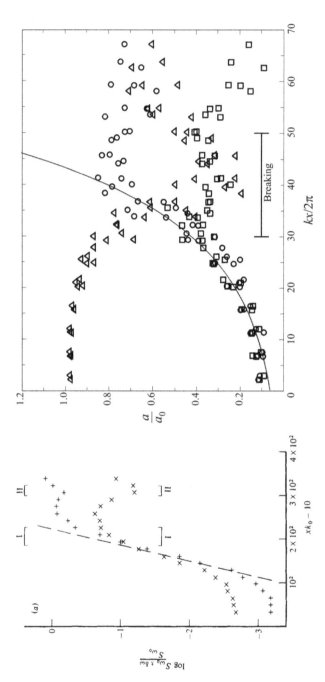

Fig. 4.17. The evolution of sideband amplitudes and the carrier wave amplitude of an unstable wave train. Adapted from Melville [1982], and from Tulin and Waseda [1999].

the energy "lost" due to spilling wave breakers will be shifted to two lower frequency modes. Therefore, the process is "conservative" as the total energy is not reduced due to the spilling breaker. The transfer of energy occurs only along the direction of that wave mode. This is based on the idea that ocean waves consist of wave groups, and at each point and time, waves are propagating coherently in one direction only [Donelan et al., 1996]. The model successfully reproduced realistic sea state [Donelan et al., 2012] but seems to require further improvement during a hurricane condition [Liu et al., 2017]. An important takeaway from this study is that the successful estimation of the wave field in terms of wave height and period does not prove that the modeled physical parameterization is correct. The result that the nonlinear wave–wave interaction is not needed is rather surprising and should be investigated further. Donelan and Magnussen [2017] go beyond spectral representation and show that freak wave occurrence probability can be explained by the random superposition of wave groups. What is not evident at this point is how these wave groups retain coherence when nonlinearity is disregarded. Even if weak nonlinearity is retained, the balance of linear diffraction and nonlinearity is yet to be understood [Babanin and Waseda, 2015]. Presence of coherent wave groups in the ocean is an outstanding question.

4.5.3. *Nonlinear source term: Beyond the third generation*

Hasselmann's kinetic integral (4.12) based on quasi-Gaussian closure has been re-investigated recently and the generalized kinetic equations as well as the evolution equation for the skewness and kurtosis were derived [Janssen, 2003; Gramstad and Stiassnie, 2013; Annenkov and Shrira, 2014, 2016, 2018]. The transient behavior at the initial stage when detuned resonance results in deviation from Gaussianity is of interest in relation to the freak wave generation. The kurtosis increases at a dynamic timescale $O((ak)^{-2})$ but eventually reduces and asymptotes to zero at a kinetic timescale $O((ak)^{-4})$

[Janssen, 2019]. Attempt has been made to include such non-resonant nonlinear processes in third-generation wave models [Gramstad and Babanin, 2016]. Tanaka [2001] on the other hand, conducted a direct numerical simulation with HOSM and obtained a nonlinear transfer function resembling Hasselmann's source term in a much shorter timescale $O(10T_p)$. This somewhat controversial result arises from the averaging of the spectrum in the $\boldsymbol{k}$-plane rather than averaging in time.

Regardless of these generalizations and alternative interpretations of Hasselmann's kinetic theory, third-generation wave models are still extensively used in operational settings. With increased computational power, the full Boltzmann integral or the Web–Resio–Tracy numerical schemes are used in realistic configurations [Liu et al., 2017; Ponce de León et al., 2018]. Compared to the DIA scheme (see Chapter 5), the energetic part of the spectrum becomes much narrower when the Hasselmann integral is evaluated with the WRT method [Ponce de León et al., 2018]. Since spectral narrowness is the key for the modulational instability to work, further study with the exact solution to the Hasselmann integral is warranted to understand the freak wave generation mechanism. Of course, the wind field is rapidly changing in time and the use of the generalized kinetic equation in a realistic setting is desired as well [Annenkov and Shrira, 2018].

4.5.4. *Nonlinear evolution under external forcing*

To conclude this chapter on nonlinear processes, the possible coupling of external forcing and resonant interactions will be discussed. The third-generation wave model solves the Action balance equation:

$$\frac{\partial N}{\partial t} + \nabla \cdot ((\boldsymbol{c}_g + \boldsymbol{U})N) = \frac{1}{\sigma}(S_{\mathrm{nl}} + S_{\mathrm{in}} + S_{\mathrm{ds}}). \tag{4.34}$$

The underlying assumption is that the source terms are independent of each other, and the current $\boldsymbol{U}$ does not affect the source terms.

Toba [1988] has suggested that waves, and wind and wave turbulence are not independent of each other. Therefore, the source terms are not independent of each other as well. Indeed, if one assumes a proportionality of wind input and dissipation, the Toba's 3/2 law can be retrieved [Waseda et al., 2001].

Now, how can current affect the source terms? The formal derivation of resonant interactions under the influence of a shear current and air pressure is presented in Zhou [1996] assuming $O(|\boldsymbol{U}|/|\boldsymbol{c}_g|) = O(ak)$ and weak pressure $p'/\rho C^2 < O((ak)^2)$. The nonlinear interaction coefficient of the Zakharov equation is modified while the dispersion relation remains unaffected. The derived modified Zakharov equation was used to study the modulated wave train, and was shown that while recurrence is unaffected by the current, the wind input enhanced spectral downshifting. Hence $S_{\rm in}$ and $S_{\rm nl}$ are not independent.

Experimentally, Waseda et al. [2015] have studied four-wave resonance under the influence of a background current field. Because of the large variation of the current field, the waves experienced random perturbation by the current, and as a result, the resonance interaction was suppressed. In this case, the current magnitude was $O(|\boldsymbol{U}|/|\boldsymbol{c}_g|) = 1$ and therefore the dispersion relation was modified. Moreover, because of the spatial current gradient, the waves refracted as well. Consequently, the resonance condition was randomly perturbed. The net effect was incorporated in the resonance detuning term in Eq. (4.9), $\exp(-i(\Delta_{1234} + \Delta_{kU})t)$, where Δ_{kU} is a random variable. As the relative magnitude of Δ_{kU} increases, mimicking the increase of the current variance, the resonant growth was suppressed (Fig. 4.18). The same effect influenced the development of the spectral tail of the random directional waves. As the background random current field strengthened, the spectral tail deviated from the ω^{-5} equilibrium and became steeper (Fig. 4.19). This result was further confirmed by additional experiments by Rapizo et al. [2016]. Impact of external forcing to the resonant interaction is an outstanding problem.

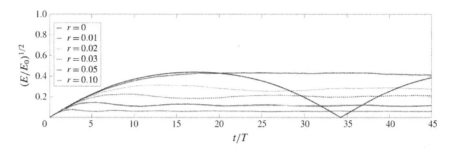

Fig. 4.18. Evolution of the amplitude of the initially non-existing member of the DIA quartet numerically estimated by solving Zakharov's equation. The solid line corresponds to exact resonant interaction whereas the other lines correspond to cases with randomly perturbed resonance detuning parameters. After Waseda et al. [2015].

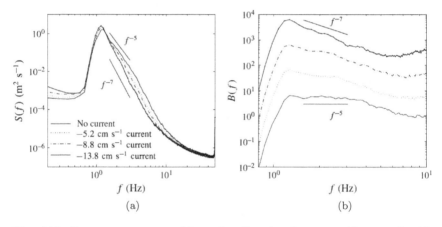

Fig. 4.19. Frequency spectra of irregular directional waves without and with opposing current, -5.2, -8.8 and -13.8 cm s^{-1}. The unperturbed spectrum is JONSWAP with a directional spread of $G(\theta) \propto \cos^{10} \theta$. Right figure is the saturated spectrum $B(f) = S(f) \times f^5$. After Waseda et al. [2015].

References

Adcock, T.A.A., Taylor, P.H., Yan, S., Ma, Q.W. and Janssen, P.A.E.M. (2011). Did the Draupner wave occur in a crossing sea? *Proc. R. Soc. A*, **467**(2134), 3004–3021.

Akhmediev, N.N. and Korneev, V.I. (1986). Modulation instability and periodic solutions of the nonlinear Schrödinger equation, *Theor. Math. Phys.* **69**(2), 1089–1093.

Akhmediev, N., Soto-Crespo, J.M. and Ankiewicz, A. (2009). Extreme waves that appear from nowhere: On the nature of rogue waves, *Phys. Lett. A*, **373**(25), 2137–2145.

Alber, I.E. (1978). The effects of randomness on the stability of two-dimensional surface wavetrains, *Proc. R. Soc. Lond. A*, **363**(1715), 525–546.

Annenkov, S.Y. and Shrira, V.I. (2014). Evaluation of skewness and kurtosis of wind waves parameterized by JONSWAP spectra, *J. Phy. Oceanogr.*, **44**(6), 1582–1594.

Annenkov, S.Y. and Shrira, V.I. (2016). Modelling transient sea states with the generalised kinetic equation, in *Rogue and Shock Waves in Nonlinear Dispersive Media*, Springer, Cham, pp. 159–178.

Annenkov, S.Y. and Shrira, V.I. (2018). Spectral evolution of weakly nonlinear random waves: Kinetic description versus direct numerical simulations, *J. Fluid Mech.*, **844**, 766–795.

Babanin, A.V. and Waseda, T. (2015). Diffraction and instability of short-crested limited-length one-dimensional coherent wave trains, in *ASME 2015: 34th Int. Conf. Ocean, Offshore and Arctic Engineering*, American Society of Mechanical Engineers, pp. V003T02A023-V003T02A023.

Benjamin, T.B. (1967). Instability of periodic wavetrains in nonlinear dispersive systems, *Proc. R. Soc. Lond. A*, **299**(1456), 59–76.

Benjamin, T.B. and Feir, J.E. (1967). The disintegration of wave trains on deep water Part 1. Theory, *J. Fluid Mech.*, **27**(3), 417–430.

Benney, D.J. (1962). Non-linear gravity wave interactions, *J. Fluid Mech.*, **14**(4), 577–584.

Bitner-Gregersen, E.M. and Toffoli, A. (2012). On the probability of occurrence of rogue waves, *Nat. Hazards Earth Syst. Sci.*, **12**(3), 751–762.

Bitner-Gregersen, E.M., Fernandez, L., Lefèvre, J.M., Monbaliu, J. and Toffoli, A. (2014). The North Sea Andrea storm and numerical simulations, *Nat. Hazards Earth Syst. Sci.*, **14**(6), 1407–1415.

Bliven, L.F., Huang, N.E. and Long, S.R. (1986). Experimental study of the influence of wind on Benjamin–Feir sideband instability, *J. Fluid Mech*, **162**, 237–260.

Cavaleri, L., Barbariol, F., Benetazzo, A., Bertotti, L., Bidlot, J.R., Janssen, P. and Wedi, N. (2016). The Draupner wave: A fresh look and the emerging view, *J. Geophys. Res.: Oceans*, **121**(8), 6061–6075.

Cavaleri, L., Benetazzo, A., Barbariol, F., Bidlot, J.R. and Janssen, P.A.E.M. (2017). The Draupner event: The large wave and the emerging view, *Bull. Am. Meteorol. Soc.*, **98**(4), 729–735.

Cartwright, D.E. and Longuet-Higgins, M.S. (1956). The statistical distribution of the maxima of a random function, *Proc. R. Soc. Lond. A*, **237**(1209), 212–232.

Chabchoub, A., Hoffmann, N.P. and Akhmediev, N. (2011). Rogue wave observation in a water wave tank, *Phys. Rev. Lett.*, **106**(20), 204502.

Chabchoub, A., Hoffmann, N., Onorato, M. and Akhmediev, N. (2012a). Super rogue waves: Observation of a higher-order breather in water waves, *Phys. Rev. X*, **2**(1), 011015.

Chabchoub, A., Hoffmann, N., Onorato, M., Slunyaev, A., Sergeeva, A., Pelinovsky, E. and Akhmediev, N. (2012b). Observation of a hierarchy of up to fifth-order rogue waves in a water tank, *Phys. Rev. E*, **86**(5), 056601.

Chabchoub, A., Onorato, M. and Akhmediev, N. (2016). Hydrodynamic envelope solitons and breathers, in *Rogue and Shock Waves in Nonlinear Dispersive Media*, Springer, Cham, pp. 55–87.

Choi, W. (1995). Nonlinear evolution equations for two-dimensional surface waves in a fluid of finite depth, *J. Fluid Mech.*, **295**, 381–394.

Christou, M. and Ewans, K. (2014). Field measurements of rogue water waves, *J. Phys. Oceanogr.*, **44**(9), 2317–2335.

Clamond, D., Francius, M., Grue, J. and Kharif, C. (2006). Long time interaction of envelope solitons and freak wave formations, *Eur. J. Mech. B/Fluids*, **25**(5), 536–553.

Dias, F., Brennan, J., de León, S.P., Clancy, C. and Dudley, J. (2015). Local analysis of wave fields produced from hindcasted rogue wave sea states, in *ASME 2015: 34th Int. Conf. Ocean, Offshore and Arctic Engineering*. American Society of Mechanical Engineers, pp. V003T02A020-V003T02A020.

Dommermuth, D.G. and Yue, D.K. (1987). A high-order spectral method for the study of nonlinear gravity waves, *J. Fluid Mech.*, **184**, 267–288.

Donelan, M., Longuet-Higgins, M.S. and Turner, J.S. (1972). Periodicity in whitecaps, *Nature*, **239**(5373), 449.

Donelan, M.A., Drennan, W.M. and Magnusson, A.K. (1996). Nonstationary analysis of the directional properties of propagating waves, *J. Phys. Oceanogr.*, **26**(9), 1901–1914.

Donelan, M.A., Curcic, M., Chen, S.S. and Magnusson, A.K. (2012). Modeling waves and wind stress, *J. Geophys. Res.: Oceans*, **117**(C11), C00J23.

Dudley, J.M., Genty, G., Dias, F., Kibler, B. and Akhmediev, N. (2009). Modulation instability, Akhmediev Breathers and continuous wave supercontinuum generation, *Opt. Express*, **17**(24), 21497–21508.

Dysthe, K.B. (1979). Note on a modification to the nonlinear Schrödinger equation for application to deep water waves, *Proc. R. Soc. Lond. A*, **369**(1736), 105–114.

Fontaine, E. (2013). A theoretical explanation of the fetch- and duration-limited laws, *J. Phys. Oceanogr.*, **43**(2), 233–247.

Fuchs, J., Regas, D., Waseda, T., Welch, S. and Tulin, M.P. (1999). Correlation of hydrodynamic features with LGA radar backscatter from breaking waves, *IEEE Trans. Geosci. Remote Sensing*, **37**(5), 2442–2460.

Fujimoto, W. and Waseda, T. (2016). The relationship between the shape of freak waves and nonlinear wave interactions, in *ASME 2016: 35th Int. Conf. Ocean, Offshore and Arctic Engineering*. American Society of Mechanical Engineers, pp. V003T02A003-V003T02A003.

Fujimoto, W., Waseda, T. and Webb, A. (2018). Impact of the four-wave quasi-resonance on freak wave shapes in the ocean, *Ocean Dynam.*, **69**(1), 101–121.

Goda, T. (1976). On the statistics of wave groups, *Rep. Port Harbour Res. Inst.*, **15**(3), 3–19.

Gramstad, O. and Trulsen, K. (2007). Influence of crest and group length on the occurrence of freak waves, *J. Fluid Mech.*, **582**, 463–472.

Gramstad, O. and Stiassnie, M. (2013). Phase-averaged equation for water waves, *J. Fluid Mech.*, **718**, 280–303.

Gramstad, O. and Babanin, A. (2016). The generalized kinetic equation as a model for the nonlinear transfer in third-generation wave models, *Ocean Dynam.*, **66**(4), 509–526.

Hasselmann, K. (1962). On the non-linear energy transfer in a gravity-wave spectrum Part 1. General theory, *J. Fluid Mech.*, **12**(4), 481–500.

Hasselmann, S. and Hasselmann, K. (1985). Computations and parameterizations of the nonlinear energy transfer in a gravity-wave spectrum. Part I: A new method for efficient computations of the exact nonlinear transfer integral. *J. Phy. Oceanogr.*, **15**(11), 1369–1377.

Hatori, M. (1984). Nonlinear properties of laboratory wind–waves at energy containing frequencies. Part 1. Probability density distribution of surface elevation, *J. Oceanogr. Soc. Japan*, **40**(1), 1–11.

Hatori, M. (1984). Nonlinear properties of laboratory wind–waves at energy containing frequencies. Part 2. Detailed structures of power spectral and their evolution with fetch, *J. Oceanogr. Soc. Japan*, **40**(1), 12–18.

Haver, S. (2004). A possible freak wave event measured at the Draupner Jacket January 1 1995, in *Rogue waves*, Vol. 2004, pp. 1–8.

Houtani, H., Waseda, T. and Tanizawa, K. (2018). Experimental and numerical investigations of temporally and spatially periodic modulated wave trains, *Phys. Fluids*, **30**(3), 034101.

Houtani, H., Waseda, T., Fujimoto, W., Kiyomatsu, K. and Tanizawa, K. (2018). Generation of a spatially periodic directional wave field in a rectangular wave basin based on higher-order spectral simulation, *Ocean Eng.*, **169**, 428–441.

Hwung, H.H., Chiang, W.S., Yang, R.Y. and Shugan, I.V. (2011). Threshold model on the evolution of Stokes wave sideband instability, *Eur. J. Mech. B/Fluids*, **30**(2), 147–155.

Janssen, P.A. (2003). Nonlinear four-wave interactions and freak waves, *J. Phys. Oceanogr.*, **33**(4), 863–884.

Janssen, P. (2004). *The Interaction of Ocean Waves and Wind*. Cambridge University Press.

Janssen, P.A.E.M. and Bidlot, J. (2009). On the extension of the freak wave warning system and its verification, European Centre for Medium-Range Weather Forecasts.

Janssen, P.A. (2009). On some consequences of the canonical transformation in the Hamiltonian theory of water waves, *J. Fluid Mech.*, **637**, 1–44.

Janssen, P.A. (2015). How rare is the Draupner wave event? European Centre for Medium-Range Weather Forecasts.

Janssen, P.A. and Janssen, A.J. (2019). Asymptotics for the long-time evolution of kurtosis of narrow-band ocean waves, *J. Fluid Mech.*, **859**, 790–818.

Kartashova, E. (2010). *Nonlinear Resonance Analysis: Theory, Computation, Applications*, Cambridge University Press.

Kartashova, E. and Shugan, I.V. (2011). Dynamical cascade generation as a basic mechanism of Benjamin–Feir instability, *Europhys. Lett.*, **95**(3), 30003.

Kinsman, B. (1984). Wind–waves: Their Generation and Propagation on the Ocean Surface, Courier Corporation.

Kharif, C., Pelinovsky, E. and Slunyaev, A. (2009). Quasi-linear wave focusing, in *Rogue Waves in the Ocean*. Springer, Berlin, pp. 63–89.

Kita, Y., Waseda, T. and Webb, A. (2018). Development of waves under explosive cyclones in the Northwestern Pacific, *Ocean Dynam.*, **68**(10), 1403–1418.

Krasitskii, V.P. (1994). On reduced equations in the Hamiltonian theory of weakly nonlinear surface waves, *J. Fluid Mech.*, **272**, 1–20.

Lake, B.M., Yuen, H.C., Rungaldier, H. and Ferguson, W.E. (1977). Nonlinear deep-water waves: Theory and experiment. Part 2. Evolution of a continuous wave train, *J. Fluid Mech.*, **83**(1), 49–74.

Lake, B.M. and Yuen, H.C. (1978). A new model for nonlinear wind–waves. Part 1. Physical model and experimental evidence, *J. Fluid Mech.*, **88**(1), 33–62.

Lamont-Smith, T., Fuchs, J. and Tulin, M.P. (2003). Radar investigation of the structure of wind waves, *J. Oceanogr.*, **59**(1), 49–63.

Levi, C., Welch, S., Fontaine, E. and Tulin, M.P. (1998). Experiments on the ringing response of an elastic cylinder in breaking wave groups, in *Proc. 13th Int. Workshop on Water Waves and Floating Bodies*, Alphen aan den Rijn, The Netherlands, p. 79.

Liu, Q., Babanin, A., Fan, Y., Zieger, S., Guan, C. and Moon, I.J. (2017). Numerical simulations of ocean surface waves under hurricane conditions: Assessment of existing model performance, *Ocean Model.*, **118**, 73–93.

Lo, E. and Mei, C.C. (1985). A numerical study of water-wave modulation based on a higher-order nonlinear Schrödinger equation, *J. Fluid Mech.*, **150**, 395–416.

Lo, E.Y. and Mei, C.C. (1987). Slow evolution of nonlinear deep water waves in two horizontal directions: A numerical study, *Wave Motion*, **9**(3), 245–259.

Longuet-Higgins, M.S. and Smith, N.D. (1966). An experiment on third-order resonant wave interactions, *J. Fluid Mech.*, **25**(3), 417–435.

Magnusson, A.K. and Donelan, M.A. (2013). The Andrea wave characteristics of a measured North Sea rogue wave, *J. Offshore Mech. Arct. Eng.*, **135**(3), 031108.

Masson, D. (1993). On the nonlinear coupling between swell and wind–waves, *J. Phys. Oceanogr.*, **23**(6), 1249–1258.

McGoldrick, L.F., Phillips, O.M., Huang, N.E. and Hodgson, T.H. (1966). Measurements of third-order resonant wave interactions, *J. Fluid Mech.*, **25**(3), 437–456.

Mei, C.C., Stiassnie, M. and Yue, D.K.P. (1989). Theory and Applications of Ocean Surface Waves: Part 1: Linear Aspects; Part 2: Nonlinear Aspects.

Melville, W.K. (1982). The instability and breaking of deep-water waves, *J. Fluid Mech.*, **115**, 165–185.

Mitsuyasu, H., Tasai, F., Suhara, T., Mizuno, S., Ohkusu, M., Honda, T. and Rikiishi, K. (1975). Observations of the directional spectrum of ocean waves using a cloverleaf buoy, *J. Phys. Oceanogr.*, **5**(4), 750–760.

Mori, N., Liu, P.C. and Yasuda, T. (2002). Analysis of freak wave measurements in the Sea of Japan, *Ocean Eng.*, **29**(11), 1399–1414.

Mori, N. and Janssen, P.A. (2006). On kurtosis and occurrence probability of freak waves, *J. Phys. Oceanogr.*, **36**(7), 1471–1483.

Mori, N., Onorato, M. and Janssen, P.A. (2011). On the estimation of the kurtosis in directional sea states for freak wave forecasting, *J. Phys. Oceanogr.*, **41**(8), 1484–1497.

Mori, N. (2012). Freak waves under typhoon conditions, *J. Geophy. Res Oceans*, **117**(C11).

Nieto-Borge, J.C., Lehner, S., Niedermeier, A. and Schulz-Stellenfleth, J. (2004). Detection of ocean wave groupiness from spaceborne synthetic aperture radar, *J. Geophy. Res.: Oceans*, **109**(C7).

Nieto-Borge, J.C., Reichert, K. and Hessner, K. (2013). Detection of spatio-temporal wave grouping properties by using temporal sequences of X-band radar images of the sea surface, *Ocean Model.*, **61**, 21–37.

Nolte, K.G. and Hsu, F.H. (1972). Statistics of ocean wave groups, in *Offshore Technology Conf.*, 1–3 May, Housto Texas.

Onorato, M., Osborne, A.R., Serio, M. and Bertone, S. (2001). Freak waves in random oceanic sea states, *Phys. Rev. Lett.*, **86**(25), 5831.

Onorato, M., Osborne, A.R., Serio, M., Cavaleri, L., Brandini, C. and Stansberg, C.T. (2004). Observation of strongly non-Gaussian statistics for random sea surface gravity waves in wave flume experiments, *Phys. Rev. E*, **70**(6), 067302.

Onorato, M., Waseda, T., Toffoli, A., Cavaleri, L., Gramstad, O., Janssen, P.A.E.M. and Serio, M. (2009a). Statistical properties of directional ocean waves: the role of the modulational instability in the formation of extreme events, *Phys. Rev. Lett.*, **102**(11), 114502.

Onorato, M., Cavaleri, L., Fouques, S., Gramstad, O., Janssen, P.A., Monbaliu, J. and Toffoli, A. (2009b). Statistical properties of mechanically generated surface gravity waves: A laboratory experiment in a three-dimensional wave basin, *J. Fluid Mech.*, **627**, 235–257.

Onorato, M., Resitori, S. and Baronio, F. (eds.). (2016). *Rogue and Shock Waves in Nonlinear Dispersive Media*, Vol. 926, Springer.

Osborne, A.R. (2010). *Nonlinear Ocean Waves and the Inverse Scattering Transform*, Academic Press.

Osborne, A.R., Resio, D., Costa, A., Ponce de León, S. and Chirivi. E. (2017). Highly nonlinear wind waves in Currituck Sound: Dense breather turbulence in random ocean waves, in *1st International Workshop on Waves, Storm Surges and Coastal Hazards*, September 10, 2017, Liverpool, UK.

Phillips, O.M. (1958). The equilibrium range in the spectrum of wind-generated waves, *J. Fluid Mech.*, **4**(4), 426–434.

Phillips, O.M. (1960). On the dynamics of unsteady gravity waves of finite amplitude Part 1. The elementary interactions, *J. Fluid Mech.*, **9**(2), 193–217.

Phillips, O.M. (1977). *The Dynamics of the Upper Ocean*. Cambridge University Press.

Ponce de León, S., Osborne, A.R. and Guedes Soares, C. (2018). On the importance of the exact nonlinear interactions in the spectral characterization of rogue waves, in *ASME 2018: 37th Int. Conf. Ocean, Offshore and Arctic Engineering*, Madrid, Spain.

Rapizo, H., Waseda, T., Babanin, A.V. and Toffoli, A. (2016). Laboratory experiments on the effects of a variable current field on the spectral geometry of water waves, *J. Phys. Oceanogr.*, **46**(9), 2695–2717.

Rasmussen, J.H. and Stiassnie, M. (1999). Discretization of Zakharov's equation, *Eur. J. Mech. B/Fluids*, **18**(3), 353–364.

Ribal, A., Stiassnie, M., Babanin, A. and Young, I. (2012). On the instability of wave-fields with JONSWAP spectra to inhomogeneous disturbances, and the consequent long-time evolution, in *EGU General Assembly Conf. Abstracts*, Vol. 14, p. 3279.

Rye, H. (1975). Wave group formation among storm waves, in *Coastal Engineering 1974*, pp. 164–183.

Romero, L. (2008). Airborne observations and numerical modeling of fetch-limited waves in the Gulf of Tehuantepec, Ph.D. dissertation, UC San Diego.

Sand, S.E., Hansen, N.O., Klinting, P., Gudmestad, O.T. and Sterndorff, M.J. (1990). Freak wave kinematics, in *Water Wave Kinematics*, Springer, Dordrecht, pp. 535–549.

Sedivy, D.G. and Thompson, W.C. (1978). Ocean wave group analysis, No. NPS-68SETH78091, Naval Postgraduate School, Monterey, CA.

Shemer, L., Jiao, H., Kit, E. and Agnon, Y. (2001). Evolution of a nonlinear wave field along a tank: Experiments and numerical simulations based on the spatial Zakharov equation, *J. Fluid Mech.*, **427**, 107–129.

Shemer, L., Kit, E. and Jiao, H. (2002). An experimental and numerical study of the spatial evolution of unidirectional nonlinear water-wave groups. *Phys. Fluids*, **14**(10), 3380–3390.

Shugan, I., Kuznetsov, S., Saprykina, Y., Hwung, H.H., Yang, R.Y. and Chen, Y.Y. (2018). The permanent downshifting at later stages of Benjamin–Feir instability of waves, *Pure Appl. Geophys.*, **176**(1), 483–500.

Slunyaev, A., Pelinovsky, E. and Soares, C.G. (2014). Reconstruction of extreme events through numerical simulations, *J. Offshore Mech. Arct. Eng.*, **136**(1), 011302.

Smith, M.J., Poulter, E.M. and McGregor, J.A. (1996). Doppler radar measurements of wave groups and breaking waves, *J. Geophys. Res.: Oceans*, **101**(C6), 14269–14282.

Socquet-Juglard, H., Dysthe, K., Trulsen, K., Krogstad, H.E. and Liu, J. (2005). Probability distributions of surface gravity waves during spectral changes, *J. Fluid Mech.*, **542**, 195–216.

Stiassnie, M. and Kroszynski, U.I. (1982). Long-time evolution of an unstable water-wave train, *J. Fluid Mech.*, **116**, 207–225.

Stiassnie, M. (1984). Note on the modified nonlinear Schrödinger equation for deep water waves, *Wave Motion*, **6**(4), 431–433.

Su, M.Y. (1982). Evolution of groups of gravity waves with moderate to high steepness, *Phys. Fluids*, **25**(12), 2167–2174.

Sverdrup, H.U. and Munk, W.H. (1946). Empirical and theoretical relations between wind, sea, and swell, *Eos, Trans. Amer. Geophys. Union*, **27**(6), 823–827.

Tamura, H., Waseda, T., Miyazawa, Y. and Komatsu, K. (2008). Current-induced modulation of the ocean wave spectrum and the role of nonlinear energy transfer, *J. Phys. Oceanogr.*, **38**(12), 2662–2684.

Tamura, H., Waseda, T. and Miyazawa, Y. (2009). Freakish sea state and swell-windsea coupling: Numerical study of the Suwa-Maru incident, *Geophys. Res. Lett.*, **36**(1).

Tamura, H., Waseda, T. and Miyazawa, Y. (2010). Impact of nonlinear energy transfer on the wave field in Pacific hindcast experiments, *J. Geophys. Res.: Oceans*, **115**(C12).

Tanaka, M. (2001). Verification of Hasselmann's energy transfer among surface gravity waves by direct numerical simulations of primitive equations, *J. Fluid Mech.*, **444**, 199–221.

Tanaka, M. (2001). A method of studying nonlinear random field of surface gravity waves by direct numerical simulation, *Fluid Dyn. Res.*, **28**(1), 41.

Tayfun, M.A. (1980). Narrow-band nonlinear sea waves, *J. Geophys. Res.: Oceans*, **85**(C3), 1548–1552.

Tayfun, M.A. and Fedele, F. (2007). Wave-height distributions and nonlinear effects, *Ocean Eng.*, **34**(11–12), 1631–1649.

Toba, Y. (1973). Local balance in the air-sea boundary processes. II. Partition of wind stress to waves and current, *J. Oceanogr. Soc. Japan*, **29**(5), 70–75.

Toba, Y. (1973). Local balance in the air-sea boundary processes. III. On the spectrum of wind waves, *J. Oceanogr. Soc. Japan*, **29**(5), 209–220.

Toffoli, A., Onorato, M., Bitner-Gregersen, E., Osborne, A.R. and Babanin, A.V. (2008). Surface gravity waves from direct numerical simulations of the Euler equations: A comparison with second-order theory, *Ocean Eng.*, **35**(3–4), 367–379.

Toffoli, A., Bitner-Gregersen, E., Onorato, M. and Babanin, A.V. (2008). Wave crest and trough distributions in a broad-banded directional wave field, *Ocean Eng.*, **35**(17–18), 1784–1792.

Tolman, H.L. (2004). Inverse modeling of discrete interaction approximations for nonlinear interactions in wind–waves, *Ocean Model.*, **6**(3–4), 405–422.

Tolman, H.L. (2009). User manual and system documentation of WAVEWATCH III^{TM} version 3.14, Technical note, MMAB Contribution, **276**, 220.

Tørum, A. and Gudmestad, O.T. (eds.). (2012). *Water Wave Kinematics*, Vol. 178. Springer Science & Business Media.

Tracy, B.A. and Resio, D.T. (1982). Theory and calculation of the nonlinear energy transfer between sea waves in deep water (No. WIS-11), Army Engineer Waterways Experiment Station Vicksburg MS Hydraulics Lab.

Trulsen, K. and Dysthe, K.B. (1996). A modified nonlinear Schrödinger equation for broader bandwidth gravity waves on deep water, *Wave Motion*, **24**(3), 281–289.

Trulsen, K. and Dysthe, K.B. (1997). Frequency downshift in three-dimensional wave trains in a deep basin, *J. Fluid Mech.*, **352**, 359–373.

Trulsen, K., Borge, J.C.N., Gramstad, O., Aouf, L. and Lefèvre, J.M. (2015). Crossing sea state and rogue wave probability during the Prestige accident, *J. Geophy. Res.: Oceans*, **120**(10), 7113–7136.

Tulin, M. (1994). Breaking of ocean waves and downshifting, in *Proc. Waves and Nonlinear Processes in Hydrodynamics: A Symp.* in *Honor of Professor Enok Palm on His 70th Birthday*, Oslo, Norway.

Tulin, M.P. (1996). Breaking of ocean waves and downshifting, in *Waves and Nonlinear Processes in Hydrodynamics*, Springer, Dordrecht, pp. 177–190.

Tulin, M.P. and Waseda, T. (1999). Laboratory observations of wave group evolution, including breaking effects. *J. Fluid Mech.*, **378**, 197–232.

van Vledder, G.P. (2006). The WRT method for the computation of non-linear four-wave interactions in discrete spectral wave models, *Coastal Eng.*, **53**(2–3), 223–242.

Waseda, T. (1997). Laboratory study of wind-and mechanically-generated water waves, Ph.D. dissertation, UC Santa Barbara.

Waseda, T. and Tulin, M.P. (1999). Experimental study of the stability of deepwater wave trains including wind effects, *J. Fluid Mech.*, **401**, 55–84.

Waseda, T., Toba, Y. and Tulin, M.P. (2001). Adjustment of wind waves to sudden changes of wind speed, *J. Oceanogr.*, **57**(5), 519–533.

Waseda, T., Kinoshita, T. and Tamura, H. (2009). Evolution of a random directional wave and freak wave occurrence, *J. Phys. Oceanogr.*, **39**(3), 621–639.

Waseda, T., Kinoshita, T. and Tamura, H. (2009). Interplay of resonant and quasi-resonant interaction of the directional ocean waves, *J. Phys. Oceanogr.*, **39**(9), 2351–2362.

Waseda, T., Hallerstig, M., Ozaki, K. and Tomita, H. (2011). Enhanced freak wave occurrence with narrow directional spectrum in the North Sea, *Geophy. Res. Lett.*, **38**(13), L13605.

Waseda, T., Shinchi, M., Nishida, T., Tamura, H., Miyazawa, Y., Kawai, Y. and Taniguchi, K. (2011). GPS-based wave observation using a moored oceanographic buoy in the deep ocean, in *Proc. 21st Int Offshore Polar Engineering Conf.*, Maui, HI, USA, pp. 19–24.

Waseda, T., Tamura, H. and Kinoshita, T. (2012). Freakish sea index and sea states during ship accidents, *J. Mar. Sci. Technol.*, **17**(3), 305–314.

Waseda, T., In, K., Kiyomatsu, K., Tamura, H., Miyazawa, Y. and Iyama, K. (2014). Predicting freakish sea state with an operational third-generation wave model, *Nat. Hazards Earth Syst. Sci.*, **14**(4), 945–957.

Waseda, T., Kinoshita, T., Cavaleri, L. and Toffoli, A. (2015). Third-order resonant wave interactions under the influence of background current fields, *J. Fluid Mech.*, **784**, 51–73.

Webb, D.J. (1978). Non-linear transfers between sea waves, *Deep Sea Res.*, **25**(3), 279–298.

Welch, S., Levi, C., Fontaine, E. and Tulin, M.P. (1999). Experimental study of the ringing response of a vertical cylinder in breaking wave groups, *Int. J. Offshore Polar Eng.*, **9**(04).

Werle, B.O. (1995). Sea backscatter, spikes and wave group observations at low grazing angles, in *Record of the IEEE 1995 Int. Radar Conf.*, IEEE. pp. 187–195.

West, B.J., Brueckner, K.A., Janda, R.S., Milder, D.M. and Milton, R.L. (1987). A new numerical method for surface hydrodynamics, *J. Geophy. Res.: Oceans*, **92**(C11), 11803–11824.

Xiao, W., Liu, Y., Wu, G. and Yue, D.K. (2013). Rogue wave occurrence and dynamics by direct simulations of nonlinear wave-field evolution, *J. Fluid Mech.*, **720**, 357–392.

Young, I.R. (1999). *Wind Generated Ocean Waves*, Vol. 2, Elsevier.

Yuen, H.C. and Ferguson Jr, W.E. (1978). Relationship between Benjamin–Feir instability and recurrence in the nonlinear Schrödinger equation, *Phy. Fluids*, **21**(8), 1275–1278.

Yuen, H.C. and Lake, B.M. (1982). Nonlinear dynamics of deep-water gravity waves, *Adv. Appl. Mech.*, **22**, 67–229.

Zakharov, V.E.E. and Filonenko, N.N. (1966). Energy spectrum for stochastic oscillations of the surface of a liquid, *Dok. Akad. Nauk* **170**(6), pp. 1292–1295.

Zakharov, V.E. (1968). Stability of periodic waves of finite amplitude on the surface of a deep fluid, *J. Appl. Mech. Tech. Phys.*, **9**(2), 190–194.

Zhou, Z. (1996). Nonlinear instability of wavetrain under influences of shear current with varying vorticity and air pressure, *Acta Mech. Sinica*, **12**(1), 24–38.

Chapter 5

Phase-Averaged Wave Models

W. Erick Rogers

Naval Research Laboratory, Stennis Space Center, MS 39529, USA

5.1. Introduction

In simplest terms, a phase-averaged wave model is one that does not treat waves individually but instead uses the wave *spectrum* as the prognostic variable. These models are sometimes called "spectral models" for this reason, though in truth, a phase-averaged model can be monochromatic, and a phase-resolving model can be used to ultimately predict spectra. We do not intend to cover the topic of phase-averaged wave modeling comprehensively herein. This is already treated excellently and in greater detail by longer works such as Komen *et al.* [1994], Young [1999], Cavaleri *et al.* [2007] and Holthuijsen [2007]. Our goal here is to offer an alternative perspective and include topics and frank discussion which are not typically included. Emphasis is on the wave models SWAN [Booij *et al.*, 1999] and WAVEWATCH III® ("WW3", Tolman [1991] WW3DG [2016]), since this author has more first-hand experience with these two models than any other model. These two are good examples of phase-averaged models since they are open source, relatively modern (e.g., using message passing interface (MPI) parallelization and including modern physics parameterizations) and enjoy widespread, international use. Both models originally borrowed heavily from features of a common ancestor, the WAM model [WAMDIG, 1988], which was a widely shared community model. Today, development of WAM is fractured, e.g., the European Centre for Medium-Range Weather

Forecasts continue to develop a proprietary, fully modernized version: ECWAM. Herein, we use the term "3GWAM" to refer to the "third generation" of wave models, which includes SWAN, WW3, WAM(s) and a handful of other models. In its formal definition, the "third generation" label implies that it is a model which explicitly includes relevant physical processes and does not make *a priori* assumptions about spectral shape [Komen *et al.*, 1994].

The governing equation of phase-averaged models is the wave energy balance equation (most versions of WAM) or wave action balance equation (SWAN, WW3). This can be represented as

$$\frac{\partial N}{\partial t} + \nabla \cdot \vec{C_g} N = \frac{S_{\text{tot}}}{\sigma}, \tag{5.1}$$

where

$$S_{\text{tot}} = S_{\text{in}} + S_{\text{ds}} + S_{\text{nl}} + S_{\text{bot}} + S_{\text{db}} + S_{\text{tr}} + S_{\text{sc}} + S_{\text{ice}} + S_{\text{ref}}. \tag{5.2}$$

In (5.2), the individual process is represented as: wind input, S_{in}; wave dissipation, S_{ds}; nonlinear wave–wave interactions (quadruplets), S_{nl}; wave–bottom interactions, S_{bot}; depth-induced breaking, S_{db}; triad wave–wave interactions, S_{tr}; bottom scattering, S_{sc}; wave-attenuation due to ice, S_{ice} and reflection of waves due to shorelines and icebergs, S_{ref}.

The models operate by integrating this equation. There are five independent variables: wave frequency (or alternatively, wave number), wave direction (or alternatively, the Cartesian form of wave number), geographic position (meters, or degrees latitude/longitude), and time. Thus, some possible forms of the prognostic variable (spectral density) are: $F(f, \theta, x, y, t)$, $N(f, \theta, \phi, \lambda, t)$, $N(k, \theta, \phi, \lambda, t)$, and $N(k_x, k_y, \phi, \lambda, t)$.

We introduce phase-averaged models by contrasting with phase-resolving models of the previous chapter.

Advantages

(1) There is no requirement of having a minimum number of computational "points per wavelength", which is particularly important if shorter waves are being modeled, e.g., the high-frequency

tail. It is challenging to efficiently represent many length scales simultaneously in a phase-resolving model.
(2) Geographic resolution is flexible, which implies that a model design can conform to the resolution of expected scale of variability, as appropriate. These scales tend to follow model input; we discuss this below.

Disadvantages

(1) Phase-averaged models tend to include more parameterization. This is especially true for source terms. Though, phase-resolving models typically do require their own parameterizations for anything that is not explicitly computed (breaking, capillary waves, atmosphere, turbulence, viscosity, compressibility, etc.).
(2) Phase-averaged models tend to utilize a larger number of assumptions built into the governing equation, e.g., that the sea surface is Gaussian. For kinematics, phase-averaged models usually rely on linear wave theory, while some phase-resolving models do not.
(3) Though today's phase-averaged models offer an enormous number of output variables computed from the spectra or the source terms, the phase-resolving model can, in principle, provide even more, e.g., some explicitly model wave shape and wave grouping. Freak wave prediction is possible using a phase-averaged wave model, but this relies on associations and correlations rather than explicit modeling [e.g., Janssen, 2002, 2003].

A careful reader will have noticed that computation time was not mentioned above. Though some phase-resolving models are exceedingly expensive to run, there are sufficient counter examples (fast phase-resolving models and slow phase-averaged models), that blanket statements are impossible. Similarly, though phase-resolving modeling tends to be more academic, and phase-averaged modeling more suitable for operational use, one can find counter-examples.

3GWAM output has many uses and forms. One common use is in operational nowcasting and forecasting of wave conditions, and

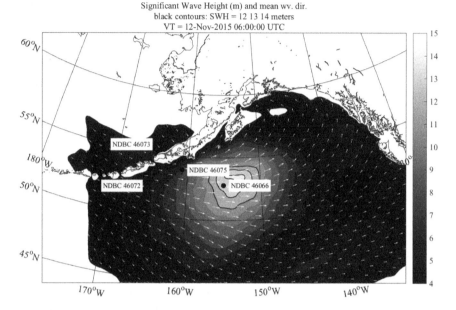

Fig. 5.1. Significant wave height (shading, in meters) and mean wave direction (arrows) from a WAVEWATCH III hindcast, south of Alaska. The three contours indicate wave heights of 12, 13, and 14 meters.

in that context, perhaps the most common product is the map of significant wave height H_s. An example is given in Fig. 5.1.

5.2. Kinematics

Here, we use "kinematics" to refer to processes described using the advection term of the governing equation, $\nabla \cdot \vec{C}_g N$. A brief list is given here:

(1) Refraction by bathymetry and currents. Obviously, this creates a change in wave direction, and perhaps less obviously, can change the energy level locally via focusing and defocusing. This occurs when the phase velocity is non-uniform along wave crests (perpendicular to the axis of propagation).
(2) Shoaling by bathymetry, and analogous behavior in cases of horizontally sheared currents. This occurs with the group velocity (plus currents) is non-uniform along the axis of propagation.

(3) Anything that causes the dispersion relation to deviate from its conventional form will necessarily produce behavior analogous to (1) and (2) above. Two examples are the effect of a viscous mud layer at the seafloor (e.g., Dalrymple and Liu [1978]) and the effect of a viscous ice cover at the surface [e.g., Keller 1998]. At the time of writing, these kinematic effects are limited to academic applications.
(4) Advection in frequency space as waves encounter horizontal shear in currents (Chapter 4).
(5) Diffraction has been implemented in SWAN via modifications to the advection terms [Holthuijsen *et al.*, 2003]. However, there are difficulties with this approach and it is our opinion that this process is best predicted using a phase-resolving model (Chapter 6), especially in the vicinity of engineering structures.

5.3. Source Terms

Though most of the source terms have already been described in Chapters 2–4, we briefly list the source terms available in 3GWAMs here, and in contrast to those prior descriptions, the following includes discussion of the compromises, approximations, and other artifice of numerical modeling.

Exponential wind input. Some methods used today are (1) the parameterization of Komen *et al.* [1984], loosely based on the experimental work of Snyder *et al.* [1981], (2) the parameterization of Janssen [1991], loosely based on the quasi-laminar theory of Miles [1957], (3) models based on fitting to a boundary layer model, e.g., Tolman and Chalikov [1996], and (4) the parameterization of Donelan *et al.* [2006], largely empirical but drawing inspiration from the sheltering theory of Jeffreys [1924]. There are large differences between these methods, even in fundamental features such as the directional distribution, and the integrated atmosphere-wave momentum flux. See Chapter 2.

Linear wind input. Exponential wind input requires energy to increase energy, so it does not act on a calm surface. In wave models, this problem can be addressed in one of two ways. One is to "seed" the model state with some very small initial energy. The other approach is to implement a linear wind input source term. The Phillips resonance mechanism (Chapter 2) is linear, so it is commonly used. Both SWAN and WW3 have linear wind input which is nominally based on this mechanism, following the work of Tolman [1992] which is, in turn, based on Cavaleri and Malanotte-Rizzoli [1981]. In truth, the connection to the Phillips resonance is practically non-existent, particularly with respect to the directional resonance at $\theta_r = \pm \cos^{-1}(C/U)$. However, it can be argued that the accuracy of the source term is inconsequential, because it has little impact on the modeling outcome beyond the first stages of wave growth.

Steepness-limited breaking (whitecapping). Breaking is strongly phase-dependent, so a spectral source term is necessarily highly parameterized. Even so, this source term has seen remarkable improvement since the early, simple formulation by Komen *et al.* [1984]. The primary shortcoming of the early formulation was non-physical features such as the breaking of swell, and the unintended non-physical influence of windsea on swell and vice versa [see Rogers *et al.*, 2003]. The key breakthrough was the effective incorporation of the concept of a spectral threshold, below which no breaking occurs, and above which breaking increases nonlinearly [e.g., Babanin and van der Westhuysen, 2008; Ardhuin *et al.*, 2010]. Also important was the recognition that multiple physical mechanisms can contribute to breaking. Modern formulations tend to include at least two of the following: (1) breaking caused by instability of waves at the frequency in question (independent of other frequencies), (2) dissipation caused by turbulence and instability associated with the breaking of waves that are longer than the frequency in question, and (3) breaking associated with the "concertina effect", in which short waves are modulated by the orbital velocities of long waves.

Non-breaking dissipation. With the threshold mechanism for wave breaking in modern wave models, it becomes necessary to include an additional mechanism to produce the slow dissipation which is implied from observations of swell propagating across long distances [e.g., Ardhuin *et al.*, 2009]. One postulated cause for this is friction at the near-surface atmospheric boundary layer, e.g., with energy lost to turbulent eddies in the air [Ardhuin *et al.*, 2010]. Another postulated cause is generation of turbulence in the water as wave orbital velocities interact with background turbulence [Babanin, 2006]. Though the causes are different, the equations are nearly equivalent [Babanin, 2011]. A third cause may be interpreted as a re-imagining of the first. Here, rather than treating the losses to the atmosphere using a friction model, they are computed by including a "negative wind input". This wind input term is computed the same way as the corresponding positive wind input, but reduced by some factor, taking the paradigm used in the experimental study of Donelan [1999].

Nonlinear interactions. This refers to energy- and momentum-conserving redistribution of energy within the spectrum. In shallow water, the fast near-resonant triad interactions are the most important, but these have historically been challenging to represent in phase-averaged models [Cavaleri *et al.*, 2007]. In the open ocean, the resonant four-wave interactions become important. These interactions are weak, requiring many wavelengths to effect significant change to the spectrum, but over the large scales of the open ocean, they become very important. Solution of these interactions has been a focus of much research for decades, and solution methods range from the fast and approximate DIA ["Discrete Interaction Approximation", Hasselmann *et al.*, 1985] to exact methods which can be computationally slower by two orders of magnitude [see Tolman, 2011]. One exact method is Hasselmann and Hasselmann [1985], and there are a number of intermediate methods, e.g., van Vledder [2012] and Tolman [2013]. The implication of the reduced accuracy of DIA is discussed in Chapter 4.

Surf breaking. This source term is sometimes referred to as depth-limited breaking. In principle, breaking in deep water and shallow water could be represented with a single unifying source function [e.g., Filipot and Ardhuin, 2012], but in practice, this is seldom done, and they are treated separately. The deep-water breaking formulations typically do not act fast enough to produce realistic profiles of waveheight across a beach profile, and so the surf breaking [e.g., Battjes and Janssen, 1978] is included as an engineering solution. The source term acts as a "safety valve" to prevent extremely large wave heights in the nearshore, keeping in mind that in the case without dissipation, energy goes to infinity as depth approaches zero, due to shoaling.

Bottom friction. In finite depths, wave orbital motion interacts with the rigid seafloor, and that near-bottom boundary layer produces dissipation of the wave energy, e.g., Madsen *et al.* [1988]. These source terms tend to be highly parameterized, and typically suffer from insufficient operational knowledge of the seafloor roughness, and variability of the same. However, the models at least capture the appropriate qualitative behavior, since the spectral calculation is based on orbital motion, which is readily calculated.

Fluid mud (non-rigid seafloor). When waves propagate in shallow and intermediate depths with a muddy seafloor, they exert time-varying pressure, which creates a two-layer wave system, and energy lost to viscosity in the mud layer implies a damping of the surface waves. This damping can be very strong, especially in shallower depths. This can be estimated using mathematical models, e.g., Dalrymple and Liu [1978], and Ng [2000]. Models have been implemented in SWAN and WW3 (Winterwerp *et al.* [2007]; Rogers and Holland [2009]; Rogers and Orzech [2013]). Unfortunately, inputs are notoriously difficult to define, so uses of these parameterizations tend to be academic. Viscosity is difficult to estimate. Mud thickness can be obtained from core samples, but since it is always inhomogeneous and often non-stationary, sparse sampling has limited value to the modeler. Also, the thickness of the mud is not in fact a useful variable. Rather, the model needs an estimate of the fluidized mud layer thickness; the immobile fraction of a mud layer does not contribute to dissipation.

Dissipation by vegetation. Marine vegetation can also dissipate wave energy. Wave motion in sea grass has been modeled as flow in and around narrow cylinders. The method of Dalrymple *et al.* [1984] has been implemented in SWAN [Suzuki *et al.*, 2011].

Dissipation by sea ice. Traditionally, the effect of sea ice on waves has been represented in wave models by deactivating grid points, i.e., treating them as land. This approach has been used with WAM, SWAN, and WW3. More recently, dissipation by sea ice has been implemented in at least two models, ECWAM and WW3, using physics parameterizations [Doble and Bidlot, 2013; Rogers and Orzech, 2013; Tolman *et al.*, 2014; WW3DG, 2016]. There are three major groups of methods: (1) mathematical models which consider the dissipation that occurs within the ice layer, (2) mathematical models which consider the dissipation that occurs below the ice layer, and (3) fully empirical parameterizations. In the first case, the losses are associated with friction. For the case of ice in a continuous sheet, there can be internal friction (hysteresis). For the case of suspensions of ice, e.g., brash and frazil ice, there can be losses from pumping of liquid water through the gaps between ice. Below the ice, the losses are associated with the boundary layer friction, and the boundary layer may be laminar or turbulent, just as with friction at the seafloor. There are further parallels: like seafloor bottom friction dissipation, all three methods of predicting dissipation by sea ice require information that is usually difficult to obtain: ice type and ice characteristics (rheology). Parameterizations and settings which provide good skill for one ice type cannot be expected to have skill with other ice types. However, there are causes for optimism. First, there has been significant progress recently to produce new, more accurate observational datasets for use in model development, e.g., Thomson *et al.* [2018] and Rogers *et al.* [2018]. Second, as with bottom friction, qualitative behavior can already be adequately represented in models, as there is a robust and intuitive increase in dissipation rate with wave frequency, e.g., Fig. 5.2. This feature did not exist in 3GWAMs prior to 2010. Selection of the power dependency of this dissipation rate vs. frequency relation is not fully resolved, but is coming into better focus with new research [Meylan *et al.*, 2018].

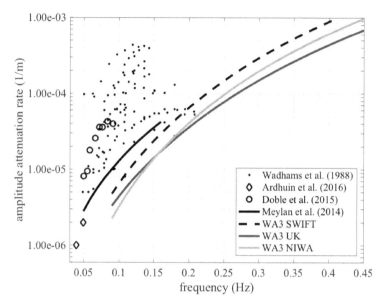

Fig. 5.2. Comparison of the frequency distributions of dissipation of wave energy by sea ice. Wadhams *et al.* [1988], Ardhuin *et al.* [2016], Doble *et al.* [2015], and Meylan *et al.* [2014] are observational studies. The "WA3" lines are based on model-data inversion using method similar to that used by Rogers *et al.* [2016, 2018], from Wave Array 3 [Thomson *et al.*, 2015] of the ONR-supported "Sea State" field experiment. "SWIFT", "UK", and "NIWA" are buoy types.

Source: SWIFT buoy data are from Jim Thomson and Madison Smith (UW/APL). UK buoy data are from Martin Doble (Polar Scientific) and Peter Wadhams (Cambridge University). NIWA buoy data are from Alison Kohout (NIWA).

Scattering by sea ice. When wavelength is comparable to the floe size, wave–ice interaction enters a "scattering regime" [Bennetts and Squire, 2012], in which the process should not be neglected. Wave energy may also be scattered in the case of propagation of swell through continuous ice, as the waves encounter sharp variations in ice thickness [Squire *et al.*, 2009]. This is a non-dissipative (conservative) process which is represented in WW3 using redistribution of energy within the spectrum [WW3DG, 2016; Ardhuin *et al.*, 2016].

Scattering by irregular seafloor. Bragg resonance with topographical features can result in scattering. This can be either forward or backward scattering. Ardhuin and Herbers [2002] found that for a case on the North Carolina shelf, forward scattering predominates, and leads to significant broadening of the spectrum. This process has been included in WW3, but like some other source terms in this list, application tends to be for academic purposes: it requires a spectrum of the seafloor and can be expensive to compute.

Reflection. Reflection from steep coastlines [Ardhuin and Roland, 2012] and icebergs [Ardhuin et al., 2011] has been implemented in WW3. In the former study, the authors find that errors in directional spread are improved by including this effect. Reflected waves have a somewhat unique role whereby they produce partial standing waves, which penetrate to extreme depths and produce seismic noise [Longuet-Higgins, 1950].

Deep-water physics packages. Collectively, the wind input source function(s), steepness-limited breaking, and non-breaking dissipation are known as "source term (ST) packages". In WW3, these include: ST1 [Komen et al., 1984; Booij et al., 1999], ST2 [Tolman and Chalikov, 1996], ST3 [Bidlot et al., 2005], ST4 [Ardhuin et al., 2010], and ST6 [Rogers et al., 2012; Zieger et al., 2015; Rapizo et al., 2017; Liu et al., 2019]. SWAN includes several other source term packages, such as ST1, ST6, and one by van der Westhuysen et al. [2007]. Other source terms are selected independently from the ST packages.

These three source terms are grouped together because their calibration is interdependent, and together they largely control the rate of growth and decay of wave energy in deep, open water (i.e., where depth-dependent processes and ice are absent). The four-wave nonlinear interactions are of course also very important in this situation, especially for spectral shape [e.g., Young and van Vledder, 1993; Resio et al., 2016], but as mentioned above, it is selected independently from the ST packages, so, for example, ST6 can be applied in conjunction with any solver for the four-wave nonlinear interaction source term in SWAN and WW3. This leads to the obvious question: does a change in nonlinear solver necessitate a retuning of the source term package? The answer is not simple. Tolman [2011] asserts that any change to the nonlinear solver does require a recalibration of

ST2. Results of Perrie *et al.* [2013] imply that this is also true in ST4 if DIA is replaced with a more accurate solver. Rogers and van Vledder [2013] find that change to the nonlinear solver does *not* require recalibration of ST6 in SWAN; however, it should be kept in mind that there was little or no swell in that study. Liu *et al.* [2019] find that a change to the nonlinear solver again does not require recalibration of wind input and breaking terms of ST6 in WW3 but they do find it necessary to recalibrate the swell dissipation in a global application of ST6. This outcome is unsurprising: DIA is known to push too much energy to the low-frequency side of the spectral peak [e.g., Rogers and van Vledder, 2013] and in a global application, this has implications for the calibration of the swell dissipation.

Thorough comparison of the ST packages is beyond the scope of this chapter, this can be found in Stopa *et al.* [2016], van Vledder *et al.* [2016], Liu *et al.* [2019] and other publications. However, we present here two idealized comparisons. Figure 5.3 shows the fetch and duration wave height growth curves for ST2, ST3, ST4, and ST6 with wind speed $U_{10} = 15\,\mathrm{ms}^{-1}$. These simulations are initialized from rest. The fetch and duration values reported by Moskowitz [1964] for fully developed conditions are marked on each plot, though it may be assumed that observed cases did *not* start from rest. The significant waveheight predicted by the Pierson and Moskowitz [1964] ("PM") parametric model for fully developed conditions is 5.5 m at this wind speed. Vertical contours indicate duration-limited conditions. Horizontal contours indicate fetch-limited conditions. Such growth curves are a useful indicator of model behavior. For example, ST2 tends to be slower than WAM4 during initial growth while overpredicting energy as windsea transitions to swell [Tolman, personal communication], and this is reflected in the figures. The plots indicate strong similarity in ST3, ST4, and ST6; especially the latter two. Notably, the models are unanimous that the fetch/duration values reported by Moskowitz [1964] *do not* correspond to fully developed conditions. In real ocean cases of very large fetch/duration, that fetch/duration is defined by the storm itself, and we postulate that the apparent cessation of growth observed by Moskowitz was associated with the inevitable slackening of winds, rather than being

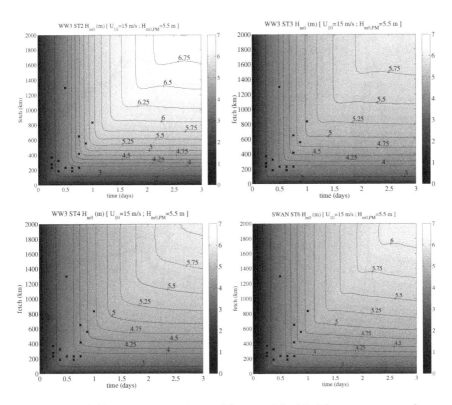

Fig. 5.3. Fetch/duration comparisons of four models. Model source term package is indicated in title above each plot. See text for explanation.

associated with the approach towards an asymptotic limit under constant winds. The latter is, of course, the traditional interpretation of the PM spectrum, e.g., Booij *et al.* [1999].

Of course, even when wave height is identical between two models, the spectrum may be very different. Figure 5.4 compares frequency distributions of energy and the three primary deep-water source functions (S_{in}, S_{ds}, S_{nl4}). These are for the same wind speed, with unlimited fetch, and 15 hours duration. In terms of spectral shape, the more modern models (ST4, ST6) are in good agreement, while the older ST2 is an outlier, with a lower dominant wave period. Wind input and whitecapping in ST6 is much stronger in the high frequencies than ST2 and ST4. In the context of whitecapping, this implies that

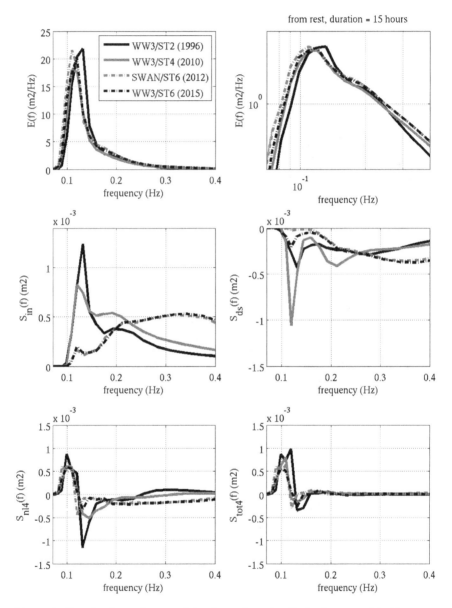

Fig. 5.4. Results from four models, for a duration-limited case, with 10-meter wind speed of $15\,\mathrm{m\,s^{-1}}$ after 15 hours duration, starting from rest. $E(f)$ is energy density. $S_{\mathrm{in}}(f)$, $S_{\mathrm{ds}}(f)$, $S_{\mathrm{nl4}}(f)$ are source terms for wind input, whitecapping, and four-wave nonlinear interactions, respectively. $S_{\mathrm{tot}}(f)$ is the summation of the three source terms, and is the growth/decay rate, since advection terms are zero (unlimited fetch).

ST6 is closest to consistency with Pushkarev and Zakharov [2016], who argue that there is no dissipation near the spectral peak. The strong difference in frequency distribution of wind input has implications for coupling to other models, since conversion from wind input to wave-supported stress (i.e., from energy flux to momentum flux) involves a factor $1/C$ where C is the phase velocity.

5.4. Computational Grids

3GWAMs are solved on grids in both spectral and geographic space. The general objective with model resolution is to sufficiently resolve the variability that would be exhibited by the model if it was solved without discretization (the latter being generally impossible, of course). This required resolution is primarily dictated by the model forcing, e.g., the scales of variation of the driving winds, the currents, the ice, or the bathymetry/coastline.

Frequency space is defined on a logarithmic spacing, implying smaller steps for lower frequencies. When the DIA is used for four-wave nonlinear interactions, the spacing is usually according to $f_i = 1.1 f_{i-1}$, where f_i indicates a bin in the frequency grid. The increment factor 1.1 is preferred because it was used in the original development and calibration of the DIA. For general cases, e.g., a global wave model, the starting frequency has traditionally been around 0.04 Hz, but it can be set significantly higher in restricted basins (e.g., lakes and enclosed seas) with defined climatology for expected spectral range. More recently, lower starting frequencies have been adopted in global wave models, e.g., 0.035 Hz, recognizing that extreme extratropical storms can produce energy at these frequencies. Even when the low-frequency bands (0.035–0.40 Hz) contribute little to the wave height within the storm, the swell energy that radiates outward at these frequencies can be important at remote locations, especially when energy at higher frequencies is absent or suppressed [e.g., Ardhuin et al., 2016]. The high-frequency limit is typically between 0.5 Hz and 1.0 Hz with modern physics packages, e.g., 0.6–0.7 Hz is used in WW3 by Ardhuin et al. [2010]. Beyond this limit of the prognostic frequency grid, a diagnostic tail is added; this can be either fixed [e.g., 1.0 Hz in Booij et al. [1999] and generally recommended for SWAN] or a flexible limit based on local wind speed and/or mean period [e.g., Tolman, 1991]. Of course, if a 3GWAM is being used to model only swell in restricted domains, forced at lateral boundaries, one may safely use a narrower frequency range.

For the directional grid, uniform spacing is used, typically around $10°$, which implies 36 directional bins to represent waves from any direction. Though again, for modeling of only swell, with restricted coastal domains and forcing at the boundaries, one may elect to omit some directions (e.g., those directed offshore) to reduce computation time.

Geographic grids exist in three main categories: structured regular, structured irregular, and unstructured. Further, they can be in a Cartesian system (x, y) or a spherical (latitude/longitude) system (ϕ, λ). "Structured" is used here to indicate that the grid is logically rectangular. A structured irregular grid in Cartesian system would have all grid points defined on x $(i = 1 \cdots n, j = 1 \cdots m)$, y $(i = 1 \cdots n, j = 1 \cdots m)$ where n and m are the number of rows and columns (thus "logically rectangular"). "Regular" is used here to indicate uniform spacing, e.g., $(\Delta x, \Delta y)$. A Cartesian structured regular grid also has points defined on $x(i, j)$, $y(i, j)$, but because of uniform spacing, this simplifies to $x(i)$, $y(j)$. Unstructured grids are not logically rectangular; here grid points are defined in a list, $x(1 \cdots n)$, $y(1 \cdots n)$, where n is the number of nodes. Unstructured grids can be composed of triangular grid cells, but this is just one type. Another example of an unstructured grid is one which is made by taking a regular grid and deleting some grid points, e.g., to have larger spacing offshore or larger longitude spacing $\Delta\lambda$ at extreme latitudes.

Selection of grid type often involves weighing the cost of grid creation against the cost of running an inefficient model. Regular grids are the easiest to create, and in the context of SWAN, can be changed by simply modifying a few numbers in an input file. However, they can be inefficient. For example, high resolution may be applied to describe a coastline and bathymetry, but this same high resolution is then applied in offshore portions of the same grid, where it is not needed. A global grid with regular latitude/longitude spacing implies highly non-uniform spacing in terms of real distances, which in the case of conditionally stable propagation, forces use of an inefficiently small time step (a small fraction of grid cells dictate a small time step for the entire grid). Irregular and unstructured grids can address these inefficiencies. For example, in the global context, we can use regular spacing at low latitudes, Lambert conformal projections at mid-latitudes, and polar stereographic projections at high latitudes to create modeling systems which permit large time steps for propagation. For modeling coastlines, the unstructured grids are most efficient, since high resolution can be applied where it is needed, and coarse resolution applied offshore, where wave conditions are more homogeneous. The cost, of course, is that irregular grids and unstructured grids — especially the latter — can require more effort to create and manage. Creation of unstructured grids usually requires specialized software, and the creation of boundary forcing for grids with irregular boundaries can be troublesome. Also, there exists a caveat for the case of unstructured grids which apply high resolution where needed: these grids usually require an unconditionally stable propagation scheme, since otherwise, the smallest grid cell will dictate the propagation time step.

Tolman [2008] created an ingenious approach for WW3 which combines the simplicity of regular grids with the efficiency of irregular and unstructured grids. This uses a regular grid, but with offshore portions masked out, meaning that those locations are computed in a host grid which provides boundary forcing. This idea unfortunately creates a new challenge of defining irregular, internal boundaries not oriented with the four cardinal directions. Tolman [2008] implemented this idea alongside a "mosaic approach" of two-way nesting, which automates the boundary management, though not the grid creation.

The strategy of "applying high resolution only where it is needed" is problematic in cases where the determining factors are non-stationary: if it is desirable to resolve high resolution details of the dynamic model inputs (e.g., surface currents, winds), then adaptive (non-stationary) grid methods should be used, e.g., Popinet et al. [2010]. Such technology is still relatively new.

5.5. Numerics

Discretization in numerical models results in numerical error, and coarser discretization generally implies larger error. However, it is not the resolution itself that increases error, but rather the resolution relative to the curvature (and other spatial derivatives) of the thing being propagated: in our case, spectral density. Also important: the numerical scheme used and the Courant number, $\mu = C\Delta t/\Delta x$, where C is the speed at which energy is advected, usually taken as the sum of the group velocity and the mean current. Different schemes have different dependency on resolution and Courant number. For example, the first-order scheme of SWAN has less error with smaller Courant numbers. The first- and third-order schemes of WAM and WW3 have opposite behavior: they have less error with larger Courant number, though they become unstable beyond a value near 1.0.

Numerical error falls into two broad categories: diffusion error and phase error. Again, different numerical schemes include these in different proportions. Diffusion tends to smooth a swell field as it propagates, so that maxima are reduced and minima increased, and also it tends to bleed energy into regions which it would not propagate into using an exact method (e.g., ray tracing). One example of the latter is excessive propagation of wave energy into sheltered regions behind islands, and this error can become so large that it increases RMS error in comparisons against observations [Rogers et al., 2002]. Phase error is the deviation of the numerical propagation speed from the true (intended) propagation speed for individual Fourier components of the numerical solution [Petit, 2001]. When these components separate during propagation, they can manifest as oscillations in the solution if they are not damped by diffusion, and these are known as "wiggles". These wiggles tend to be most severe in higher-order implicit schemes such as the one used in SWAN. Thus, a user is faced with the dilemma of choosing between larger numerical error with no wiggles (the diffusive first-order implicit scheme) versus numerical error that is significantly smaller but also more visibly non-physical. Further, unfortunately, the selection is not case-dependent, since the same thing that creates diffusion in the first-order scheme (curvature of the wave field) also creates wiggles in the higher-order scheme. Thus the selection is largely based on a user's preference.

The above dilemma is an outcome of the use of implicit schemes in SWAN which is in turn a result of the requirement of unconditional stability, which is in turn a result of SWAN's primary design criterion: efficient modeling at high resolution. WAM and WW3 take a different approach. These models were originally designed for larger-scale modeling, which implies coarser resolution (e.g., greater than 2 km), so conditional stability is not objectionable. This allows use of explicit schemes which have higher overall accuracy and speed than their counterparts in SWAN.

WW3 is not without problems, however. It separates x- and y-propagation into separate sequential steps, in contrast to SWAN which solves for propagation in both dimensions with a single operation. In extreme canonical cases, this "splitting error" results in a distortion of features: initially round features become more square in appearance after propagation.

The higher-order schemes of both SWAN and WW3 have numerical diffusion that is so small that it produces yet another problem: the "Garden Sprinkler Effect", whereby the spectral discretization causes distinct features to become visible in the geographic distribution of swell fields after propagation. This is most severe when spectral discretization is coarse. With first-order schemes, the effect is mostly masked by diffusion. In both SWAN and WW3, methods exist to counteract the effect [Booij and Holthuijsen, 1987; Tolman, 2002].

The problems mentioned above are not necessarily apparent in skill scores, particularly in the open ocean where the curvature of the wave field tends to be smaller. Rather they are first evident as aesthetic problems, e.g., in spatial distribution of swell energy. In a time series comparison against observations, diffusion will manifest as a reduction in the variance of the model time series, but again, this does not necessarily affect traditional skill scores such as RMS error when larger errors (e.g., from wind forcing) dominate.

SWAN has an additional method for improving the efficiency of high resolution modeling: stationary computations. Here, the assumption is made that the waves are in steady state, having achieved balance with the local winds and boundary conditions. This is more appropriate for small regions, e.g., 25 km × 25 km, and error becomes significant for larger areas, e.g., with waves reacting too quickly to changes in wind, or swells arriving too early [Rogers et al., 2007].

The time step also plays a role in numerical error. Similar to the x/y propagation issue above, WW3 separates the computations of propagation from computation of source terms, where SWAN solves this simultaneously. The impact of the split time stepping in WW3 is largely in the control of the user. By using infrequent reconciliation of source term integration and propagation, computations can be accelerated, but with a loss of accuracy.

SWAN has a time step problem which is arguably more severe. It applies a "limiter" such that large changes to the spectrum (from one time step to the next) are prevented. This improves model stability. However, it means that the model solution can become dependent on the time step size [Tolman, 2002]. This implies that in cases of rapid wave growth, e.g., a sudden squall following relative calm, the growth will be underpredicted by SWAN during the initial hours. However, like with the case of propagation error, the problem is most evident in idealized cases. In hindcasts of real scenarios, this author has not observed significant sensitivity to time step size. WW3 solves the limiter problem by dynamically reducing the time step size according to the strength of the net source terms [Tolman, 1992].

5.6. Inputs

The non-uniform input permitted for 3GWAMs include: wind at 10 meters, wind stress, bathymetry, surface currents, air–sea temperature difference, water levels, bottom friction parameter(s), sea ice concentration, sea ice thickness, other sea ice parameters, iceberg parameters, mud parameters, and vegetation parameters. Of course, there is significant variation between models and no single model includes every item in this list. Moreover, there are differences between models with respect to which fields are permitted to be non-stationary. Just as wave models can be created for many different scales (global models, surf zone models, etc.) and different conditions (swell propagation through a tropical archipelago, windsea in an ice-infested sea, waves under a tropical cyclone, swells arriving at a muddy coastline, etc.), the crucial input variables will be different. At global scale and larger regional scales, the accuracy of the wind input is often the primary factor determining wave model skill, even more so than the model physics parameterizations [Cardone *et al.*, 1996; Bidlot *et al.*, 2002; Rogers *et al.*, 2005]. This implies that, in those situations, the wave modeler can only influence the second-order errors. Further, the wind problem is not unique. Bathymetry error, coupled with shoreline- and depth-dependent physics, can cause severe difficulties. Some coastlines, especially in the military context, may be inaccessible for traditional bathymetric surveys. At an exposed sandy coastline, the beach profile is seasonal and determined by storm events [CERC, 1973]. A third example is the situation in and near the ice edge, where even modest errors in the ice edge position can make accurate wave forecasting impossible [e.g., Rogers *et al.*, 2018], and that position can change much even during the short (4–24 hours) window between satellite overpasses.

Notably, there is no explicit method in these models for providing them with gridded information about bottom type or ice type. For example, a model of the northern Gulf of Mexico is not explicitly instructed regarding where the southern Louisiana coastline is muddy or the Florida panhandle is sandy. However, extending SWAN or WW3 to allow multiple sediment types in a single grid would be only a modest technical challenge, since corresponding dissipation mechanisms already exist in the models. The larger technical challenge is to provide the model with non-stationary and non-uniform information required by these mechanisms, such as the depth of fluidization of the mud layer and the state of a sandy seafloor as it transitions between a flat surface where the roughness is associated with the sand grains, and a rippled surface, where the roughness is associated with the bedforms. In both the mud and sand cases, it is of course a coupled problem, with the wave orbital motion impacting the state of the seafloor [Winterwerp et al., 2012; Smith et al., 2011], but application to deterministic modeling is still relatively academic. Similarly, with respect to ice type, many parameterizations for dissipation by sea ice exist in WW3 [WW3DG, 2016; Rogers et al., 2018]. Some are purely empirical, while others are solutions to boundary value problems or otherwise based on theoretical calculations. Regardless, parameterizations or settings with parameterizations are sometimes associated with specific ice types, e.g., grease ice, thick frazil ice, pancake ice, sheet ice, or floes from broken sheet ice. The intuitive conclusion is that if we know the ice type, we can more accurately estimate the dissipation of wave energy. Unfortunately, the analogy to seafloor interaction still holds: estimation of sea ice type is still not a reality for operational use. We can imagine that this situation will improve soon, as satellite technology and process models improve. Again, like the seafloor, it is a coupled problem, both with ice breakup by waves [e.g., Williams et al., 2013a,b; Collins et al., 2015; Boutin et al., 2018] and the role of waves in ice formation [Shen et al., 2001, 2004; Doble, 2009; Thomson et al., 2017].

Two issues with wind forcing are worth further discussion. One is the modification of the effective wind speed by surface currents, such that the wind speed used for source term calculation is reduced when the wind and currents are aligned, which is of course not uncommon. However, a wave modeler must be judicious when including this effect. In the real ocean, such a scenario results in lower drag on the atmosphere and thus an increase in wind speed. As such, it may be problematic to apply in cases where the atmospheric model does not also include the currents. The second issue is air–sea stability. Wave modelers often assume a stable atmospheric boundary layer, since it is inconvenient to provide the air–sea temperature differences to the model, the effect on wave growth is usually small, and the available stability corrections are not routinely evaluated for effectiveness. However, as our models steadily improve, small errors, previously unnoticeable, may now be otherwise. Figure 5.5 compares

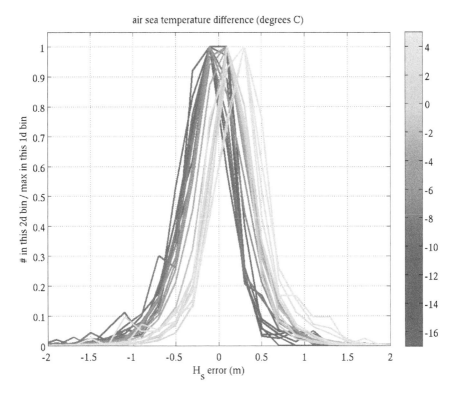

Fig. 5.5. Histograms of wave height error from a WAVEWATCH III hindcast. Observed wave height is from buoys. Color scaling indicates air–sea temperature differences (°C) taken from buoy observations.
Sources: The WW3 hindcast was performed by Yalin Fan (NRL). Buoy co-locations were performed by Yalin Fan and Silvia Gremes (University of New Orleans).

model wave height errors against air–sea temperature differences for a hindcast with the NRL global model, run without this stability information. The correlation is especially remarkable if one considers that over much of the ocean, swell dominates, implying that wave height is not determined by the local winds. The result suggests that a well-crafted stability correction would provide significant benefit.

For wave model grids that are not global and not in enclosed seas, boundary forcing must be provided. This is most often in the form of directional spectra. Ideally, the geographic resolution of the boundary forcing should approximately match the resolution of the model providing the forcing, so that computed spatial variability of the wave field is not lost during the nesting process. The transition from open water to coastal regions can reveal errors that are not as apparent offshore. Errors in spectral shape offshore can become errors in significant wave height nearshore. For example, accurate prediction of the blocking of swell by islands demands high accuracy directional distribution [Rogers et al., 2007]. Also, strongly frequency-dependent reactions to bathymetry enhance the impact of spectral error, even for validations that focus primarily on significant wave height.

5.7. Validation

Validation establishes the degree to which output from a model can be trusted. It quantifies the skill of the wave model, and (if done carefully) provides attribution or evidence for the cause of error, so that future predictions can be improved. The most conventional wave parameters included in validation are significant wave height H_s and dominant wave period, for which peak period T_p is sometimes used. However, experienced wave modelers avoid T_p because the model frequency resolution is often coarse relative to the natural variability of peak period during a wave event, resulting in step-wise time series. They instead use a metric for dominant wave period that is calculated via integration of the spectrum (discussed below). Experienced modelers include the 10-meter wind speed U_{10} in even the most basic validation, since atmospheric forcing is always the first suspect when windsea is poorly predicted. Also, it is always a good idea to compute bulk parameters using integration over a common frequency range. For example, if the buoy spectra stop at 0.4 Hz, the model integration should also stop at 0.4 Hz. This requires additional work organizing and managing spectra, but it is most often worthwhile. We summarize our three recommendations for the most elementary validation: (1) include wind speed, (2) use a wave period metric based on spectral integration, and (3) use consistent limits on frequency integration.

The moments of the non-directional spectra are particularly useful in validation: $m_n = \int E(f)f^n df$. In the spectral context, significant wave height H_s is defined using the zero-moment wave height: $H_{m0} = 4\sqrt{m_0}$. The third moment m_3 is proportional to surface Stokes drift in deep water [e.g., Ardhuin et al., 2009], and the fourth moment m_4 is proportional to mean square slope in deep water [e.g., Liu et al., 2000; Li et al., 2013]. These higher moments are particularly useful when studying the impact of source terms on the spectral tail [e.g., dissipation by sea ice, Rogers et al., 2016]. However, for the same reason, they are sensitive to measurement errors which are often larger in the spectral tail, so model-data misfit must be interpreted carefully. The m_2 moment is related to mean square of particle velocities (orbital motion). The m_{-1} moment, which places more emphasis on lower frequencies, is proportional to energy flux in deep water. Mean periods are defined using ratios of moments, e.g., $T_{m,-1,0}$ is calculated using m_{-1} and m_0 and is a superior alternative to T_p as a metric for dominant period. Additional emphasis can be placed on energy at the spectral peak in the period metric by using the moments on $E^4(f)$ rather than those of $E(f)$ [Collins and Rogers, 2017]. The period $T_{m,0,1}$ is perhaps the most commonly used definition of "mean period". The mean period $T_{m,0,2}$ places more emphasis on the shorter waves and in deep water is close to the "zero up-crossing period" T_z calculated from sea surface time series, just as H_{m0} is close to $H_{1/3}$, being, respectively, the spectral and time series definitions of H_s.

A group of metrics for non-directional spectra is related to frequency width or its inverse (narrowness). The so-called "peakedness" parameters fall into this category. The metrics are often used for evaluation of simulations of idealized wave growth, establishing basic model behaviors [e.g., Hasselmann et al., 1985], and in fact such comparisons are simple enough that they are sometimes performed qualitatively, without defining a metric [e.g., Alves et al., 2002]. Validation with width/narrowness parameters using scatter plots and skill scores is quite challenging to interpret, since one part of a time series may be windsea only, so the width parameter will reflect the width of the windsea, and another part of a time series may be a mixed sea/swell scenario, and the same parameter indicates the separation and relative size of the two systems. This can be addressed by restricting usage to unimodal spectra, e.g., by looking at windsea-dominated cases, Rogers and van Vledder [2013], or by isolating the system of interest. The spectral narrowness is associated with groupiness and probability of freak waves [e.g., Janssen, 2002, 2003]. Many metrics have been proposed, e.g., Longuet-Higgins [1957], Goda [1985] and Babanin and Soloviev [1998]. An excellent review can be found in Saulnier et al. [2011]. We do not recommend the use of metrics which require fitting to an idealized spectrum, e.g., the JONSWAP peakedness parameter γ, since it is indirect and impedes reproducibility.

The bulk parameters described above are calculated from the entire wave spectrum. A highly effective alternative to this is calculation of energy within a handful of frequency bands, e.g., one for low frequencies, two for middle frequencies, and one for highest frequencies. Frequency bands can either be fixed in frequency (f) space, e.g., Rogers et al. [2005], or in normalized frequency (f/f_p) space, e.g., Rogers and Wang [2007].

Directional accuracy is sometimes included in validations. Most often, this comes in one of two forms: (1) qualitative side-by-side comparisons of directional spectra [e.g., WAMDIG, 1988] or (2) quantitative evaluation of time series of a "mean direction" or "peak direction" metric [e.g., Moon et al., 2003]. When the ground truth comes from a directional buoy, the best approach is to use the "directional moments" computed from the Fourier coefficients describing directional distribution at each wave frequency, e.g., Kuik et al. [1988]. It is not advisable to apply a directional estimator (e.g., Maximum Likelihood Estimator) to the buoy data when skill statistics are the ultimate goal; this adds an unnecessary step, and some estimators will even modify the directional moments. Mean wave direction and directional spread are the low order moments. Shape parameters skewness and kurtosis are higher order moments, though quantitative comparisons with these are rare, and with good reason. Thorough quantitative comparison with the lower-order moments are already difficult: since they are frequency dependent, the modeler must adopt a strategy for condensing into manageable form. Skill scores are more meaningful if the parameters are not computed by integrating/averaging over all frequencies. One useful approach is to compute the parameters integrating frequencies near the spectral peak [Ardhuin et al., 2003]. Another approach is to analyze the skill for several frequency bands [Rogers and Wang, 2007]. A third approach is to focus on a short-time series and present analysis on the full frequency distribution [Romero and Melville, 2010].

Any validation effort should be mindful of the agency of the model. For example, if observations indicate a major influence of a physical process on a wave parameter, and the model captures most of this variability in the parameter via its parameterization of the process, with, say, 15% error in the parameter, that may be a success. In another case, with very low spatial/temporal variability of the same parameter, 15% error may be intolerable. In another case, errors in wind forcing (or ice, bathymetry, etc.) may make accuracy of the model's parameterizations irrelevant. In a final example, if a swell prediction model is compared to observations only a short distance from the offshore boundary, then it is primarily a validation of the boundary forcing rather than the parameterizations of the model.

Validation should also be mindful of measurement error. This is especially important when looking at spectra, higher moments of spectra, or directional information. We should also be careful when dealing with new measurement types (e.g., new buoy designs). One approach is to attempt to estimate measurement uncertainty, e.g., Alves et al. [2002] include a quantification of statistical uncertainty due to sampling variability. However, there are other types of measurement error, more difficult to estimate. For example, Thomson et al. [2015] find a major impact of biofouling of a Datawell buoy on the higher moments of the non-directional spectrum. O'Reilly et al. [1996] argue that NDBC (National Data Buoy Center) buoys have problems with directional spreading estimates, relative to Datawell buoys. The strength (or lack thereof) of the noise filter used during processing may affect the spectrum. In the largest wave experiment during the ONR-funded Sea State field experiment, there are major discrepancies between data from two drifting buoy types, deployed in the same area, and this had major impact on interpretation of comparisons against WW3 hindcasts [Rogers et al., 2018]. In recent unpublished work, we have found that the response function used in 3-meter NDBC discuss buoys is non-optimal in the higher frequencies, which can result in incorrect conclusions about the magnitude (and even the sign!) of model bias in low-pass mean square slope (proportional to m_4).

It is uncommon for validation exercises to address phase error in the time series. However, this is an issue of practical importance. For example, if the forecast swell amplitude is accurate, but is early or late, this will affect operational planning. This is especially true for the case of older swells, since there is a longer duration for phase errors to accumulate. It may be caused by numerical error, incorrect frequency distribution, or physical processes that are omitted from the modeling. For a good example of this type of model evaluation, the reader is referred to Jiang et al. [2016].

Most wave validation studies evaluate the skill of model hindcasts or analyses. However, in an operational context, the forecast skill is obviously of interest. For forecasts that are further in the future (longer horizons), the error is larger, and this is primarily caused by larger errors in the wind forcing, and in the case of the Marginal Ice Zone, the ice forcing. Good examples of evaluations of forecast skill are Bidlot and Holt [1999], Bidlot et al. [2002], Tolman et al. [2002], and Bernier et al. [2016].

Finally, a validation exercise requires selection of statistical measures (e.g., bias, RMS error, correlation, scatter index) for each parameter. The measures used by Cardone et al. [1996] are a good starting point. In cases where only one to three bulk parameters (e.g., wave height, wind speed, and dominant period) are included, it is not difficult to include several statistical parameters. In other cases, there is risk of overwhelming the reader with numbers. In recent years, the Taylor diagram (e.g., Tolman et al. [2013]; Zieger et al. [2015]; Stopa et al. [2016]) has become a popular method of presenting several statistical measures visually, rather than in a table. Correlation is a popular choice as a statistic, but it can be highly misleading, especially when used out of context. For example, a model with severe bias may have perfect correlation. A second example: when standard deviation of a parameter is small (e.g., common in a short-time series of dominant wave period), a model may exhibit high skill but with very poor correlation (failure to follow the minor changes in observational time series). Similarly, measures which involve a normalization by the observed values can favor models with negative bias and result in misleading statistics; Mentaschi *et al.* [2013] recommend an alternative method of normalization.

5.8. Other Challenges

A wave modeler is faced with a number of difficulties. We have already discussed what is arguably the primary challenge: accurate input fields. This difficulty is heightened in cases where the model solution is required near rapid spatial and/or temporal changes of the fields, or if the variable is inherently difficult to determine (e.g., ice type from satellite).

Another challenge is in the spectral tail. Traditionally, e.g., with ECWAM with Janssen [1991] physics and WW3 with ST2 physics, the challenge is obviated by attaching the parametric tail (see Chapter 4) at a relatively low frequency, e.g., 0.30–0.45 Hz. However, recently there has been an increase in efforts to explicitly model a larger frequency range, and with greater attention to accuracy in this portion of the spectrum (ST4 and ST6). An argument can be made that this is not the energy-containing region of a typical ocean spectrum, and for most applications, the benefit of modeling "more of the spectrum" is intangible. However, one can imagine certain benefits. If we compute source terms in the tail more accurately than a parametric extension, this is useful for coupling to other models, e.g., wave-to-ocean momentum flux from breaking. Also, if the prognostic portion of the tail (e.g., 0.4–1.0 Hz) is more accurate than a parametric tail in the same range, then it is reasonable to assume that where the parametric tail *is used* (beyond 1.0 Hz in this example), there will also be a benefit to accuracy. This will facilitate comparison to some satellite products, e.g., radiometers that estimate a mean square slope including very high frequencies.

The broad objective just mentioned is more accurate tail level. A more specific objective is the accurate prediction of the point of transition of the non-directional energy spectrum from $E(f) \propto f^{-4}$ to $E(f) \propto f^{-5}$. Authors have proposed dependencies of this transition point on inverse wave age U/C_p [Forristal, 1981; Kahma and Calkoen, 1992] or a combination of wave age and wind speed [Babanin, 2011], where U is wind speed and C_p is the phase velocity of the peak frequency. Progress in prediction of the tail level has been substantial in recent years, as evidenced by the skill of m_4 prediction in Fig. 5.6. Progress in prediction of transition frequency has been slower, but in WW3, new results are promising [Liu et al., 2019].

The efficient coupling between multiple models (atmosphere, ocean, ice, wave, sediment, hydrology, aerosol) presents a number of major challenges which are primarily technical in nature. The first problem is re-gridding. It is not necessarily a good idea to run the wave model on the same grid that is used by another model. Computational costs are dissimilar between models on a per-grid-cell basis, making it necessary to run the wave model at a coarser resolution than the ocean model. Running the wave model on a decimated version of the ocean model is not necessarily wise either. For example, we have found that WW3 is inefficient when run on the "tripole" grids used by the Navy's ocean models because these grids [e.g., Barron et al., 2007] have highly non-uniform resolution. In the case of coastal coupling, the motivations for high resolution are similar (resolving bathymetry, coastlines, coastal currents, etc.) but not identical. Thus, different grids are often used, which necessitates rapid re-gridding of fields during runtime. Export (or import) of fields from (or to) WW3 is especially difficult in cases where the multi-grid feature of WW3 is used. At the time of writing, this challenge has not been fully addressed. Another challenge is purely technical: building the code to allow all of these models to compile together and run as a single executable, so that the fields can be exchanged through system memory. Lastly (also purely technical), in an operational environment, there must be careful coordination and scheduling of required tasks for each model, e.g., field exchange, data assimilation, pre- and post-processing.

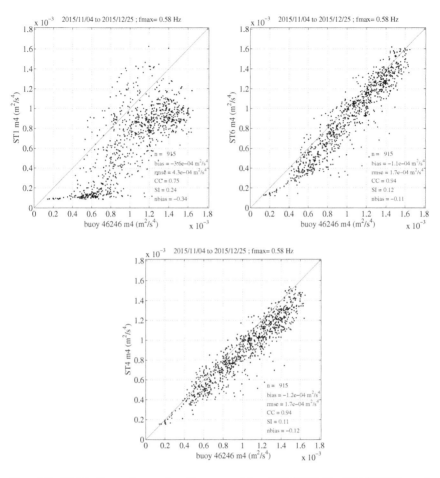

Fig. 5.6. Validation of m_4 for three models: ST1/SWAN, ST4/WW3, and ST6/SWAN, using Waverider buoy CDIP 166, at Ocean Station Papa, owned and operated by APL/UW. Time period is 0400 UTC 4 November to 1200 UTC December 25, 2015. The m_4 parameter is proportional to the contribution to the mean square slope of the sea surface by frequency components up to 0.58 Hz. [These figures are taken from a poster presented by the author in 2017]

Various forms of observational data have been assimilated into wave models, either operationally or as technical demonstrations: Synthetic Aperture Radar [SAR, Breivik et al., 1998; Abdalla et al., 2006; Aouf et al., 2006]; altimetry [Lionello et al., 1992; Komen et al., 1994;

Wittmann and Cummings, 2004]; buoy data [Voorrips et al., 1997; Portilla, 2009; Veeramony et al., 2010; Orzech et al., 2013]; and coastal radar [Panteleev et al., 2015]. Some use optimal interpolation to modify the model initialization and others use one of several approaches to three- or four-dimensional variational (3D-Var or 4D-Var) assimilation, in which the latter improves the simulation dynamically. One often-cited problem with ocean-scale data assimilation is the tendency of windsea to quickly "forget" the data assimilation, relaxing to a state that is in balance with the local winds [Komen et al., 1994]; this has been addressed in some cases through modification of the wind forcing [e.g., Bauer et al., 1996]. Quality control is always a challenge, especially when the observational methods are less mature or robust (e.g., SAR and WERA radar, in our experience). Altimetry, on the other hand, is very robust, but is impaired by the fact that it only provides the total energy of the wave spectrum. These observations can be used to adjust the entire spectrum up or down, but in cases of both windsea and swell components, this is problematic, and one can easily imagine situations where one component is adjusted such that the bias of this component is actually made worse. When skill is quantified using only wave height, such assimilation may improve the skill of short-term forecasts while yielding mixed outcomes for the longer forecast horizons.

5.9. Common Mistakes

This section is specifically for readers that are new to phase-averaged wave modeling. We review some common mistakes. Of course, several of these mistakes are not specific to this type of modeling.

Familiarize with the manual. Though most users' manuals are over 100 pages long, it is always worthwhile to at least scan the contents before starting, and repeat the process periodically. It is not uncommon for a user to become aware too late of key instructions or advice.

Avoid excessive numbers of grid points. This problem is especially common with regular grids. There is a temptation to model a large region at high resolution. This is usually unwise. The user should instead consider irregular or unstructured grids, or nesting. If a model implementation runs too slowly, the user should first ask if the number of grid points is the primary problem, before looking for nonstandard shortcuts in the model physics, spectral grid, or numerics.

Use correct bounds of integration. As noted in Section 5.7, during validation, the bulk parameters should be calculated using consistent frequency range, model versus observations. Doing this incorrectly usually has only minor consequence for wave height and dominant period, but it is often severe for parameters computed from higher moments of the spectrum, e.g., $T_{m,02}$ and m_4. Also, it can have severe consequences for *all* parameters in cases where the climatological dominant period is not much lower than the buoy's maximum frequency, e.g., a large buoy deployed in a lake.

Check for blunders. Of course, the possibilities of human blunder are virtually limitless, but one example is to perform visual spot checks of input fields (bathymetry, ice, wind, etc.) by *outputting* these fields from the wave model and plotting them, thus verifying that they were not read in upside down, missing a factor 100, offset by 180° longitude, etc.

Be suspicious. We have already discussed how one should consider measurement error, especially when dealing with higher spectral moments or new measurement platforms. Similarly, notwithstanding the review process, the published literature can contain errors in methods, reasoning, and interpretation. A good example is the interpretation of idealized model results. Such experiments are useful, but one should avoid naïve extrapolation from these results to realistic applications. A thorough study is supported by results from both types of simulations (idealized and realistic).

References

Abdalla, S., Bidlot, J.R. and Janssen, P.A.E.M. (2006). Global Validation and Assimilation of ENVISAT ASAR Wave Mode Spectra, in *Proc. SeaSAR 2006*, January 23–26, 2006, Frascati, Italy (ESA SP-613, April 2006). [Retrieved from https://earth.esa.int/workshops/seasar2006/proceedings/papers/s1_7_abd.pdf on May 22, 2019]

Alves, J.H.G.M., Greenslade, D.J.M. and Banner, M.L. (2002). Impact of a saturation-dependent dissipation source function on operational hindcasts of wind-waves in the Australian region, *Glob. Atm.-Ocean Syst.*, **8**, 239–267.

Aouf, L., Lefevre, J.-M. and Hauser, D. (2006). Assimilation of directional wave spectra in the wave model WAM: An impact study from synthetic observations in preparation for the SWIMSAT satellite mission, *J. Atmos. Ocean. Techno.*, **23**, 448–463.

Ardhuin, F. and Herbers, T.H.C. (2002). Bragg scattering of random surface gravity waves by irregular seabed topography, *J. Fluid Mech.*, **451**, 1–33.

Ardhuin, F. and Roland, A. (2012). Coastal wave reflection, directional spread and seismoacoustic noise sources, *J. Geophys. Res.: Oceans*, **117**, doi:10.1029/2011JC007832.

Ardhuin, F., O'Reilly, W.C., Herbers, T.H.C. and Jessen, P.F. (2003). Swell transformation across the continental shelf. Part I: Attenuation and directional broadening, *J. Phys. Oceanogr.*, **33**, 1921–1939.

Ardhuin, F., Chapron, B. and Collard, F. (2009). Observation of swell dissipation across oceans, *Geophys. Res. Lett.*, **36**, L06607, doi:10.1029/2008GL037030.

Ardhuin, F., Rogers, W.E., Babanin, A.V., Filipot, J.-F., Magne, R., Roland, A., van der Westhuysen, A., Queffeulou, P., Lefevre, J.-M., Aouf, L. and Collard, F. (2010). Semiempirical dissipation source functions for ocean waves. Part I: Definitions, calibration and validations, *J. Phys. Oceanogr.*, **40**, 1917–1941.

Ardhuin, F., Stutzmann, E., Schimmel, M. and Mangeney, A. (2011). Ocean wave sources of seismic noise, *J. Geophys. Res.: Oceans*, **116**(C9), C006952, doi:10.1029/2011JC006952.

Ardhuin, F., Sutherland, P., Doble, M. and Wadhams, P. (2016). Ocean waves across the Arctic: Attenuation due to dissipation dominates over scattering for periods longer than 19 s, *Geophys. Res. Lett.*, **43**, 5775–5783, doi:10.1002/2016GL068204.

Babanin, A.V. (2006). On a wave-induced turbulence and a wave-mixed upper ocean layer, *Geophys. Res. Lett.*, **33**, L20605, doi:10.1029/2006GL027308.

Babanin, A.V. (2011). *Breaking and Dissipation of Ocean Surface Waves*, Cambridge University Press.

Babanin, A.V. and van der Westhuysen, A.J. (2008). Physics of "saturation-based" dissipation functions proposed for wave forecast models, *J. Phys. Oceanogr.*, **38**, 1831–1841.

Babanin, A.V. and Soloviev, Yu.P. (1998). Field investigation of transformation of the windwave frequency spectrum with fetch and the stage of development, *J. Phys. Oceanogr.*, **28**, 563–576.

Barron, C.N., Kara, A.B., Rhodes, R.C., Rowley, C. and Smedstad, L.F. (2007). Validation Test Report for the 1/8° Global Navy Coastal Ocean Model Nowcast/Forecast System. NRL Tech Report NRL/MR/7320–07-9019, p. 144. [Available from www7320.nrlssc.navy.mil/pubs.php.]

Battjes, J.A. and Janssen, J.P.F.M. (1978). Energy loss and set-up due to breaking of random waves, in *Proc. 16th Int. Conf. Coastal Eng.*, ASCE, pp. 569–587.

Bauer, E., Hasselmann, K., Young, I.R. and Hasselmann, S. (1996). Assimilation of wave data into the wave model WAM using an impulse response function method, *J. Geophys. Res.: Oceans*, **101**, 3801–3816.

Bennetts, L.G. and Squire, V.A. (2012). On the calculation of an attenuation coefficient for transects of ice-covered ocean, *Proc. R. Soc. A*, **468**(2137), 136–162, doi:10.1098/rspa.2011.0155.

Bernier, N.B., Alves, J.-H.G.M., Tolman, H., Chawla, A., Peel, S., Pouliout, B., Belanger, J.-M., Pellerin, P., Lepine, M. and Roch, M. (2016). Operational wave prediction system at Environment Canada: Going global to improve regional forecast skill, *Weather Forecast.*, **31**, 353–370.

Bidlot, J.R. and Holt, M.W. (1999). Numerical wave modeling at operational weather centres, *Coast. Eng.*, **37**(3–4), 409–429.

Bidlot, J.R., Holmes, D.J., Wittmann, P.A., Lalbeharry, R. and Chen, H.S. (2002). Intercomparison of the performance of operational ocean wave forecasting systems with buoy data, *Weather Forecast.*, **17**(2), 287–310.

Bidlot, J., Janssen, P.A.E.M. and Abdalla, S. (2005). A revised formulation for ocean wave dissipation in CY25R1. Technical Report Memorandum R60.9/JB/0516, Research Department, ECMWF, Reading, UK, 35 pp.

Booij, N., Ris, R.C. and Holthuijsen, L.H. (1999). A third-generation wave model for coastal regions, Part 1: Model description and validation, *J. Geophys. Res.: Oceans*, **104**(C4), 7649–7666.

Booij, N. and Holthuijsen, L.H. (1987). Propagation of ocean waves in discrete spectral wave models, *J. Comput. Phys.: Oceans*, **68**, 307–326.

Boutin, G., Ardhuin, F., Dumont, D., Sévigny, C., Girard-Ardhuin, F. and Accensi, M. (2018). Floe size effect on wave-ice interactions: Possible effects, implementation in wave model, and evaluation, *J. Geophys. Res.: Oceans*, **123**, 4779–4805, https://doi.org/10.1029/2017JC013622.

Breivik, L.A., Reistad, M., Schyberg, H., Sunde, J., Krogstad, H. and Johnsen, H. (1998). Assimilation of ERS SAR wave spectra in an operational wave model, *J. Geophys. Res.: Oceans*, **103**(C4), 7887–7900, https://doi.org/10.1029/97JC02728.

Cardone, V.J., Jensen, R.E., Resio, D.T., Swail, V.R. and Cox, A.T. (1996). Evaluation of contemporary ocean wave models in rare extreme events: The "Halloween Storm" of October 1991 and the "Storm of the Century" of March 1993, *J. Atmos. Ocean. Technol.*, **13**, 198–230.

Cavaleri, L. and Malanotte-Rizzoli, P. (1981). Wind-wave prediction in shallow water: Theory and applications, *J. Geophys. Res.: Oceans*, **86**(C11), 10961–10975.

Cavaleri, L. and 25 co-authors (2007). Wave modelling — The state of the art, *Prog. Oceanogr.*, **75**, 603–674.

Collins, C.O., Rogers, W.E., Marchenko, A. and Babanin, A.V. (2015). In situ measurements of an energetic wave event in the Arctic marginal ice zone, *Geophys. Res. Lett.*, **42**(6), 1863–1870, doi:10.1002/2015GL063063.

Collins, C.O. and Rogers, W.E. (2017). A source term for wave attenuation by sea ice in WAVEWATCH III®: IC4, NRL Report NRL/MR/7320-17-9726, 25 pp. [Available from www7320.nrlssc.navy.mil/pubs.php.]

Dalrymple, R.A. and Liu, P.L.F. (1978). Waves over soft muds: A two-layer fluid model, *J. Phys. Oceanogr.*, **8**, 1121–1131.

Dalrymple, R.A., Kirby, J.T. and Hwang, P.A. (1984). Wave diffraction due to areas of energy dissipation, *J. Waterw. Port C-ASCE*, **110**, 67–79.

Doble, M.J. (2009). Simulating pancake and frazil ice growth in the Weddell Sea: A process model from freezing to consolidation, *J. Geophys. Res.: Oceans*, **114**(C9), C09003, doi:10.1029/2008JC004935.

Doble, M.J. and Bidlot, J.-R. (2013). Wavebuoy measurements at the Antarctic sea ice edge compared with an enhanced ECMWF WAM: Progress towards global waves-in-ice modeling, *Ocean Model.*, **70**, 166–173, doi:10.1016/j.ocemod.2013.05.012.

Doble, M.J., De Carolis, G., Meylan, M.H., Bidlot, J.-R. and Wadhams, P. (2015). Relating wave attenuation to pancake ice thickness, using field measurements and model results, *Geophys. Res. Lett.*, **42**, 4473–4481, doi:10.1002/2015GL063628.

Donelan, M.A. (1999). Wind-induced growth and attenuation of laboratory waves, in *Wind-Over-Wave Couplings: Perspectives and Prospects*, eds. S.G. Sajadi, N.H. Thomas, and J.C.R. Hunt, Clarendon Press, pp. 183–194.

Donelan, M.A., Babanin, A.V., Young, I.R. and Banner, M.L. (2006). Wave follower field measurements of the wind input spectral function. Part II. Parameterization of the wind input, *J. Phys. Oceanogr.*, **36**, 1672–1688.

Filipot, J.-F. and Ardhuin, F. (2012). A unified spectral parameterization for wave breaking: From the deep ocean to the surf zone, *J. Geophys. Res.: Oceans*, **117**(C11), doi:10.1029/2011JC007784.

Forristall, G.Z. (1981). Measurements of a saturation range in ocean wave spectra, *J. Geophys. Res.: Oceans*, **86**(C9), 8075–8084.

Goda, Y. (1985). *Random Sea and Design of Maritime Structures*, University Tokyo Press.

Hasselmann, S. and Hasselmann, K. (1985). Computations and parameterizations of the nonlinear energy transfer in a gravity-wave spectrum. Part I. A new method for efficient computations of the exact nonlinear transfer integral, *J. Phys. Oceanogr.*, **15**, 1369–1377.

Hasselmann, S., Hasselmann, K., Allender, J.H. and Barnett, T.P. (1985). Computations and parameterizations of the nonlinear energy transfer in a gravity-wave spectrum. Part II: Parameterizations of the nonlinear energy transfer for application in wave models, *J. Phys. Oceanogr.*, **15**, 1378–1391.

Holthuijsen, L.H. (2007). *Waves in Oceanic and Coastal Waters*, Cambridge University Press, Cambridge, UK.

Holthuijsen, L.H., Herman, A. and Booij, N. (2003). Phase-decoupled refraction–diffraction for spectral wave models, *Coast. Eng.*, **49**, 291–305.

Janssen, P.A.E.M. (1991). Quasi-linear theory of wind-wave generation applied to wave forecasting, *J. Phys. Oceanogr.*, **21**, 1631–1642.

Janssen, P.A.E.M. (2002). Nonlinear four wave interactions and freak waves, ECMWF tech. report., Reading, UK, 35 pp. [Retrieved May 22, 2019 from http://www.ecmwf.int/publications/.]

Janssen, P.A.E.M. (2003). Nonlinear four-wave interactions and freak waves, *J. Phys. Oceanogr.*, **33**, 863–884.

Jeffreys, H. (1924). On the formation of waves by wind, *Proc. R. Soc. A*, **107**, 189–206.

Jiang, H., Babanin, A.V. and Chen, G. (2016). Event-based validation of swell arrival time, *J. Phys. Oceanogr.*, **46**, 3563–3569.

Kahma, K.K. and Calkoen, C.J. (1992). Reconciling discrepancies in the observed growth of wind-generated waves, *J. Phys. Oceanogr.*, **22**, 1389–1405.

Keller, J.B. (1998). Gravity waves on ice-covered water, *J. Geophy. Res.: Oceans*, **103**(C4): 7663–7669.

Komen, G.J., Hasselmann, S. and Hasselmann, K. (1984). On the existence of a fully developed wind-sea spectrum. *J. Phys. Oceanogr.*, **14**, 1271–1285.

Komen, G.J., Cavaleri, L., Donelan, M., Hasselmann, K., Hasselmann, S. and Janssen, P.A.E.M. (1994). *Dynamics and Modelling of Ocean Waves*, Cambridge University Press.

Kuik, A.J., van Vledder, G. and Holthuijsen, L.H. (1988). A method for the routine analysis of pitch-and-roll buoy wave data, *J. Phys. Oceanogr.*, **18**, 1020–1034.

Li, S., Zhao, D., Zhou, L. and Liu, B. (2013). Dependence of mean square slope on wave state and its application in altimeter wind speed retrieval, *Int. J. Remote Sens.*, **34**(1), 264–275.

Lionello, P., Gunther H. and Janssen, P.A.E.M. (1992). Assimilation of altimeter data in a global third-generation wave model, *J. Geophys. Res.: Oceans*, **97**(C9), 14453–14474.

Liu, Q., Rogers, W.E., Babanin, A.V., Young, I.R., Romero, L., Zieger, S., Qiao, F. and Guan, C. (2019). Observation-based source terms in the third-generation wave model WAVEWATCH III: Updates and verification, *J. Phys. Oceanogr.*, **49**, 489–517.

Liu, Y., Su, M.-Y., Yan, X.-H. and Liu, W.T. (2000). The mean-square slope of ocean surface waves and its effects on radar backscatter, *J. Atmos. Ocean. Technol.*, **17**, 1092–1105.

Longuet-Higgins, M.S. (1950). A theory of the origin of microseisms, *Philos. Trans. R. Soc. Lond. A*, **243**(857), 1–35.

Longuet-Higgins, M.S. (1957). The statistical analysis of a random moving surface, *Philos. Trans. R. Soc. Lond. A*, **249**, 321–387.

Madsen, O.S., Poon, Y.K. and Graber, H.C. (1988). Spectral wave attenuation by bottom friction: Theory, in *Proc. 21st Int. Conf. Coastal Eng.*, ASCE, New York, pp. 492–504.

Mentaschi, L., Besio, G., Cassola, F. and Mazzino, A. (2013). Problems in RMSE-based wave model validations, *Ocean Model.*, **72**, 53–58.

Meylan, M., Bennetts, L.G. and Kohout, A.L. (2014). *In situ* measurements and analysis of ocean waves in the Antarctic marginal ice zone, *Geophys. Res. Lett.*, **41**, 5046–5051, doi:10.1002/2014GL060809.

Meylan, M.H., Bennetts, L.G., Mosig, J.E.M., Rogers, W.E., Doble, M.J. and Peter, M.A. (2018). Dispersion relations, power laws, and energy loss for waves in the marginal ice zone, *J. Geophys. Res.: Oceans*, **123**(5), 3322–3335, https://doi.org/10.1002/2018JC013776.

Miles, J.W. (1957). On the generation of surface waves by shear flows, *J. Fluid Mech.*, **3**, 185–204.

Moon, I., Ginis, I., Hara, T., Tolman, H.L., Wright, C.W. and Walsh, E.J. (2003). Numerical simulation of sea surface directional wave spectra under hurricane wind forcing, *J. Phys. Oceanogr.*, **33**, 1680–1706.

Moskowitz, L. (1964). Estimates of the power spectrums for fully developed seas for wind speeds of 20 to 40 knots, *J. Geophys. Res.*, **69**(24), 5161–5179.

Ng, C.O. (2000). Water waves over a muddy bed: A two-layer Stokes boundary layer model, *Coast. Eng.*, **40**(3), 221–242.

O'Reilly, W.C., Herbers, T.H.C., Seymour, R.J. and Guza, R.T. (1996). A comparison of directional buoy and fixed platform measurements of Pacific swell, *J. Atmos. Ocean. Technol.*, **13**, 231–238.

Orzech, M.D., Veeramony, J. and Ngodock, H. (2013). A variational assimilation system for nearshore wave modeling, *J. Atmos. Oceanic Technol.*, **30**(5), 953–970, doi:10.1175/JTECH-D-12-00097.1.

Panteleev, G., Yaremchuk, M. and Rogers, W.E. (2015). Adjoint-free variational data assimilation into a regional wave model, *J. Atmos. Ocean. Technol.*, **32**(7), 1386–1399, doi:10.1175/JTECH-D-14-00174.1.

Perrie, W., Toulany, B., Resio, D.T., Roland, A. and Auclair, J.-P. (2013). A two-scale approximation for wave-wave interactions in an operational wave model, *Ocean Model.*, **70**, 38–51.

Petit, H.A.H. (2001). Diffusion and dispersion of numerical schemes for hyperbolic problems. Communications on Hydraulic and Geotechnical Engineering, T.U. Delft Faculty of Civil Eng. and Geosciences Report No. 01–04, p. 217.

Pierson, W.J. and Moskowitz, L. (1964). A proposed spectral form for fully developed wind seas based on the similarity theory of S. A. Kitaigorodskii, *J. Geophys. Res.*, **69**(24), 5181–5190.

Popinet, S., Gorman, R.M., Rickard, G.J. and Tolman, H.L. (2010). A quadtree-adaptive spectral wave model, *Ocean Model.*, **34**, 36–49.

Portilla, J. (2009). Buoy data assimilation in nearshore wave modeling, Ph.D. dissertation, Katholieke Universiteit Leuven, Heverliee, Belgium, p. 193.

Pushkarev, A. and Zakharov, V. (2016). Limited fetch revisited: Comparison of wind input terms, in surface wave modeling, *Ocean Model.*, **103**, 18–37.

Rapizo, H., Babanin A.V., Provis, D. and Rogers, W.E. (2017). Current-induced dissipation in spectral wave models, *J. Geophys. Res.: Oceans*, **122**(3), 2205–2225, doi:10.1002/2016JC012367.

Resio, D.T., Vincent, L. and Ardag, D. (2016). Characteristics of directional wave spectra and implications for detailed-balance wave modeling, *Ocean Model.*, **103**, 38–52.

Rogers, W.E. and Holland, K.T. (2009). A study of dissipation of wind-waves by mud at Cassino Beach, Brasil: Prediction and inversion, *Cont. Shelf Res.*, **29**, 676–690.

Rogers, W.E. and Orzech, M.D. (2013). Implementation and testing of ice and mud source functions in WAVEWATCH III. NRL Memorandum Report, NRL/MR/7320-13-9462, p. 31. [Available from www7320.nrlssc.navy.mil/pubs.php.]

Rogers, W.E. and Van Vledder, G.Ph. (2013). Frequency width in predictions of windsea spectra and the role of the nonlinear solver, *Ocean Model.*, **70**, 52–61, dx.doi.org/10.1016/j.ocemod.2012.11.010.

Rogers, W.E. and Wang, D.W. (2007). Directional validation of wave predictions, *J. Atmos. Ocean. Technol.*, **24**, 504–520.

Rogers, W.E., Kaihatu, J.M., Booij, N., Holthuijsen, L.H. and Petit, H. (2002). Diffusion reduction in an arbitrary scale third generation wind wave model, *Ocean Eng.*, **29**, 1357–1390.

Rogers, W.E., Hwang, P.A. and Wang, D.W. (2003). Investigation of wave growth and decay in the SWAN model: Three regional-scale applications. *J. Phys. Oceanogr.*, **33**, 366–389.

Rogers, W.E., Wittmann, P.A., Wang, D.W., Clancy, R.M. and Hsu, Y.L. (2005). Evaluations of global wave prediction at the fleet numerical meteorology and oceanography center, *Weather Forecast.*, **20**, 745–760.

Rogers, W.E., Kaihatu, J.M., Hsu, L., Jensen, R.E., Dykes, J.D. and Holland, K.T. (2007). Forecasting and hindcasting waves with the SWAN model in the Southern California Bight, *Coast. Eng.*, **54**, 1–15.

Rogers, W.E., Babanin, A.V. and Wang, D.W. (2012). Observation-consistent input and whitecapping dissipation in a model for wind-generated surface waves: Description and simple calculations, *J. Atmos. Ocean. Technol.*, **29**(9), 1329–1346.

Rogers, W.E., Thomson, J., Shen, H.H., Doble, M.J., Wadhams, P. and Cheng, S. (2016). Dissipation of wind waves by pancake and frazil ice in the autumn Beaufort Sea, *J. Geophys. Res.: Oceans*, **121**(11), 7991–8007, doi:10.1002/2016JC012251.

Rogers, W.E., Posey, P., Li, L. and Allard, R.A. (2018). Forecasting and hindcasting waves in and near the marginal ice zone: Wave modeling and the ONR 'Sea State' field experiment, NRL Report NRL/MR/7320-18-9786, 179 pp. [Available from www7320.nrlssc.navy.mil/pubs.php.]

Romero, L. and Melville, W.K. (2010). Numerical modeling of fetch-limited waves in the Gulf of Tehuantepec, *J. Phys. Oceanogr.*, **40**, 466–486.

Saulnier, J.-P., Clement, A., Falcao, A.F.O., Pontes, T., Prevesto, M. and Ricci, P. (2011). Wave groupiness and spectral bandwidth as relevant parameters for the performance assessment of wave energy converters, *Ocean. Eng.*, **38**, 130–147.

Shen, H.H., Ackley, S.F. and Hopkins, M.A. (2001). A conceptual model for pancake-ice formation in a wave field, *Ann. Glaciol.*, **33**, 361–367.

Shen, H.H., Ackley, S.F. and Yuan, Y. (2004). Limited diameter of pancake ice, *J. Geophys. Res.: Oceans*, **109**(C12), C002123, https://doi.org/10.1029/2003JC002123.

Smith, G.A., Babanin, A.V. and Riedel, P. (2011). Introduction of a new friction routine into the SWAN model that evaluates roughness due to bedform and sediment size changes, *Coast. Eng.*, **58**, 317–326.

Snyder, R.L., Dobson, F.W., Elliot, J.A. and Long, R.B. (1981). A field study of wind generation of ocean waves, *J. Fluid Mech.*, **102**, 1–59.

Squire, V.A., Vahghan, G.L. and Bennetts, L.G. (2009). Ocean surface wave evolvement in the Arctic Basin, *Geophys. Res. Lett.*, **36**, L22502, doi:10.1029/2009GL040676.

Stopa, J.E., Ardhuin, F., Babanin, A. and Zieger, S. (2016). Comparison and validation of physical wave parameterizations in spectral wave models, *Ocean Model.*, **103**, 2–17, doi:10.1016/j.ocemod.2015.09.003.

Suzuki, T., Zijlema, M., Burger, B., Meijer, M.C. and Narayan, S. (2011). Wave dissipation by vegetation with layer schematisation in SWAN, *Coast. Eng.*, **59**, 64–71.

Thomson, J. (2015). ONR Sea State DRI Cruise Report: R/V Sikuliaq, Fall 2015 (SKQ201512S). Retrieved from http://www.apl.washington.edu/project/project.php?id=arctic_sea_state on March 14, 2019, p. 45.

Thomson, J., Talbert J., de Klerk, A., Brown, A., Schwendeman, M., Goldsmith, J., Thomas, J., Olfe, C., Cameron, G. and Meinig, C. (2015). Biofouling effects on the response of a wave measurement buoy in deep water, *J. Atmos. Ocean. Technol.*, **32**, 1281–1286, doi: 10.1175/JTECH-D-15-0029.1.

Thomson, J., Ackley S., Shen, H.H. and Rogers, W.E. (2017). The balance of ice, waves, and winds in the arctic autumn. *Earth & Space Science News*, Eos.org, pp. 30–34.

Thomson, J. et al. (2018). Overview of the Arctic sea state and boundary layer physics program, *J. Geophys. Res.: Oceans*, **123**(12), 8674–8687, doi:10.1002/2018JC013766.

Tolman, H.L. (1991). A Third-generation model for wind-waves on slowly varying, unsteady, and inhomogeneous depths and currents, *J. Phys. Oceanogr.*, **21**(6), 782–797.

Tolman, H.L. (1992). Effects of numerics on the physics in a third-generation wind–wave model, *J. Phys. Oceanogr.*, **22**, 1095–1111.

Tolman, H.L. (2002). *User Manual and System Documentation of WAVEWATCH-III*, version 2.22, Tech. Note 222, NOAA/NWS/NCEP/MMAB, 133 pp.

Tolman, H.L. (2008). A mosaic approach to wind-wave modeling, *Ocean Model.*, **25**, 35–47.

Tolman, H.L. (2011). The impact of nonlinear interaction parameterization on practical wind wave models, in *Proc. 12th Int. Workshop on Wave Hindcasting and Forecasting*, Hawaii, USA, p. 10.

Tolman, H.L. (2013). A generalized multiple discrete interaction approximation for resonant four-wave interactions in wind wave models, *Ocean Model.*, **70**, 11–24.

Tolman, H.L. and Chalikov, D. (1996). Source terms in a third-generation wind wave model, *J. Phys. Oceanogr.*, **26**, 2497–2518.

Tolman, H.L., Balasubramaniyan, B., Burroughs, L.D., Chalikov, D.V., Chao, Y.Y., Chen, H.S. and Gerald, V.M. (2002). Development and implementation of wind-generated ocean surface wave models at NCEP, *Weather Forecast.*, **17**, 311–333.

Tolman, H.L., Banner, M.L. and Kaihatu, J.M. (2013). The NOPP operational wave model improvement project, *Ocean Model.*, **70**, 2–10.

Tolman, H.L. and the WAVEWATCH III Development Group (2014). *User Manual and System Documentation of WAVEWATCH III*, version 4.18, Tech. Note 316, NOAA/NWS/NCEP/MMAB, p. 282 + Appendices.

US Army Corps of Engineers (1973). *Shore Protection Manual*, CERC, US Army Corps of Engineers, Tech. Rep. No. 4.

Van Vledder, G.Ph. (2012). Efficient algorithms for non-linear four-wave interactions, in *Proc. ECMWF Workshop on Ocean Waves*, June 25–27, 2012.

Van Vledder, G.Ph., Hulst, S.Th.C. and McConochie, J.D. (2016). Source term balance in a severe storm in the Southern North Sea, *Ocean Dynam.*, **66**, 1681–1697.

van der Westhuysen, A.J., Zijlema, M. and Battjes, J.A. (2007). Nonlinear saturation-based whitecapping dissipation in SWAN for deep and shallow water, *Coast. Eng.*, **54**(2), 151–170.

Veeramony, J., Walker, D. and Hsu, L. (2010). A variational data assimilation system for nearshore applications of SWAN, *Ocean Model.*, **35**, 206–214, doi:10.1016/j.ocemod.2010.07.008.

Voorrips, A.C., Makin, V.K. and Hasselmann, S. (1997). Assimilation of wave spectra from pitch-and-roll buoys in a North Sea wave model, *J. Geophys. Res.: Oceans*, **102**(C3), 5829–5849.

Wadhams, P., Squire, V.A., Goodman, D.J., Cowan, A.M. and Moore, S.C. (1988). The attenuation rates of ocean waves in the marginal ice zone, *J. Geophys. Res.*, **93**(C6), 6799–6818.

WAMDIG. (1988). The WAM model — a third generation ocean wave prediction model, *J. Phys. Oceanogr.*, **18**, 1775–1810.

WAVEWATCH III Development Group (WW3DG) (2016). *User Manual and System Documentation of WAVEWATCH III*, version 5.16. Tech. Note 329, NOAA/NWS/NCEP/MMAB, College Park, MD, USA, p. 326 + Appendices.

Williams, T.D., Bennetts, L.G., Squire, V.A., Dumont, D. and Bertino, L. (2013a). Wave-ice interactions in the marginal ice zone. Part 1: Theoretical foundations, *Ocean Model.*, **71**, 81–91, doi:10.1016/j.ocemod.2013.05.010.

Williams, T.D., Bennetts, L.G., Squire, V.A., Dumont, D. and Bertino, L. (2013b). Wave-ice interactions in the marginal ice zone. Part 2: Numerical implementation and sensitivity studies along 1D transects of the ocean surface, *Ocean Model.*, **71**, 91–101, doi:10.1016/j.ocemod.2013.05.011.

Winterwerp, J.C., de Graaff, R.F., Groeneweg, J. and Luijendijk, A.P. (2007). *Coast. Eng.*, **54**, 249–261.

Winterwerp, J.C., de Boer, G.J., Greeuw, G. and van Maren, D.S. (2012). Mud-induced wave damping and wave-induced liquefaction, *Coast. Eng.*, **64**, 102–112.

Wittmann, P.A. and Cummings, J.A. (2004). Assimilation of altimeter wave measurements into Wavewatch III, in *Proc. 8th Int Workshop on Wave Hindcasting and Forecasting and 2nd Coastal Hazards Symp.*, North Shore, Oahu, Hawaii, November 14–19, 2004, p. 11.

Young, I.R. (1999). *Wind Generated Ocean Waves*. Elsevier.

Young, I.R. and Van Vledder, G. (1993). A review of the central role of nonlinear interactions in wind-wave evolution, *Philos. Trans. R. Soc. Lond. A*, **342**, 505–524.

Zieger, S., Babanin, A.V., Rogers, W.E. and Young, I.R. (2015). Observation-based source terms in the third-generation wave model WAVEWATCH, *Ocean Model.*, **96**, doi:10.1016/j.ocemod.2015.07.014.

Chapter 6

Phase-Resolving Models

Dmitry Chalikov

*Shirshov Institute of Oceanology Russian Academy of Science,
Moscow, Russia*

Melbourne University, Melbourne, Australia

6.1. Introduction

Phase-resolving modeling of sea waves is the mathematical modeling of surface waves including explicit simulations of surface elevation and velocity field evolution. Compared with the *spectral wave modeling* (see Chapter 5), phase-resolving modeling is more general since it reproduces a real visible physical process and is based on the well-formulated full equations. Phase-resolving models usually operate with a large number of degrees of freedom. In general, this method is more complicated and requires more computational resources. The simplest approach for such modeling is the calculation of a wave field evolution based on the linear equations. Such an approach allows reproducing the main effects of the linear wave transformation due to superposition of wave modes, reflections, refraction, etc. This approach is useful for many technical applications, although it cannot reproduce the nonlinear nature of waves and the transformation of the wave field due to nonlinearity. Another example of a relatively simple object is a case of shallow-water waves. Nonlinearity can be taken into account in the more sophisticated models derived from the fundamental fluid mechanics equations with some simplifications. The most popular approach is based on the nonlinear

Schrödinger equation of different orders [see Dysthe, 1979] obtained by expansion of the surface wave displacement. This approach is also used for solving the problem of *freak waves* (see Chapter 4). The main advantage of a simplified approach is that it allows the reduction of a three-dimensional (3D) problem to a two-dimensional (2D) problem (or two-dimensional problem to one-dimensional (1D) problem). However, it is not always clear which of the non-realistic effects are eliminated or included in the model after simplifications. This is why the most general approach being developed over the past years is based on the initial 2D or 3D equations (still potential). All the tasks based on these equations can be divided into two groups: the periodic and non-periodic problems. An assumption of periodicity considerably simplifies construction of numerical models though such a formulation can be applied to cases when the condition of periodicity is acceptable, e.g., when the domain is considered as a small part of a large uniform area. For limited domains with no periodicity, the problem becomes more complicated since the Fourier presentation cannot be used directly.

From the point of view of physics, the problem of phase-resolving modeling can be divided into two groups: adiabatic and non-adiabatic modeling. A simple adiabatic model assumes that the process develops with no input or output of energy. Being not completely free of limitations, such a formulation allows investigation of the wave motion on the basis of true initial equations. This includes the effects of energy input and its dissipation and is always connected with the assumptions that generally contradict an assumption of potentiality, i.e., new terms added to the equations should be referred to as pure phenomenological. This is why treatment of a non-adiabatic approach is often based on quite different constructions.

All phase-resolving models use the methods of computational mathematics and inherit all their advantages and disadvantages, i.e., on the one side, the possibility of a detailed description of the processes, on the other side, a range of specific problems connected with computational stability, space and time resolution. Mathematical modeling produces tremendous volumes of information, the processing of which can be more complicated than the modeling itself.

Phase-resolving wave modeling takes considerable computer time since it normally uses a surface-following coordinate system, which considerably complicates the equations. The most time-consuming part of the model is an elliptic equation for velocity potential usually solved iteratively. Fortunately, for a 2D problem, this problem is completely eliminated by use of conformal coordinates reducing the problem to a 1D system of equations which can be solved with high accuracy. For a 3D problem, the reduction to a 2D form is evidently impossible; hence, the solution of a 3D elliptical equation for velocity potential becomes an essential part of the entire problem. This equation is quite similar to the equation for pressure in a non-potential problem. It follows that the 3D Euler equations, being more complicated, still can be solved with acceptable computer time.

There is a vast volume of literature devoted to the numerical methods developed for investigation of wave processes over the past decades. It includes a finite difference method (FDM) [Engsig-Karup et al., 2012], a finite volume method (FVM) [Causon et al., 2010], a finite element method (FEM) [Ma and Yan, 2010; Greaves, 2010], a boundary integral element method (BIEM) [Grue and Fructus, 2010] and spectral methods [Ducrose et al., 2007; Touboul and Kharif, 2010; Bonnefoy et al., 2010]. These include a smoothed particle hydrodynamics (SPH) method [Dalrymple et al., 2010], a large eddy simulation (LES) method [Issa et al., 2010; Lubin and Caltagirone, 2010], a moving particle semi-implicit (MPS) method [Kim et al., 2014], a constrained interpolation profile (CIP) method [Zhao, 2016], a method of fundamental solutions (MFS) [Young et al., 2010] and a meshless local Petrov–Galerkin (MLPG) method [Ma, 2010]. Numerous solution techniques have been developed for different mathematical models and formulations. Some of them were developed for a fully nonlinear potential theory; some are based on the higher-order Boussinesq equations [Madsen and Fuhrman, 2010; Zou et al., 2010]; some deal with Navier–Stokes equations; some are suitable for a single-fluid flow, while others consider two or multiple fluids [Causon et al., 2010]; some adopt a Lagrangian formulation; some are built on a familiar Eulerian formulation of the mesh-based methods; and others use a mixed Lagrangian–Eulerian formulation or an arbitrary Lagrangian–Eulerian (ALE) formulation. There are many situations when a fully nonlinear model should be applied. Most of the models were designed for engineering applications such as overturning waves, broken waves, waves generated by landslides, freak waves, solitary waves, tsunamis, violent sloshing waves, interaction of extreme waves with beaches, interaction of steep waves with fixed structures or with freely-responding floating structures. The references given above make up less than 1% of the publications on those topics. In this chapter, the main attention is devoted to methods developed for the investigation of the physics of wave motion, i.e., a finite difference method, a spectral method and a boundary element method.

6.2. Governing Equations

Let us consider the principal 3D equations for potential waves written in Cartesian coordinates, i.e., Laplace equation for velocity potential:

$$\Phi_{xx} + \Phi_{yy} + \Phi_{zz} = 0, \tag{6.1}$$

and two boundary conditions at a free surface $\eta = \eta(x, y, t)$, i.e., the kinematic condition:

$$\eta_t + \eta_x \varphi_x + \eta_y \varphi_y - \Phi_z = 0, \tag{6.2}$$

and Bernoulli integral:

$$\varphi_t + \frac{1}{2}(\varphi_x^2 + \varphi_y^2 + \Phi_z^2) + \eta + p = 0, \tag{6.3}$$

where (x, y, z) are the Cartesian coordinate system, t is time; $\eta(x, y, t)$ describes shape of the free surface; Φ is a 3D velocity potential and ϕ is a value of Φ at the surface η; p is the external pressure created by a flow above the surface and normalized using the density of water. The subscripts denote partial differentiation with respect to the corresponding coordinate. Equations (6.1)–(6.3) are written in a non-dimensional form with the following scales: length L, where $2\pi L$ is a dimensional period in the horizontal, time $L^{1/2}g^{-1/2}$ and velocity potential $L^{3/2}g^{-1/2}$ (g is gravitational acceleration). The capillarity is not taken into account. Accounting for the surface tension effect is quite straightforward. Equations (6.2) and (6.3) contain a vertical derivative of the potential on the surface, defined from a 3D solution of the Laplace equation. The main difficulty is that the boundary conditions for the Laplace equation are given at a variable surface $\eta(x,y,t)$, which excludes the possibility to directly use a standard solution for the Laplace equation.

6.3. Two-Dimensional Phase-Resolving Model

In the two-dimensional model, the terms with derivatives over y are eliminated, so Eqs. (6.1)–(6.3) describe the motion in a plane $(x-z)$ (so-called *unidirected waves*). If the periodicity over x is valid, the most effective approach is introduction of the conformal coordinates [Chalikov and Sheinin, 1998]:

$$x(\xi,\zeta) = \xi + x_0(\tau) + \sum_{-M \leq k < M, k \neq 0} \eta_{-k}(\tau) \frac{\cosh(\zeta + H)}{\sin kH} \vartheta_k(\xi), \quad (6.4)$$

$$z(\xi,\zeta) = \zeta + \eta_0(\tau) + \sum_{-M \leq k < M, k \neq 0} \eta_k(\tau) \frac{\sinh k(\zeta + H)}{\sin kH} \vartheta_k(\xi), \quad (6.5)$$

where x and z are Cartesian coordinates, ξ and ζ are conformal surface-following coordinates, τ is time, H is depth, M is the number of Fourier modes, η_k are coefficients of the Fourier expansion of the free surface $\eta(\xi,\tau)$ with respect to a new horizontal coordinate ξ:

$$\eta(\xi,\tau) = h(x(\xi,\zeta=0,\tau),t=\tau) = \sum_{-M \leq k \leq M} \eta_k(\tau) \vartheta_k(\xi). \quad (6.6)$$

Here ϑ_k denotes the following functions:

$$\vartheta_k(\xi) = \begin{cases} \cos k\xi, & k \geq 0, \\ \sin k\xi, & k < 0. \end{cases} \quad (6.7)$$

Due to the conformity, the Laplace equation retains its form in (ξ,ζ) coordinates, which is the main advantage of the transformation (6.4) and (6.5). It is shown in Chalikov and Sheinin [1998] that the potential wave equations can be represented in the new coordinates as follows:

$$\Phi_{\xi\xi} + \Phi_{\zeta\zeta} = 0, \tag{6.8}$$

$$z_\tau = x_\xi \tilde{g} + z_\xi \tilde{f}, \tag{6.9}$$

$$\varphi_\tau = \tilde{f}\varphi_\xi - \frac{1}{2}J^{-1}(\Phi_\xi^2 - \Phi_\zeta^2) - z, \tag{6.10}$$

where (6.9) and (6.10) are written for surface $\zeta = 0$ (so that $z = \eta$ as represented by expansion (6.6)), J is the Jacobian of the transformation, $\tilde{g}$ is an auxiliary function:

$$\tilde{g} = (J^{-1}\Phi_\zeta)_{\zeta=0}, \tag{6.11}$$

and f is connected with $\tilde{g}$ by the Hilbert transform which for $k \neq 0$ can be defined in Fourier space as follows:

$$\tilde{f}_k = \tilde{g}_{-k}, \quad \tilde{g}_k = -\tilde{f}_{-k}\tanh(kH), \tag{6.12}$$

by Eqs. (6.4) and (6.5).

The boundary condition assumes vanishing of the vertical velocity

$$\Phi_\zeta(\xi, \zeta = H, \tau) = 0. \tag{6.13}$$

The solution of the Laplace equation (6.8) with a boundary condition (6.13) yields the Fourier expansion, which reduces the system (6.8)–(6.10) to a 1D problem:

$$\Phi = \sum_{-M \leq k \leq M} \phi_k(\tau) \exp(k\zeta) \vartheta_k(\xi), \tag{6.14}$$

where ϕ_k are Fourier coefficients of the surface potential $\Phi(\xi, \zeta = 0, \tau)$. Equations (6.9)–(6.12) constitute a closed system of prognostic equations for surface functions

$$z(\xi, \zeta = 0, \tau) = \eta(\xi, \tau) \quad \text{and} \quad \Phi(\xi, \zeta = 0, \tau). \tag{6.15}$$

The problem of the numerical scheme validation for the wave model was discussed in Chalikov and Sheinin [1998]. The scheme was found to be very precise, i.e., a typical accuracy of solution for a sufficiently high resolution was around 10^{-10}. Such accuracy is not surprising, since the equations written in the conformal coordinates turn into the 1D evolutionary equations that can be accurately solved by the Fourier transform method using no local approximations. Both high accuracy of the solution and conservation of integral invariants are crucial for the numerical wave simulation. The ratio of the timescale for waves (wave period) and the timescale for the energy input and dissipation is of the order of 10^{-4}; hence, the wave motion is highly conservative, i.e., at timescales of the order of wave period (in the absence of breaking), the motion is actually adiabatic. The accuracy of the model was tested by simulation of long-term propagation of a steep Stokes wave ($ak = 0.42$). The shape of the Stokes wave was preserved with the accuracy of order 10^{-20}.

The model is mostly intended for simulation of a multi-mode long-term wave field evolution with a realistic spectrum. No matter how high the spectral resolution might be, the energy flux into a truncated part of the spectrum must be parameterized for the long-term simulations of nonlinear waves. Otherwise, spurious energy accumulation violating the energy conservation law at large wave numbers always corrupts the numerical solution. In the numerical solutions of the fluid mechanics equations, this effect is suppressed by introducing different types of viscosity. This effect must be taken into account in all the 2D and 3D phase-resolving wave models.

Equations (6.8)–(6.12) gave a unique example of the equations describing a natural process with computer accuracy. This approach can be applied for investigation of wave dynamics when the two-dimensionality of the spectrum does not play any important role, e.g., for freak wave dynamics [Chalikov, 2009] and nonlinear adiabatic transformation of the 1D spectrum [Chalikov, 2012]. Later, the approach developed was used by Zakharov *et al.* [2002] to demonstrate some nonlinear properties of steep waves. Chalikov and Babanin [2012] performed a detailed investigation of wave breaking onset. Conformal coordinates are so flexible that they allow us to reproduce a non-single-valued surface (which in the conformal coordinates remains a single-valued surface). Hence, the modeling is able to reproduce the initial stage of wave breaking (see Chapter 3).

It is important to emphasize that, after the appearance of inverse inclination, the solution never returns to stability. Up to this moment, the conservation of the sum of potential and kinetic energy, horizontal momentum and volume is excellent. When the surface becomes a non-single-valued surface (at the initial stage of breaking), the conservation of invariants still holds, though later a sharp increase of energy occurs and further integration becoming useless. Disintegration of the solution happens mostly due to inapplicability of potential approximation for a single-phase fluid. This effect can refer to any type of modeling. Using a conformal model, Chalikov [2005] investigated the statistical properties of a 1D wave field. It was shown that opposite to common views, the modes of the wave field are highly non-steady, i.e., their amplitudes quickly fluctuate at relatively short timescales. The phase velocity of each mode for high wave numbers is essentially higher than the linear phase velocity, as shown in Fig. 6.1. It means that the spectral "tail" consists mostly of "bound waves" of low wave number modes. Amplitude of the fluctuations grows with increase of the *rms* steepness of elevations.

A numerical simulation of the Benjamin–Feir (BF) instability was carried out in Chalikov [2007]. According to the Benjamin and Feir [1967] instability theory for the first-order Stokes waves, the amplitudes of disturbances a_k in the vicinity of the main mode with wave number $k = K$ grow exponentially, i.e., $a_k \propto \exp(\gamma_k K^{1/2} t)$. An explicit formula for β_k derived in BF instability can be represented in the following form:

$$\beta_k = \gamma_k K^{1/2}, \quad \text{where } \gamma_k = 0.5|d_k|(2(AK)^2 - d_k^2)^{1/2}, \qquad (6.16)$$

where d characterizes "distance" in Fourier space between the modes with amplitude a_k and the main mode A_K: $d_k = (k/K)^{1/2} - 1$. The function γ_k is represented as a function of the two parameters: AK and δ_k, where $\delta_k = k/K - 1$; hence $d_k = (\delta_k + 1)^{1/2} - 1$. The function $\gamma(AK, \delta)$ is shown in Fig. 6.2 by contours. The function γ is symmetric with respect to the value $d_k = 0$ and reaches a maximum value at $AK = 0.32$ and $\delta = 0.5$. Usually, sea waves have a steepness less than 0.1 where γ is of the order of 10^{-3}; hence, the timescale for development of the BF instability is of the order of 10^3.

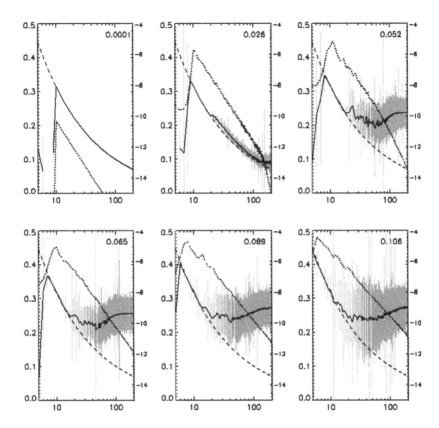

Fig. 6.1. Phase velocities of wave components as a function of wave number (solid curve) for different *rms* steepness of elevation (numbers at top right corner). Gray vertical bars correspond to *rms* of phase velocities. Dashed line is the linear dispersion relation. Dotted line is the wave spectrum (right axis).

The main advantage of a 2D model in the conformal coordinates is high accuracy. Unlike the simple surface-following coordinates (when the height is measured from the surface), the periodic conformal modeling does not impose any limitations on surface steepness. The surface can form intervals with vertical "walls", hence, the instability has a physical nature rather than a computational one. This is why a conformal model is the most convenient for investigation of local processes such as the formation of freak waves or wave breaking.

For non-periodic waves (e.g., waves in a restricted domain or waves above a non-periodic bottom), a conformal mapping does not

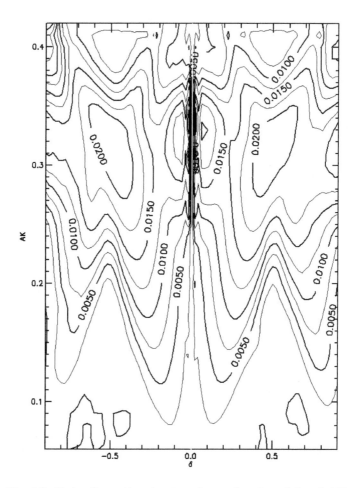

Fig. 6.2. Rate of growth γ (contours) as a function of δ and AK.

exist, and the solution of Eqs. (6.1)–(6.3) should include the solution of the Laplace equation. In this case, it is also reasonable to use semi-Lagrangian methods not necessarily based on the assumption of potentiality.

6.4. Three-Dimensional Phase-Resolving Models

A 2D approach considers a strongly idealized wave field, since even monochromatic waves in the presence of lateral disturbances quickly obtain a 2D structure. The difficulty arising is not a direct result of the increase of dimension. The fundamental complication is that the problem cannot be reduced to a 2D problem, and even for the case of a double-periodic wave field, the problem of the solution of the Laplace equation (6.1) for velocity potential arises.

The majority of the models designed for investigation of 3D wave dynamics are based on simplified equations such as the second-order perturbation methods in which higher-order terms are ignored. Overall, it is unclear which effects are missing in such simplified models.

6.4.1. *Boundary integral equation model*

The most sophisticated method is based on full 3D equations and surface integral formulations [Beale, 2001; Xue et al., 2001; Grilli et al., 2001; Clamond and Grue, 2001; Clamond et al., 2005; Fructus et al., 2005; Guyenne et al., 2006; Fochesato et al., 2006]. In this chapter, we present a new numerical model for fully nonlinear, 3D water waves, which extends an approach by Craig and Sulem [1993] originally given in the 2D setting. The model is based upon a Hamiltonian formulation [Zakharov, 1968], which allows reducing a problem of surface variables computation by introducing a Dirichlet–Neumann operator which is expressed in terms of its Taylor series expansion in homogeneous powers of surface elevation. Each term in this Taylor series can be obtained from the recursion formula and efficiently computed using fast Fourier transforms.

The main advantage of boundary integral equation methods (BIEM) is that they are accurate and can describe highly nonlinear waves. A method of solution of the Laplace equation is based on the use of Green's function, which allows us to reduce a 3D water-wave problem to a 2D boundary integral problem.

The 3D free space Green's function is defined as follows:

$$G(\boldsymbol{x},\boldsymbol{x}_l) = \frac{1}{4\pi|\boldsymbol{r}|}, \quad \frac{\partial G}{\partial n}(\boldsymbol{x},\boldsymbol{x}_l) = -\frac{1}{4\pi}\frac{\boldsymbol{r}-\boldsymbol{n}}{|\mathbf{r}|^3}, \qquad (6.17)$$

where $|\boldsymbol{r}| = |\boldsymbol{x} - \boldsymbol{x}_l|$ is a distance from the source point $\boldsymbol{x}$ to the collocation point $\boldsymbol{x}_l$ (both are on the boundary) and $\boldsymbol{n}$ is a normal vector pointing out of the fluid. The notation $\partial G/\partial n$ represents a normal derivative, i.e., $\partial G/\partial n = \nabla G \cdot \boldsymbol{n}$. The Green's second identity transforms Laplace equation (1) for velocity potential into a boundary integral equation on the boundaries of the fluid domain:

$$\alpha(\boldsymbol{x}_l)\varphi(\boldsymbol{x}_l) = \int_{\Gamma(t)} \left\{ \frac{\partial \Phi}{\partial n}(\boldsymbol{x}) G(\mathbf{x}, \boldsymbol{x}_l) - \varphi(\boldsymbol{x}) \frac{\partial G}{\partial n}(\boldsymbol{x}, \boldsymbol{x}_l) \right\} d\Gamma, \quad (6.18)$$

where $\alpha(\boldsymbol{x}_l)$ is proportional to a solid exterior angle made by the boundary at the collocation point $\mathbf{x}_l$. Kinematic and dynamic conditions (6.2) and (6.3) are used in a mixed Eulerian–Lagrangian form for tracing of a variable surface. Integral equations are solved by the BIEM. The boundary is discretized into N collocation nodes and M high-order elements which are used to interpolate in-between these nodes. Within each element, the boundary geometry and field variables are discretized using polynomial functions.

The surface integral method is well suited for simulation of wave effects connected with very large steepness, especially, for investigation of freak wave generation. These methods can be applied both to periodic and non-periodic flows. The methods do not impose any limitations on wave steepness, so they can be used for simulation of waves that even approach breaking [Grilli et al., 2001] when the surface obtains a non-single value shape. The method allows us to take into account bottom topography [Grue and Fructus, 2010] and investigate the interaction of waves with fixed structures or with freely-responding floating structures [Liu et al., 2016; Gou et al., 2010].

However, the BIEM method is quite complicated and time-consuming when applied to long-term evolution of a multimode wave field in large domains. The simulation of relatively simple wave fields illustrates an application of the method, and it is unlikely that the method can be applied to the simulation of long-term evolution of a large-scale multimode wave field with a broad spectrum. Implementation of a multipole technique for a general problem of sea wave simulation [Fochesato et al., 2006] can solve the problem, but obviously leads to considerable algorithmic difficulties.

6.4.2. *High-order scheme*

Currently, the most popular approach is a high-order scheme (HOS) model developed by Dommermuth and Yue [1987] and West *et al.* [1987]. The HOS is proposed by Zakharov [1968], where a convenient form of the dynamic and kinematic surface conditions was suggested:

$$\eta_\tau = -\eta_\xi \varphi_\xi - \eta_\vartheta \varphi_\vartheta + (1 + \eta_\xi^2 + \eta_\vartheta^2)\Phi_\varsigma, \tag{6.19}$$

$$\varphi_\tau = -\frac{1}{2}(\varphi_\xi^2 + \varphi_\vartheta^2 - (1 + \eta_\xi^2 + \eta_\vartheta^2)\Phi_\zeta^2) - \eta - p. \tag{6.20}$$

Formally, Eqs. (6.19) and (6.20) differ from Eqs. (6.2) and (6.3) by a metric factor $1 + \eta_\xi^2 + \eta_\vartheta^2$, but what is more important is that the "horizontal" derivatives over ξ and ϑ are taken along the surface $\eta(\xi, \vartheta)$. Equations (6.19) and (6.20) used by Zakharov were not intended for modeling, but rather for investigation of stability of finite amplitude waves. In fact, a system of coordinates where depth is measured from the surface was used, but the Laplace equation for velocity potential was taken in its traditional form. However, the Zakharov followers have accepted this idea literally. They used the two coordinate systems: a curvilinear surface-fitting system for surface conditions and the Cartesian system for calculation of a surface vertical velocity. The analytic solution for velocity potential in the Cartesian coordinate system is known. It is based on the Fourier coefficients on a fixed level, while the true variables are the Fourier coefficients for the potential on the free surface. Here, a problem of transition from one coordinate system to another arises. This problem is solved by expansion of the surface potential in a Taylor series in the vicinity of the surface. The accuracy of this method depends on the estimation of an exponential function $\exp(k\eta)$ with a finite number terms of the Taylor series. For small-amplitude waves and for a narrow wave spectrum, such accuracy is satisfactory. However, for the case of a broad wave spectrum that contains many wave modes, the order of the Taylor series should be high. The problem is now that the waves with high wave numbers are superposed over the surface of larger waves. Since the amplitudes of a surface potential attenuates exponentially, an amplitude of a small wave at a positive elevation increases, and on the contrary, can approach zero at negative elevations. It is clear that such a formulation of the HOS model cannot reproduce high-frequency waves, which actually reduces the nonlinearity of the model. This is why such models can be integrated for long periods using no high-frequency smoothing. In addition, the accuracy of calculation of the vertical velocity on the surface depends on the full elevation at each point. Hence, the accuracy is not uniform along a wave profile. A substantial increase of the order of the Taylor expansion can result in numerical instability due to occasional amplification of modes with high wave numbers. The authors of a surface integral method share a similar point of view [Clamond *et al.*, 2005]. We should note, however, that the comparison of a HOS method based on the West *et al.* [1987] approach using a method of surface integral for an idealized wave field [Clamond *et al.*, 2006] shows quite acceptable results. It was shown in the previous paper that a method suggested by Dommermuth *et al.* [1987] demonstrates poor divergence of the expansion for vertical velocity. The HOS model has been widely used [for example, Tanaka, 2001; Toffoli *et al.*, 2010; Touboul and Kharif, 2010] and it has shown its ability to efficiently simulate wave evolution (propagation, nonlinear wave–wave interactions, etc.) in a large-scale domain [Ducrozet *et al.*, 2007, 2012]. It is obvious that the HOS model can be used for many practical purposes. Recently, Ecole Centrale Nantes, LHEEA Laboratory (CNRC) announced that the nonlinear wave models based on HOS would be published as an open source (GPL license). Two different versions of the model are available.

(1) HOS-Ocean is a numerical model dedicated to propagation of nonlinear wave fields in the open ocean with an arbitrary constant depth in a double-periodic domain. It is based on the HOS method which enables simulation of highly nonlinear wave fields (https://github.com/LHEEA/HOS-ocean/wiki). An example of wave spectrum development is given in Fig. 6.3.

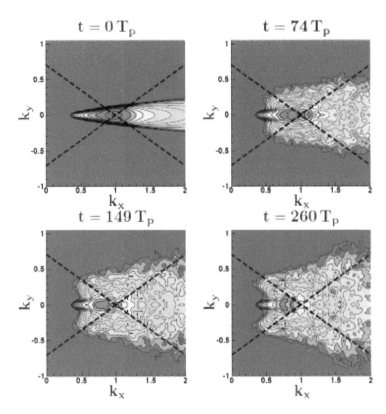

Fig. 6.3. An example of development of a 2D wave spectrum $S(k_x, k_y)$; t is the time expressed in wave peak period.

(2) HOS-NWT is a numerical wave tank based on the HOS method. It is dedicated to generation and propagation of highly nonlinear wave fields in the wave basins with an arbitrary constant depth. The model reproduces all the features of an ocean wave basin: a directional wave maker, reflective walls and an absorbing beach (see Fig. 6.4) (https://github.com/LHEEA/HOS-NWT/wiki).

6.4.3. Schemes based on direct solution of potential equations

In contrast to the HOS method based on the analytical solution of the Laplace equation in Cartesian coordinates, a group of models is based

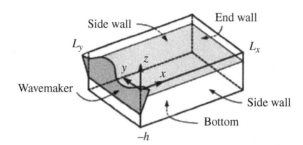

Fig. 6.4. A schematic illustrating the HOS-NWT setting. In front of the domain, initial waves are assigned. At the back side of the domain, the wave absorption is formulated.

on direct solution of the equation for velocity potential in curvilinear coordinates [Engsig-Karup et al., 2012; Chalikov et al., 2014].

It should be noted that although Eqs. (6.2) and (6.3) are written for a free surface, there are no straightforward ways to reduce the equations to a 2D problem, since for evaluation of Φ_z the Laplace equation for velocity potential φ

$$\varphi_{xx} + \varphi_{yy} + \varphi_{zz} = 0 \qquad (6.21)$$

should be solved in the domain

$$\{0 < \xi \leq 2\pi,\ 0 < \vartheta < 2\pi, H < z \leq \eta\} \qquad (6.22)$$

with a curvilinear upper boundary which is a function of x and y. The integration of the system in Cartesian coordinates is either quite inaccurate or too expensive computationally and is not efficient for time intervals that are much greater than a maximum period of a wave. Periodicity conditions over the "horizontal" coordinates x and y are assumed. This suggests that the domain is considered as a small part of an infinitely large basin. The periodicity assumption considerably simplifies the problem since it makes possible the use of the Fourier transform method.

Let us introduce a *non-stationary surface-following non-orthogonal coordinate system*:

$$\xi = x,\quad \vartheta = y,\quad \zeta = z - \eta(\xi, \vartheta, \tau),\quad \tau = t, \qquad (6.23)$$

where $\eta(x,y,t) = \eta(\xi,\vartheta,\tau)$ is a moving periodic wave surface given by the Fourier series

$$\eta(\xi,\vartheta,\tau) = \sum_{-M_x<k<M_x} \sum_{-M_y<l<M_y} h_{k,l}(\tau)\Theta_{k,l}, \qquad (6.24)$$

where M_x and M_y are the numbers of modes in directions ξ and ϑ respectively, while $\Theta_{k,l}$ is the function:

$$\Theta_{kl} = \begin{cases} \cos(k\xi + l\vartheta), & -M_x \leq k \leq M_x, \ -M_y < l < 0, \\ \cos(k\xi), & -M_x \leq k \leq 0, \ l = 0, \\ \sin(k\xi), & 0 \leq k \leq M_y, \ l = 0, \\ \sin(k\xi + l\vartheta), & -M_x \leq k \leq M_x, \ 0 < l \leq M_y. \end{cases} \qquad (6.25)$$

If the accuracy of the Fourier approximation in both directions is the same, then $\gamma = M_x/M_y$ is a ratio of the domain sides L_x/L_y in ξ and ϑ directions. Since a ratio of horizontal scales γ is taken into account in the definition of the derivative over ϑ, it is not included in the equations given below.

A vertical coordinate (6.23) is constructed for a deep water case. As seen, the vertical fluctuations of the "horizontal" coordinates ξ and ϑ do not attenuate with depth. Such fluctuations do not create any approximation problems. However, the lower boundary condition is applied at a variable level $H = \zeta + \eta$. Since all the variables in wave motion attenuate with depth exponentially, the difference between the fixed and fluctuating levels for depth $|H| \gg |\eta|$ becomes negligible. A finite depth as well as a variable depth can be taken into account by use of the sigma-coordinate transformation (see below).

The main advantage of the surface-following coordinate system (6.23) is that a variable surface η is mapped onto the fixed plane $\zeta = 0$. The 3D equations of potential waves in the system of coordinates (6.23) at $\zeta < 0$ take the following form:

$$\eta_\tau = -\eta_\xi \varphi_\xi - \eta_\vartheta \varphi_\vartheta + (1 + \eta_\xi^2 + \eta_\vartheta^2)\Phi_\varsigma, \qquad (6.26)$$

$$\varphi_\tau = -\frac{1}{2}(\varphi_\xi^2 + \varphi_\vartheta^2 - (1 + \eta_\xi^2 + \eta_\vartheta^2)\Phi_\zeta^2) - \eta - p, \qquad (6.27)$$

$$\Phi_{\xi\xi} + \Phi_{\vartheta\vartheta} + \Phi_{\zeta\zeta} = \Upsilon(\Phi), \qquad (6.28)$$

where Φ is a 3D velocity potential, p is the external pressure, and φ is a value of Φ at the surface $\zeta = 0$ while $\Upsilon(\ )$ is an operator:

$$\Upsilon(\ ) = 2\eta_\xi(\ )_{\xi\zeta} + 2\eta_\vartheta(\ )_{\vartheta\zeta} + (\eta_{\xi\xi} + \eta_{\vartheta\vartheta})(\ )_\zeta - (\eta_\xi^2 + \eta_\vartheta^2)(\ )_{\zeta\zeta}. \qquad (6.29)$$

As mentioned above, Eqs. (6.26) and (6.27) are written at a free surface whose position in the surface-following coordinate system is fixed at $\zeta = 0$. These equations formally look as 2D; however, they include a vertical derivative of the potential Φ_ζ at the surface, which should be derived from an elliptical equation (6.28) with the following boundary conditions:

$$\Phi(\zeta = 0) = \varphi, \quad \frac{\partial \Phi}{\partial \zeta}(\zeta \to -\infty) = 0. \qquad (6.30)$$

The second condition (6.30) in the numerical scheme is replaced by the condition at finite depth $\frac{\partial \Phi}{\partial \zeta}(\zeta = H) = 0$, where depth H should be large enough to be considered as infinitely large. The previous calculations with a 1D model show that such H can be defined by the relation $H = 2\pi n/k_p$ where k_p is a mode wave number with the largest amplitude, while $1 < n \leq 2$.

The 2D equations for potential waves written in the conformal coordinates have a remarkable property, i.e., the Laplace equation remains the same. This is why Fourier modes of the velocity potential can be represented through a standard expansion. It means that the potential and any of its derivatives decrease exponentially from a free surface. In a 3D case in the Cartesian coordinates, as well as in the curvilinear coordinates, this is not so. However, it would be reasonable to suggest that the exponential behavior remains dominant, while the potential can be represented as a sum of the two components, i.e., an analytic (linear) component $\bar{\Phi}$, $(\bar{\varphi} = \bar{\Phi}(\xi, \vartheta, 0))$ and an arbitrary nonlinear component $\tilde{\Phi}$, $(\tilde{\varphi} = \tilde{\Phi}(\xi, \vartheta, 0))$:

$$\varphi = \bar{\varphi} + \tilde{\varphi}, \quad \Phi = \bar{\Phi} + \tilde{\Phi}. \qquad (6.31)$$

An analytic component $\bar{\Phi}$ satisfies Laplace equation:

$$\bar{\Phi}_{\xi\xi} + \bar{\Phi}_{\vartheta\vartheta} + \bar{\Phi}_{\zeta\zeta} = 0, \tag{6.32}$$

with the known solution:

$$\bar{\Phi}(\xi, \vartheta, \zeta) = \sum_{k,l} \bar{\varphi}_{k,l} \exp(|k|\zeta)\Theta_{k,l}, \tag{6.33}$$

($\bar{\varphi}_{k,l}$ are Fourier coefficients of a surface analytical potential $\bar{\varphi}$ at $z = 0$). The solution satisfies the following boundary conditions:

$$\begin{aligned} \zeta = 0: &\quad \bar{\Phi} = \bar{\varphi}, \\ \zeta \to -\infty: &\quad \tilde{\Phi}_\zeta \to 0. \end{aligned} \tag{6.34}$$

A nonlinear component satisfies the equation:

$$\tilde{\Phi}_{\xi\xi} + \tilde{\Phi}_{\vartheta\vartheta} + \tilde{\Phi}_{\zeta\zeta} = \Upsilon(\tilde{\Phi}) + \Upsilon(\bar{\Phi}). \tag{6.35}$$

Equation (6.20) is solved with the boundary conditions:

$$\begin{aligned} \zeta = 0: &\quad \tilde{\Phi} = 0, \\ \zeta \to -\infty: &\quad \tilde{\Phi}_\zeta \to 0. \end{aligned} \tag{6.36}$$

Derivatives of the linear component $\bar{\Phi}$ are calculated analytically. The scheme combines a 2D Fourier transform method on "horizontal surfaces" and the second-order finite difference approximation on a stretched staggered grid defined by the relation $\Delta\zeta_{j+1} = \chi\Delta\zeta_j$ ($\Delta\zeta$ is a vertical step, and $j = 1$ at the surface). A stretched grid provides an increase of the accuracy of approximation for the exponentially decaying modes. The values of the stretching coefficient χ lie within the interval 1.10–1.20. A finite difference second-order approximation of Eq. (6.35) on a non-uniform vertical grid is quite straightforward. The details of the numerical scheme and its validation are given by Chalikov [2016]. The scheme was validated by simulation of a Stokes wave with steepness $ak = 0.4$. It was shown that the first several modes of Stokes waves with amplitudes as small as 10^{-4} remain practically unchanged over the entire period of calculations for about 200 periods. Simulation of a disturbed Stokes wave reproduces well-known experimental data on a "horse-shoe" structure [Su et al., 1982] and agrees with the numerical simulations based on a surface integral [Fructus et al., 2005]. The rate of growth of disturbances is in good agreement with a theory of instability of 2D Stokes waves [McLean, 1982].

6.5. Application of 3D Model

The 3D model was used to investigate physical properties of a multi-mode wave field [see Chalikov et al., 2016]. For example, it was found that the deviation from a linear dispersion relation depends on the energy of the mode, i.e., the less the energy, the stronger the deviation. These effects are demonstrated in Fig. 6.5 where spectral density of energy is plotted in coordinates $(\omega_{\text{lin}}, \omega_{\text{mod}})$ where $\omega_{\text{lin}} = k^{1/2}$ and $\omega_{\text{mod}} = \bar{\omega}$ [$\bar{\omega}$ is the actual frequency calculated by the model; see Chalikov, 2005]. Different levels of energy (normalized by its maximum) are shown by different densities of gray tone. The solid curve

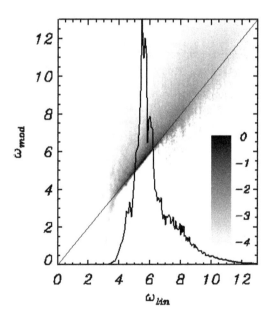

Fig. 6.5. Distribution of spectral energy $\log_{10} \bar{S}$ ($\bar{S}$ is a spectral density normalized by its maximum) in the coordinates $(\omega_{\text{lin}}, \omega_{\text{mod}})$ where $\omega_{\text{lin}} = k^{1/2}$ and ω_{mod} is an actual phase velocity. Solid line shows the spectral density averaged over directions.

shows the spectral energy distribution averaged over equal values of ω_{lin}, i.e., over directions. As can be seen, the modes with large energy obey a linear dispersion relation, i.e., large energy is concentrated along the straight line $\omega_{\text{lin}} = \omega_{\text{mod}}$, while starting approximately from $0.1 S_p$ (S_p is a peak value of spectrum), the calculated frequency ω_{mod} is mostly larger than the linear frequency ω_{lin}. This effect was discovered experimentally, being reproduced in numerical models and explained in Lake and Yuen [1978] and Chalikov and Sheinin [1998]. In reality, surface waves are nonlinear, each wave being constructed from a carrying mode and so-called *bound waves*. These waves are not real waves; they are just shorter modes moving with a speed of the main mode. Besides, the wave field contains free small-amplitude waves whose phase velocity is close to the linear phase velocity $c = \omega/k$. Consequently, at each wave number, free waves and bound waves coexist. Their averaged calculated frequency is larger than the linear frequency ω_{lin}. This effect is more pronounced if the total nonlinearity is large, while the energy of free waves remains small.

It was found that statistical properties of the solution depend essentially on the initial set of phases. This effect is illustrated in Fig. 6.6 where elevation spectra are presented for 50 different runs, calculated for the same initial spectrum but different sets of the initial phases (ensemble modeling). Gray curves correspond to different runs; while solid curve corresponds to the averaged over ensemble spectrum and dashed curves correspond to dispersion.

As seen, the difference between the spectra is very large, which means that the evolution of spectrum depends on a set of the initial phases, i.e., the individual evolution of amplitude of each mode is unpredictable. A most curious property of surface waves is demonstrated in Fig. 6.7. The spectrum assigned as the initial conditions is smooth (top panel). However, after just several peak wave periods, the spectrum starts transforming, i.e., sharp peaks and deep holes appear. Finally, a continuous spectrum transforms into a nearly discrete one consisting of individual peaks.

It is tempting to explain this phenomenon on the basis of a wave–wave resonance mechanism, i.e., the resolution is not high enough

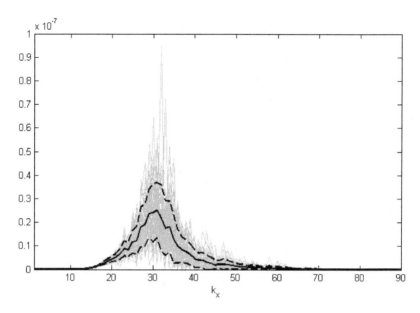

Fig. 6.6. The wave spectra obtained upon reaching the 94th peak wave period corresponding to runs starting from the same wave spectrum but with different random sets of initial phases. Thick curve corresponds to the spectrum averaged over the ensemble; dashed curves characterize dispersion of the data.

to cover all possible resonant combinations of wave numbers and frequencies. This explanation, however, should be based on the assumption that an exact dispersion relation is valid. In reality, the phase velocity of each wave mode is fluctuating due to many reasons, such as nonlinearity, Doppler effects, presence of bound waves, etc. Consequently, the resonant conditions can get blurred over a finite area, and therefore, such an explanation is not valid. If it were, the spectrum would have been continuous. Moreover, if the resolution were a problem, then, following its increase, the spectrum would have converged into a continuous spectrum similar to that at the top panel of Fig. 6.7, which does not happen. Note that similar results were obtained using a simplified model based on the equations derived through expansion of the Hamiltonian up to the fourth-order [Zakharov et al., 2002]. The simplified approach allowed the authors to use a resolution several times higher than that used in the current work. However, simulation of an evolution of the initially homogeneous spectrum resulted

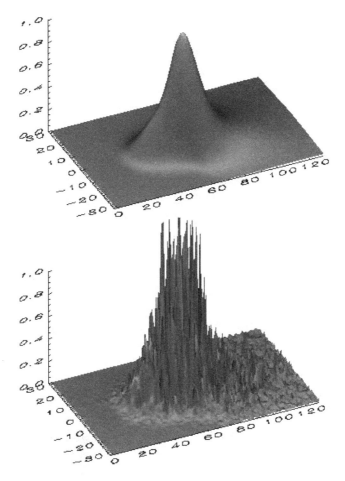

Fig. 6.7. Top panel corresponds to the initial Pierson–Moskowitz 2D wave spectrum $\log_{10}(S)$. Bottom panel corresponds to the final spectrum after integration over 318 peak wave periods.

in a strictly discrete spectrum similar to that in Fig. 6.7. It should be noted that a discretization effect can be visible in the 2D Fourier wave number space, while this effect manifests itself much more weakly in a single-point low-resolution frequency spectrum.

Another hypothesis for the wave spectrum tendency to discretize is based on consideration of a convergence problem. Actually, nonlinear interactions occur in the orbital velocity field. A change of spectral resolution results in modification of statistical characteristics of elevation and velocity fields. It is quite obvious that with an increase of spectral resolution (provided that total energy conservation is strictly valid) the statistical properties of velocity and elevation fields cannot formally come to any reasonable limit. This means that the physical mechanism that prevents homogenization of the spectrum does exist. Probably, the modes with very close wave numbers cannot exist independently; hence, a wave spectrum consists of a finite number of nonlinear modes rather than of an infinite number of linear modes. In other words, a wave field probably has a "corpuscular nature".[1] It is confirmed by Fig. 6.8 where the positions of local maxima in the wave spectrum are shown.

We refer all the points to the local maxima where the value of the spectrum exceeds the values in all eight surrounding points. To make the plot clearer, only those points where the spectral density in the central point exceeds $0.01 S_p$ (S_p is the peak spectral density) are included. The data on the left panel refer to a single spectrum, while the data on the right panel include the points for all 50 spectra.

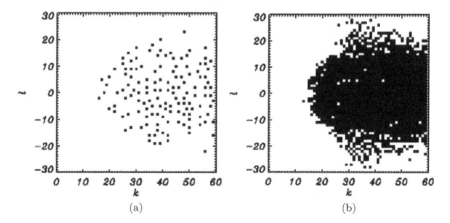

Fig. 6.8. Part (a) shows positions of local maxima in a single spectrum. Part (b) shows positions of maxima in 50 parallel runs.

---

[1]The wave spectrum looks rather like La Sagrada Familia (Gaudi) in Barcelona than the St. Mary Axe ("Gherkin") in London.

As can be seen, the points on the right panel are distributed over the wave number space more or less evenly. Note that the maxima can change their location during a single long integration, but this process is very slow, i.e., the uniformity shown on the right panel of Fig. 6.8 can be reached over thousands of peak wave periods. These results raise doubt as to whether peaks and holes can be explained by resonance mechanisms.

6.6. Freak Wave Phenomenon

Extreme waves are currently a subject of intense study. Various theoretical studies and laboratory experiments have been carried out recently [see Dysthe *et al.*, 2008; Kharif and Pelinovsky, 2003; Kharif *et al.*, 2009]. As it usual in the early stages of a research field, the formation of extreme waves has been explained by many different mechanisms. The simplest explanation of the phenomenon of extreme waves can be given by the assumption of linear superposition of waves. This idea was promoted in a number of papers [see Pelinovsky *et al.*, 2011; Shemer *et al.*, 2007; Chapter 3] where it was shown that the estimates based on this assumption give quite acceptable statistics of extreme waves. Later on, this scheme was found to be too simple (with no rigorous justification), and researchers became engaged in construction of much more complex nonlinear schemes [see Janssen, 2003] and direct numerical simulation [Chalikov, 2005; Touboul and Kharif, 2010]. It was assumed in all those schemes that a linear theory is not able to describe the appearance of extreme waves. Therefore, based on a linear theory, one often makes an additional assumption that wave energy can concentrate on specific structures of surface flows and/or on bathymetry. However, it is known that extreme waves appear both in deep water and in shallow water in the presence or absence of appropriate currents. All the processes mentioned above were studied within a framework of weakly nonlinear models such as the nonlinear Schrödinger equation, a Davey–Stewartson system, the Korteweg–de Vries equation and an equation of Kadomtsev–Petviashvili. Strictly speaking, all those equations do not belong to the equations of hydrodynamics.

Moreover, basic equations are often simplified by being reduced to the only equation for surface elevation.

The Benjamin–Feir [1967] instability is an important mechanism of development of the continuity of the wave spectrum due to the growth of new wave components; however, this process evolves relatively slowly and is probably not applicable to the description of the fast evolution of the finite-amplitude waves controlled by preservation of energy and strong nonlinearity. In the numerical study of the evolution of 1D waves, it is preferable to use precise numerical models based on the full equations of fluid mechanics. The most accurate model of this sort is a 2D model based on the conformal transformation [Chalikov, 2005]. This model was used for studying the mechanics and statistics of 1D extreme waves [Chalikov, 2009]. Many researchers believe that a restriction of the process by assuming a 1D structure reduces the generality of the results and therefore, they recommend use of a 3D model. Other researchers consider that the highest waves occur in a unidirected wave field. There are three main approaches to 3D modeling of waves — the most comprehensive approach having been developed in Chalikov et al., [2014], Touboul and Kharif [2010], and Grue and Fructus [2010]. All of the above models are based on full equations of the potential motion of the fluid with a free surface. The typical number of degrees of freedom of a model reaches 10^6, and the integration being carried out over hundreds of periods of peak wave period. The main difficulties in construction and use of such a model are: (1) the necessity of strict conservation of total energy; (2) the need for individual tracing of evolution of the multimode wave fields with a high phase velocity, which suggests use of a very small time step; (3) the need to solve a 3D elliptic equation for velocity potential at each time step. As a result, one calculation can require computation times of weeks or months. Numerical experiments generate large ensembles of 2D fields of the wave surface $\eta(x, y, t)$. The problem is to calculate the probability of wave heights.

Surprisingly, there is still some confusion in the simple problem of determining wave height. Wave height is defined both in oceanography and marine practice as a vertical distance between the closest trough and crest of a wave. The height of a harmonic wave is twice its amplitude; however, there are no harmonic waves of finite amplitude in nature. In a monochromatic train of nonlinear waves (e.g., Stokes waves), the wave height is equal to the difference between a maximum and a minimum. A real wave field consists of many nonlinear dispersing wave modes and, hence, the surface shape is changing rapidly. In this case, the concept of "wave" becomes uncertain because a wave consists of many modes with their own phases. The amplitudes of modes fluctuate as a result of nonlinear reversible interactions. The height of the wave above the average level is not a complete characteristic as the wave field is vertically asymmetric, i.e., the heights of wave crests are usually larger than the depth of troughs, which is a sign of strong nonlinearity of the process. However, the destructive power of the wave depends on its total height from trough to crest. Therefore, the only way to distinguish an extreme wave is by direct search of closely located crest/trough pairs. The qualitative definition of such closeness should be formalized.

As a rule, the spectrum of wind waves is rather narrow. This means that the wave field always contains dominant waves whose wave numbers are close to the wave number of a mode related to the peak of the spectrum. The form of this wave is changing due to superposition with other modes and due to reversible nonlinear interactions. It is reasonable to assume that the full height of the wave can be defined as the vertical distance between a maximum and a minimum in some interval of the order of the peak wavelength L_p. Since the length of a real wave varies, the interval should be slightly extended. The data processing of the numerical experiments presented in Chalikov [2009] showed that in the case of 1D waves, the total wave height H_{ct} should be determined in a moving window of length $1.5L_p$. The window itself can move with discrete increment of $0.5L_p$. The calculations have shown that the probability of extreme waves does not practically depend on the specific values of the selected parameters, if the size of the ensemble is very large. A 2D wave field can be represented as an ensemble of 1D fields oriented in the direction of general wave propagation. This interpretation is traditional, yet, not complete, as the locations of a minimum and a maximum of a real wave as well as the direction of their movement do not necessarily coincide with a general direction of wave propagation. Therefore, it seems reasonable to generalize a 1D algorithm for a 2D case by introducing a rectangular moving window with the sizes $1.5L_p \times 1.5L_p$. The window is moved over the field $\eta(x,y)$ with discrete steps of $0.5L_p$ in both directions. It is clear that if the window narrows in y-direction, then the statistics collected with a 2D algorithm converges to the 1D one.

The initial conditions were generated using the JONSWAP spectrum written as a function of wave numbers k and l in a half-plane $(-My \leq l \leq My, 0 \leq k \leq Mx)$. The angular spreading in the energy containing part of spectrum $S(k,l)$ was taken proportional to $\cos^2 \vartheta_{k,l}$ (ϑ is a direction of the kth mode). The maximum of the spectrum was associated with the wave number. The amplitudes of wave modes $h_{k,l}$ and $h_{-k,-l}$ were calculated by formulas:

$$h_{k,l} = (\Delta k \Delta l S_{k,l})^{1/2} \cos \theta_{k,l}, \quad h_{-k,-l} = (\Delta k \Delta l S_{k,l})^{1/2} \sin \theta_{k,l}, \tag{6.37}$$

where $\theta_{k,l}$ are phases of the modes distributed uniformly and randomly in the interval $(0 \leq \varphi \leq 2\pi)$. Uniform distribution of random phases is a natural assumption because the influence of "bound" waves on the statistics is negligibly small. The field of amplitudes can be further transformed by the inverse Fourier transform of the elevation field $\eta(\xi, \vartheta)$:

$$\eta(\xi, \vartheta) = \sum_{0 \leq k \leq M_x} \sum_{-M_y \leq l \leq M_y} h_{k,l} \cos(k\xi + l\vartheta) + h_{-k,-l} \sin(k\xi + l\vartheta). \tag{6.38}$$

The Fourier amplitudes f for the surface velocity potential φ were calculated using the formulas of the theory of small amplitude waves (see Chapter 1), i.e.,

$$f_{k,l} = -k^{1/2} h_{-k,-l}, \quad 0 \leq k \leq M_x, -M_y \leq l \leq M_y,$$
$$f_{-k,-l} = -|k|^{1/2} h_{k,l}, \quad 0 \leq k \leq M_x, -M_y \leq l \leq M_y.$$

This algorithm describes the generation of the initial wave field given by a superposition of linear modes with certain angular resolution.

The wave modes of this field are not affected by nonlinearity; however, a non-steady solution obtained on the basis of Eqs. (6.26)–(6.28), using the initial conditions (6.37) and (6.38) quickly develops nonlinear features, i.e., waves crests sharpen, their troughs becoming flatter. In particular, the wave field becomes closer to a set of Stokes waves rather than to linear waves. If the crests coincide, then the local energy may be focused rapidly [Chalikov and Babanin, 2012]. The slower nonlinear effects develop in the wave field such as group structure [Sanina *et al.*, 2015], Benjamin–Feir instability [Benjamin and Feir, 1967; McLean, 1982] and nonlinear processes like Hasselmann's [1962] quadruplet interactions. It is assumed that the main cause of the extreme waves whose heights far exceeds the average height of waves is the BF instability (or the so-called modulation instability). The theories based on this assumption generate the greatest distrust. It is known that development of new modes for the steepness typical for sea waves occurs according to the BF scenario for tens and hundreds of periods of the carrier wave, whereas an extreme wave develops over the time of the order of its period [Chalikov, 2007].

The aim of this study is to clarify to what degree the nonlinearity affects the generation processes of extreme waves. The most reliable answer to this question can be given by comparison of the probability of linear and nonlinear extreme waves. We call the results obtained from the "static" representation (6.38) as "the linear probability", while "the nonlinear probability" corresponds to the results of integration of the system (6.26)–(6.28). If nonlinearity plays a significant role in generation of extreme waves, then the linear probability must be evidently less than the nonlinear one; moreover, it should be small. In generation of a linear field, we used an initial JONSWAP spectrum, not the spectra obtained as a result of integration at a given moment. This is possible since the period of integration is small and the spectrum does not change significantly.

The results depend on the general distribution of the energy over the spectrum, but not on the particular details of spectrum.

The probability of surface deviation from a mean level P_H is shown in Fig. 6.9.

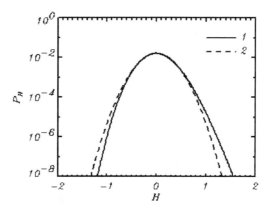

Fig. 6.9. Probability of wave surface deviations PH from a mean level, 1 — calculations for the data reproduced by a nonlinear model; 2 — calculations for an ensemble of linear waves given by formula (6.38).

Curve 1 corresponds to the probability calculated from an ensemble of wave surfaces reconstructed with the numerical integration of the system (6.26)–(6.28). Curve 2 corresponds to the similar probability calculated from the superposition of linear waves with random phases. It is evident that curve 2 shows strict symmetry of perturbations with regard to the mean level. Curve 1 confirms that the probability of positive perturbations far exceeds the probability of negative perturbations; in particular, the heights of wave crests are essentially larger than the depths of wave troughs, i.e., the nonlinear waves are more sharpened in comparison with linear waves. Firstly, this means that each mode approaches a Stokes mode and, secondly, it reflects a focusing effect for energy in coincidence of wave crests [Chalikov and Babanin, 2016].

Further, we calculated an integral probability of wave heights $\tilde{H} = H_{ct}/H_s$ from trough to crest, determined with the use of a moving rectangular window (a 2D algorithm). The size of the ensemble was 256 for one field. Thus, the total size of an ensemble including all time moments for all variants of the model was equal to 3,200,000. The integral probability of $\tilde{H}$ is presented in Fig. 6.10 (curve 1). The integral probability of $\tilde{H}$ calculated by the same 2D algorithm, but for an ensemble of linear waves is shown by curve 3. As seen from

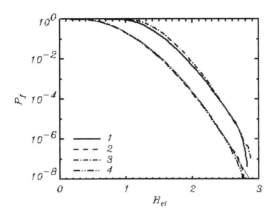

Fig. 6.10. Integral probability P_f of wave height from trough to crest: 1 and 3 — a 2D algorithm for the nonlinear and linear wave fields, respectively; 2 and 4 — the same for a 1D algorithm.

Fig. 6.10, curves 1 and 3 are very close to each other, in particular, for the high values of $\tilde{H}$. Some difference observed for $P_f < 10^{-6}$ can be explained by an insufficient volume of data. It is interesting to note that opposite to expectations, a linear model predicts the reality of appearance of very large values of dimensionless height $\tilde{H} > 2.7$. It was earlier believed that such high waves can be generated only under the influence of nonlinearity. There are some reasons to suggest that the established patterns do not significantly depend on resolution. The qualitatively similar results, i.e., coincidence of the linear and nonlinear statistics, were obtained with a single-processor version of the model with double resolution.

Despite the fact that an algorithm for the detection of extreme waves based on a moving rectangular window seems reasonable enough, some researchers believe that waves with a maximum and a minimum located in the general direction of wave propagation can also be classified as extremes. The probability of such cases can be calculated by the algorithm used above, while the dimension of a window in x-direction is taken equal to 1. The probability of the linear and nonlinear waves calculated by such a 1D algorithm is presented in Fig. 6.10 by curves 2 and 4, respectively. As expected, the linear and nonlinear 1D statistics give the same results.

The results shown in Figs. 6.9 and 6.10 allow us to come to an interesting conclusion. The nonlinearity leads to strong modifications of wave shape; such modifications are well pronounced if positive and negative deviations of wave surface are considered separately. However, they are not pronounced at all if the probability of the full trough-to-crest height of the waves is considered. For calculation of the trough-to-crest statistics, no nonlinear theory is required, thus, the probability of the trough-to-crest extreme waves can be calculated by the use of (6.38). However, the statistics of positive and negative elevations can be investigated only with the nonlinear theory which takes into account the presence of bound waves.

6.7. Non-Adiabatic Effects

The amplitudes of all modes in a spectrum fluctuate, but a non-reversible evolution of the spectrum occurs under the influence of wind input, dissipation and nonlinear interactions.

6.7.1. *Input energy*

According to linear theory [Miles, 1957] (see Chapter 2), the Fourier components of surface pressure p are connected with those of the surface elevation through the following expression:

$$p_{k,l} + ip_{-k,-l} = \frac{\rho_a}{\rho_w}(\beta_{k,l} + i\beta_{-k,-l})(h_{k,l} + ih_{-k,-l}), \quad (6.39)$$

where $h_{k,l}, h_{-k,-l}, \beta_{k,l}, \beta_{-k,-l}$ are real and imaginary parts of elevation η and the so-called β-function (i.e., Fourier coefficients at cos and sin, respectively), and ρ_a/ρ_w is a ratio of air and water densities, respectively. Hence, for derivation of the shape of the β-function, it is necessary to simultaneously measure the wave surface elevation and non-static pressure on the surface. An experimental measurement of the surface pressure is a very difficult problem since the measurements should be done very close to a moving surface, preferably, with a surface-following sensor. Such measurements are done quite seldom, especially, in the field. Measurements were carried out for the first time by a team of authors both in laboratory and field

[Snyder et al., 1981; Hsiao and Shemdin, 1983; Hasselmann and Bösenberg, 1991; Donelan et al., 2005, 2006]. The data obtained in this way allowed construction of the imaginary part of the β-function used in some versions of wave forecasting models [Rogers et al., 2012]. The second approach to evaluate the β-function is based on the results of numerical investigations of the statistical structure of the boundary layer above waves with use of the Reynolds' equations and an appropriate closure scheme. In general, this method works so well that many problems in the technical fluid mechanics are often solved using numerical models, not experimentally. This method was developed beginning from [Chalikov, 1978, 1986], followed by [Chalikov and Makin, 1991; Chalikov and Belevich, 1992; Chalikov, 1995]. The results were implemented in the WAVEWATCH model, i.e., the third-generation wave forecast model [Tolman and Chalikov, 1996] and thoroughly validated against experimental data in the course of developing WAVEWATCH-III [Tolman et al., 2014] (see Chapter 5). This method was later improved on the basis of a more advanced coupled modeling of waves and the boundary layer [Chalikov and Rainchik, 2010; hereafter CR], while the β-function used in WAVEWATCH-III was corrected and extended up to high frequencies. Direct calculation of the energy input to waves requires both real and imaginary parts of the β-function. The total energy input to waves depends on the imaginary part of the β-function, while the moments of higher-order depend both on the imaginary and real parts of β. This is why the full approximation constructed in CR was used in the current work. Note that in the range of relatively low frequencies the new method is very close to the scheme implemented in WAVEWATCH-III.

It is a traditional suggestion that both coefficients are the functions of non-dimensional frequency $\Omega = \omega_k U \cos\psi = U/c_k \cos\psi$ (where ω_k and U are the non-dimensional radian frequency and wind speed, respectively; c_k is a phase speed of the kth mode, and ψ is the angle between the wind and wave mode directions). Most of the schemes for calculations of the β-function consider a relatively narrow interval of non-dimensional frequencies Ω. In the current work, the range of frequencies covers an interval ($0 < \Omega_p < 10$), and

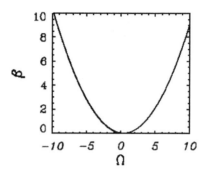

Fig. 6.11. The imaginary part of the β-function.

occasionally the values of $\Omega > 10$ can appear. This is why the function derived in Chalikov and Rainchik [2010] through coupled simulations of waves and boundary layer is used here. The wave model is based on the potential equations for flow with a free surface, extended with an algorithm for breaking dissipation (see the below description of the breaking dissipation parameterization). A wave boundary layer (WBL) model is based on the Reynolds equations closed with a $K - \varepsilon$ scheme; the solutions for air and water are matched through the interface. The β-function obtained in CR was used for the accurate evaluation of surface pressure, p. The shape of the β-function connecting the surface elevations and surface pressure has been studied up to high non-dimensional wave frequencies both in positive and negative (i.e., for wind opposite to waves) domains. The data on the β-function exhibit wide scatter, but since the volume of the data is quite large (47 long-term numerical runs allowed us to generate about 1,400,000 values of β), the shape of the β-function was defined with satisfactory accuracy up to very high non-dimensional frequencies ($-50 < \Omega < 50$). As a result, the data on the β-function in such a broad range allow us to calculate the wave drag up to very high frequencies and to explicitly divide the fluxes of energy and momentum transferred between the pressure and molecular viscosity. This method is free of arbitrary assumptions on the drag coefficient C_d, and, on the contrary, such calculations allow investigation of the nature of the wave drag [see Ting et al., 2012].

The energy and momentum can be transferred to waves only through the imaginary part of $\beta(\Omega_{k,l})$ where $\Omega_{k,l} = U(\lambda_{k,l}/2)/c_{k,l}\cos\psi_{k,l}$ is the virtual frequency of the mode with wave numbers k and l, $U(\lambda_{k,l}/2)$ is a wind speed at a height equal to half of wavelength $\lambda_{k,l}$, $c_{k,l}$ is a phase speed of the mode (always positive), and $-\pi < \psi_{k,l} \le \pi$ is an angle between the wind direction and the direction of mode propagation.

The revised approximation to the β-function is given by

$$\beta_i = \begin{cases} \beta_0 + a_0(\Omega - \Omega_0) + a_1(\Omega - \Omega_0)^2, & \Omega > \Omega_0, \\ \beta_0 - a_0(\Omega - \Omega_0) + a_1(\Omega - \Omega_0)^2, & \Omega < \Omega_0, \end{cases} \quad (6.40)$$

where $\Omega_0 = 0.0355$, $\beta_0 = -0.02$, $a_0 = 0.0228$, $a_1 = 0.0948$ (see Fig. 6.11). The imaginary part of function β is negative in the interval $(0 < \Omega < 0.7)$ where the phase velocity of waves is larger than the wind velocity; hence the momentum and energy are transferred from waves to wind. This means that the pressure at the upwind slope of the waves is less than the pressure at the downwind slope, hence the waves accelerate the wind and lose momentum and energy. For all other intervals of Ω, the flux of energy and momentum is directed from wind to waves, i.e., pressure at upwind slope of wave is larger than at the downwind slope and momentum and energy is directed from wind to waves which decreases the absolute value of the wind velocity. The flux of momentum is directed to the waves, but wave momentum is negative, hence the waves reaccelerate and their energy dissipates. In cases when the wind blows against the waves, the function is larger than in the opposite case at the same value of Ω, so, the function β is not symmetric relatively $\Omega = 0$.

6.7.2. *High-frequency dissipation*

A nonlinear flux of energy directed to the small wave numbers produces downshifting of spectrum, while an opposite flux to high wave numbers shapes of the spectral tail. This second process can produce an accumulation of energy near the "cutoff" wave number in the model. Both processes become more intensive with an increase in energy input. Growth of amplitudes at high wave numbers is followed by growth of the local steepness and numerical instability. This phenomenon is well known in numerical fluid mechanics and is eliminated by the use of a highly selective filter simulating nonlinear viscosity.

To support stability, additional terms are included in the right-hand sides of Eqs. (6.26) and (6.27):

$$\frac{\partial \eta_{k,l}}{\partial \tau} = E_{k,l} - \mu_{k,l}\eta_{k,l}, \qquad (6.41)$$

$$\frac{\partial \varphi_{k,l}}{\partial \tau} = F_{k,l} - \mu_{k,l}\varphi_{k,l}, \qquad (6.42)$$

where $E_{k,l}$ and $F_{k,l}$ are Fourier amplitudes of the right-hand sides of Eqs. (6.26) and (6.27), and $\mu_{k,l}$ is a coefficient whose dependence on k and l is chosen for absorption of energy on the periphery of spectrum. The aim of the algorithm is to support smoothness and monotonicity of the wave spectrum within a high wave number range. Since the algorithm affects amplitudes of small modes, it actually does not reduce total energy, though it efficiently prevents the development of the numerical instability. Note that any long-term calculations cannot be performed without this "tail dissipation" eliminating development of numerical instability at high wave numbers.

6.7.3. Wave breaking

The main process of wave dissipation is wave breaking. This process is taken into account in all spectral wave forecasting models similar to WAVEWATCH [see Tolman and Chalikov, 1996]. Since there are no (individual) waves in spectral models, no local criteria of wave breaking can be formulated. This is why the breaking dissipation is represented in the spectral models in a distorted form. The areal breaking occurs in relatively narrow areas of physical space; however, a spectral image of such breaking is stretched over the entire wave spectrum, while in reality the breaking decreases the height and energy of dominant waves. This contradiction occurs because waves in spectral models are assumed as linear, while actually the breaking occurs in a physical space with a nonlinear sharp crested wave usually composed of several modes.

The mechanics of wave breaking for a developed wave spectrum differs from those in a wave field represented by few modes normally considered in many theoretical and laboratory investigations. Since the breaking in laboratory conditions is initiated by special assigning of amplitudes and phases, it cannot be similar to the breaking in natural conditions. To some degree, wave breaking is similar to the development of extreme waves that appear suddenly with no pronounced prehistory [Chalikov and Babanin, 2016] (see also Chapter 3). There are no signs of the modulational instability in both phenomena, which suggests a process of energy borrowing from other modes. The evolution leading to the breaking or "freaking" seems just opposite: the full energy of the main wave remains nearly constant while the columnar energy is focused around the crest of the wave that becomes sharper and unstable. Probably, even more frequent cases of wave breaking and extreme wave appearance can be explained by a local superposition of several modes (see Chapter 3).

The instability of the interface leading to breaking is an important but poorly developed problem of fluid mechanics. In general, this essentially nonlinear process should be investigated for a two-phase flow. Such an approach was demonstrated, for example, by Iafrati [2001]. However, progress in solving this highly complicated problem has not been rapid.

A problem of breaking parameterization includes two points: (1) establishing criterion for breaking onset and (2) developing an algorithm for breaking parameterization. The problem of breaking is discussed in details in Babanin [2011] (see Chapter 3). Chalikov and Babanin [2012] have performed a numerical investigation of the processes leading to wave breaking. It was found that a clear predictor of breaking, formulated in dynamical and geometrical terms, probably does not exist. The most evident criterion of breaking is the breaking itself, i.e., the process when some part of the upper portion of a sharp wave crest falls down. This process is usually followed by separation of a detached volume of liquid into water and air phases. Unfortunately, there is no possibility to describe this process within the scope of potential theory.

Some investigators suggest using a physical velocity approaching the rate of surface movement in the same direction as a criterion for breaking onset. This is incorrect, since a kinematic boundary condition suggests that these quantities are exactly equal to each other. Consideration of a ratio of the water particle velocity and crest velocity in the direction of wave propagation [Seifet and Ducrozet, 2016] seems more reasonable. It is quite clear that the onset of breaking can be characterized by appearance of a non-single-valued section of the water surface. This stage can be investigated with a 2D model which, due to the high flexibility of the conformal coordinates, allows us to reproduce a surface with the inclination in the Cartesian coordinates exceeding 90°. (In the conformal coordinates, the dependence of elevation on a curvilinear coordinate is always single-valued.) The duration of this stage is extremely short, the calculations always being interrupted by numerical instability with sharp violation of the conservations laws (constant integral invariants, i.e., full energy and volume) and a strong distortion of the local structure of flow. Numerous numerical experiments with conformal models showed that after the appearance of a non-single value, the model never returns to stability. However, introduction of the appearance of a non-single-valued surface as a criterion of breaking instability even in a conformal model is impossible, since the behavior of the model at a critical point is unpredictable, and the run is most likely to be terminated, no matter what kind of parameterization of breaking is introduced. It means that even in a very precise conformal model, stabilization of the solution should be initiated prior to breaking.

Consideration of an exact criterion for breaking onset for the models using transformation of a coordinate type (1) is useless, since the numerical instability in such models arises not because of the breaking approach but due to the appearance of large local steepness. Multiple experiments with a direct 3D wave model show that appearance of the local steepness $\max(\partial \eta/\partial x, \partial \eta/\partial y)$ exceeding ≈ 2 (which corresponds to a slope of about 60°) is always followed by numerical instability. A decrease of time step does not make any effect. As seen, a surface with such a slope is very far from being a vertical "wall", when real breaking starts. However, an algorithm for breaking parameterization must prevent appearance of large local steepness. The situation is similar to the numerical modeling of turbulence (LES technique) where the local highly selective viscosity is used to prevent the appearance of too large local gradients of velocity. A description of breaking in direct wave modeling should satisfy the following conditions. (1) It should prevent large local gradients of elevation; in our case a breaking algorithm should prevent the onset of instability at each point of half million of grid points over more than 100 thousand time steps. (2) It should describe in a more or less realistic way the loss of the kinetic and potential energies with preservation of balance between them. (3) It should preserve the volume. It was suggested in Chalikov [2005] that an acceptable scheme can be based on a local highly selective diffusion operator with a special diffusion coefficient. Several schemes of such type were validated, and finally the following scheme was chosen:

$$\eta_\tau = E_\eta + J^{-1}\left(\frac{\partial}{\partial \xi}B_\xi \frac{\partial \eta}{\partial \xi} + \frac{\partial}{\partial \vartheta}B_\vartheta \frac{\partial \eta}{\partial \vartheta}\right), \qquad (6.43)$$

$$\varphi_\tau = F_\varphi + J^{-1}\left(\frac{\partial}{\partial \xi}B_\xi \frac{\partial \varphi}{\partial \xi} + \frac{\partial}{\partial \vartheta}B_\vartheta \frac{\partial \varphi}{\partial \vartheta}\right), \qquad (6.44)$$

where F_η and F_φ are the right-hand sides of Eqs. (6.3) and (6.4) including the terms introduced by (6.16)–(6.20). It was suggested in the first versions of the scheme that a diffusion coefficient depends on a local slope; however, such a scheme did not prove to be very reliable since it did not prevent all of the events of the numerical instability. A scheme based on the calculation of the local curvilinearity $\eta_{\xi\xi}$ and $\eta_{\vartheta\vartheta}$ turned out to be much more reliable. Calculations of 75 different runs were performed with a full 3D model in Chalikov et al. [2014] over a period of $t = 350$ (70,000 time steps). The total number of values used for the calculations of dependence in Fig. 6.12 (thick curve) is about 6 billion. The normal probability calculated with the same dispersion is shown by thin curve.

It is seen that the probability of large negative values of curvilinearity is orders of magnitude larger than the probability calculated over an ensemble of linear modes with the same spectrum.

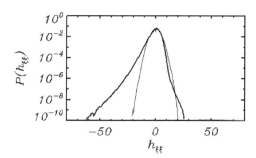

Fig. 6.12. Probability of curvilinearity $\eta_{\xi\xi}$. Thick curve calculated with a full 3D model; thin curve is the probability calculated over an ensemble of linear modes with the same spectrum.

The curvilinearity turned out to be very sensitive to the shape of the surface. This is why it was chosen as a criterion of the breaking approach. Coefficients B_ξ and B_ϑ depend nonlinearly on the curvilinearity

$$B_\xi = \begin{cases} \Delta\xi C_B \eta_{\xi\xi}^2, & \eta_{\xi\xi} < \eta_{\xi\xi}^{cr}, \\ 0, & \eta_{\xi\xi} \geq \eta_{\xi\xi}^{cr}, \end{cases} \quad (6.45)$$

$$B_\vartheta = \begin{cases} \Delta\vartheta C_B \eta_{\vartheta\vartheta}^2, & \eta_{\vartheta\vartheta} < \eta_{\xi\xi}^{cr}, \\ 0, & \eta_{\vartheta\vartheta} \geq \eta_{\xi\xi}^{cr}, \end{cases} \quad (6.46)$$

where $\Delta\xi$ and $\Delta\zeta$ are horizontal steps in x- and y-direction in a grid space, and the coefficients are $C_B = 2.0$, $\eta_{\xi\xi}^{cr} = \eta_{\vartheta\vartheta}^{cr} = -50$. The algorithm (6.43)–(6.46) does not change the volume and decreases the local potential and kinetic energy. It is assumed that the lost momentum and energy are transferred to current and turbulence [see Chalikov and Belevich, 1992]. In addition, the energy also goes to other wave modes. A choice of parameters in (6.21)–(6.24) is based on the simple consideration: a local piece of surface can closely approach the critical curvilinearity but not exceed it. Conservative values are chosen for the coefficients to provide stability of long runs.

Obviously, the suggested breaking parameterization is not the final solution of the problem. Other schemes are to be tested in the next version of the model. However, the results presented below show that the scheme is reliable and provides a realistic energy dissipation rate.

6.8. Development of Waves Under the Action of Wind

Until recently, direct modeling was used for reproduction of a quasi-stationary wave regime when the wave spectrum essentially did not change. A unique example of the direct numerical modeling of a surface wave evolution is given in Chalikov and Babanin [2014] where development of the wave field was calculated with the use of a 2D model based on the full potential equations written in the conformal coordinates. The model included algorithms for parameterization of the input and dissipation of energy (a description of similar algorithms is given below). The model reproduced the evolution of the wave spectrum under the action of wind. However, the strictly 1D (unidirected) waves are not realistic; hence, the full problem of wave evolution should be formulated on the basis of 3D equations. An example of such modeling is given in the current section. The preliminary results were published by Chalikov and Bulgakov [2017]. The new calculations were performed with the improved physics and better resolution.

The elevation and surface velocity potential fields are approximated in the current calculations by $M_x = 256$ and $M_y = 128$ modes in directions x and y. A corresponding grid includes $N_x \times N_y = (1024 \times 512)$ knots. The vertical derivatives are approximated on a vertical stretched grid $d\zeta_{j+1} = \nu d\zeta_j$ ($j = 1, 2, 3, \ldots, L_w$) where $\nu = 1.2$ and $L_w = 10$. The small number of levels used for solution of the equation for a nonlinear component of the velocity potential are possible because just a surface vertical derivative for the velocity potential $\partial \Phi / \partial \zeta (\zeta = 0)$ is required. The velocity potential mainly consists of an analytical component $\bar{\varphi}$, while a nonlinear component provides but a small correction. To reach the accuracy of solution $\varepsilon = 10^{-6}$ for Eq. (6.11), no more than two iterations are usually required.

The parameters chosen are used for solution of the problem of a wave field evolution over an acceptable time (of the order of 10 days). Initial conditions were assigned on the basis of the empirical JONSWAP spectrum [Hasselmann et al., 1973] with the spectral peak at wave number $k_p = 100$ with the angle spreading $(\cosh \psi)^{256}$.

The details of the initial conditions are of no importance because the initial energy level is quite low.

The total energy of wave motion $E = E_p + E_k$ (where E_p is potential energy, while E_k is kinetic energy) is calculated with the following formulas:

$$E_p = 0.25\overline{\overline{\eta^2}}, \quad E_k = 0.5\overline{\overline{(\varphi_x^2 + \varphi_y^2 + \varphi_z^2)}}, \qquad (6.47)$$

where single overbar denotes averaging over the ξ and ϑ coordinates, while double overbar denotes averaging over the entire volume. The derivatives in (6.47) are calculated according to the transformation (6.23). An equation for the integral energy $E = E_p + E_k$ evolution can be represented in the following form:

$$\frac{dE}{dt} = \overline{\overline{I}} + \overline{\overline{D_b}} + \overline{\overline{D_t}} + \overline{\overline{N}}, \qquad (6.48)$$

where $\overline{\overline{I}}$ is an integral input of energy from the wind (Eqs. (6.39) and (6.40)); $\overline{\overline{D_b}}$ is a rate of energy dissipation due to the wave breaking (Eqs. (6.43)–(6.46)); $\overline{\overline{D_t}}$ is a rate of energy dissipation due to filtration of the high wave number modes ("tail dissipation", Eqs. (6.41) and (6.42)); $\overline{\overline{N}}$ is the integral effect of the nonlinear interactions described by the right-hand side of the equations when surface pressure p is equal to zero. The differential form for calculation of the energy transformation can be, in principle, derived from Eqs. (6.26)–(6.28), but here a more convenient and simple method is applied. Different rates of the integral energy transformations can be calculated with the help of fictitious time steps (i.e., apart from the basic calculations). For example, the value of $\overline{\overline{I}}$ is calculated by the relation:

$$\overline{\overline{I}} = \frac{1}{\Delta t}(\overline{\overline{E^{t+\Delta t}}} - \overline{\overline{E^t}}), \qquad (6.49)$$

where $\overline{\overline{E^{t+\Delta t}}}$ is the integral energy of the wave field obtained after one time step with the right-hand side of Eq. (6.27) containing only the surface pressure calculated with Eqs. (6.39) and (6.40). For calculation of the dissipation rate due to filtration, the right-hand side of the equations contain just the terms describing high-frequency

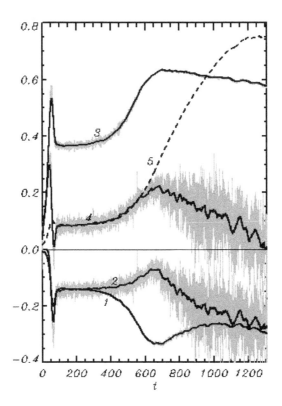

Fig. 6.13. Evolution of the integral characteristics of solution (a rate of evolution of the integral energy multiplied by 10^7) due to: 1 — tail dissipation D_t (Eqs. (6.41) and (6.42)); 2 — breaking dissipation D_b (Eqs. (6.43)–(6.46)); 3 — input of energy from wind I (Eqs. (6.39) and (6.40)); 4 — balance of energy $I + D_t + D_b$. Curve 5 shows evolution of wave energy $10^5 E$. Vertical bars of gray color show instantaneous values; thick curve shows a smoothed behavior.

dumping, while for calculation of the effects of breaking, only the terms introduced in (6.41)–(6.44) are in use.

The evolution of the characteristics calculated by (6.47) is shown in Fig. 6.13. Up to the end of the integration, the sum of all energy transition terms (tail dissipation $\overline{\overline{D_t}}$, breaking dissipation $\overline{\overline{D_b}}$ and energy input $\overline{\overline{I}}$) approaches zero (curve 4), and the energy growth E (curve 5) stops. Then the energy tends to decrease, but we are not sure about the nature of this effect. Such behavior can be explained by the fluctuating character of mutual adjustment of the input and

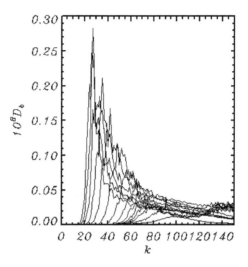

Fig. 6.14. Wave spectra $S_h(r)$ integrated over angle ψ in polar coordinates and averaged over a period of approximately 100 units of non-dimensional time t. The spectra grow and shift from right to left.

dissipation or simply by worsening of the approximation because of the downshifting process. Note that unlike the more or less monotonic behavior of tail dissipation (curve 1), the breaking dissipation is highly intermittent, which is consistent with the common views on the nature of wave breaking.

The data on the evolution of wave spectrum are shown in Fig. 6.14.

A 2D wave spectrum $S(k,l)$ $(0 \leq k \leq M_x, -M_y \leq l \leq M_y)$ averaged over 13 time intervals of length equal to $\Delta t \approx 100$ was transferred to the polar coordinates $S_p(\psi, r)$ $(-\pi/2 \leq \psi \leq \pi/2, 0 \leq r \leq M_x)$ and then averaged over angle ψ to obtain the 1D spectrum $S_h(r)$:

$$S_h(r) = \sum S_p(\psi, r) r \Delta \psi. \tag{6.50}$$

An angle $\psi = 0$ coincides with the direction of wind U, $\Delta \psi = \pi/180$.

The wave spectra $S_h(r)$ calculated by averaging over angle ψ in the polar coordinates and averaged over approximately 100 units of non-dimensional time t are presented in Fig. 6.15. The spectra

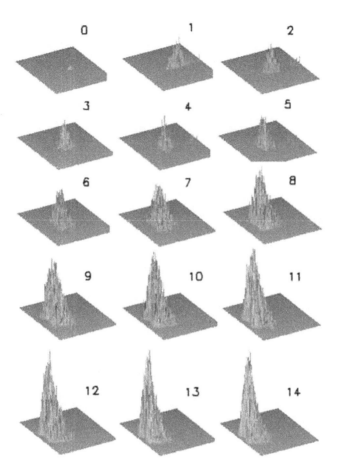

Fig. 6.15. Sequence of 3D images of $\log_{10}(S(k,l))$ where each panel corresponds to single curve in Fig. 6.14. The left side refers to wave number $l(-M_y \leq l \leq M_y)$ while the front side corresponds to $k(-M \leq k \leq M)$.

increase in energy and peak moves from high to low wave numbers, i.e., they undergo downshifting. A maximum value of $S_h(r)$ increases by as much as 152 times. According to the data in Fig. 6.13, the total energy increases 44 times. This difference is explained by the spectrum narrowing and by an overlapping effect (i.e., decrease of a high-frequency spectrum for long fetches). The 3D images of the wave spectrum $\log_{10}(S(k,l))$ are shown in Fig. 6.15.

As seen, each spectrum consists of separated peaks and holes.[2] This phenomenon was first observed and discussed by Chalikov et al. [2014]. The repeated calculations with different resolution show that such structure of the 2D spectrum is typical. It cannot be explained by a fixed combination of interacting modes, since in different runs (with the same initial conditions but a different set of phases for the modes) peaks are located in different locations in Fourier space.

Another presentation is given in Fig. 6.16 where $\log_{10}(S(\psi,r))$ averaged over the successive seven period length $\Delta t = 200$ is given. The first panel with a mark 0 refers to the initial conditions. The disturbances within the range $(125 < k < 150)$ reflect the initial adjustment of the input and dissipation at the high wave number slope of the spectrum. The pictures characterize well the downshifting and angular spreading of the spectrum due to the nonlinear interactions. The discrete structure of spectrum is also well depicted.

The evolution of the wave spectrum $S_h(r)$ integrated over angles ψ can be described with the following equation:

$$\frac{dS_h(r)}{dt} = I(r) + D_t(r) + D_b(r) + N(r), \tag{6.51}$$

where $I(r)$, $D_t(r)$, $D_b(r)$ and $N(r)$ are the spectra of the input energy, tail dissipation, breaking dissipation and the rate of the nonlinear interactions, all obtained by integration over angles ψ. All of the spectra shown below were obtained by transformation of the 2D spectra into a polar coordinate (ψ, r) and then integrated over angles ψ within the interval $(-\pi/2, \pi/2)$.

The spectra can be calculated using an algorithm similar to the algorithm (6.27) for integral characteristics. For example, the spectrum of energy input $I(k,l)$ is calculated as follows:

$$I(k,l) = (S_c^{t+\Delta t}(k,l) - S_c^t(k,l))/\Delta t, \tag{6.52}$$

---

[2]Opposite to our expectations, the wave spectrum looks more like La Sagrada Familia (Gaudi) in Barcelona than the St. Mary Axe (Gherkin) in London.

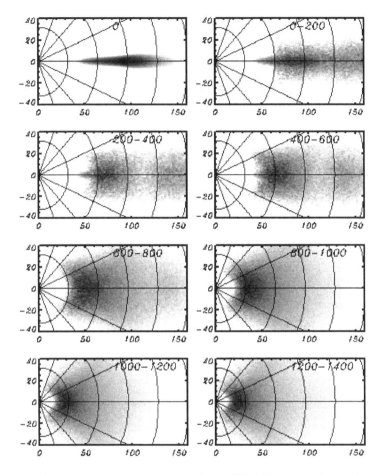

Fig. 6.16. A sequence of the 2D images of $\log_{10}(S(k,l))$ averaged over the consequent seven period length $\Delta t = 200$. The numbers indicate the period of averaging (first panel marked 0, refers to the initial conditions). The horizontal and vertical axes correspond to wave numbers k and l, respectively.

where $S_c(k_x, k_y)$ is a spectrum of columnar energy calculated by the relation:

$$S_c(k,l) = \frac{1}{2}\left(h_{k,l}^2 + h_{-k,-l}^2 \right.$$
$$\left. + \int_{-H}^{0} (u_{k,l}^2 + u_{-k,-l}^2 + v_{k,l}^2 + v_{-k,-l}^2 + w_{k,l}^2 + w_{-k,-l}^2)d\zeta \right),$$
(6.53)

where grid values of the velocity components u, v, w are calculated by the relations:

$$u = \varphi_\xi + \varphi_\zeta \eta_\xi, \quad v = \varphi_\vartheta + \varphi_\zeta \eta_\vartheta, \quad w = \varphi_\zeta, \quad (6.54)$$

and $u_{k,l}$, $v_{k,l}$ and $w_{k,l}$ are their Fourier coefficients.

For calculation of $I(k_x, k_y)$, the fictitious time steps Δt are made only with a term responsible for the energy input, i.e., surface pressure p. A spectrum $I(k,l)$ was averaged over the periods $\Delta t \approx 100$, then transformed into a polar coordinate system and integrated in Fourier space over angles ψ within the interval $(-\pi/2, \pi/2)$.

The evolution of input spectra is, in general, similar to that of wave spectra and shown in Fig. 6.17. Note that a maximum of the input spectra is located at the maximum of the wave spectra since the input depends mainly on the spectral density, while the dependence on frequency is less important.

The algorithm (6.41)–(6.43) was applied for calculation of the dissipation spectra due to dumping of a high wave number part of the spectrum [tail dissipation, Chalikov and Bulgakov, 2017] as well as for the calculation of a spectrum of breaking dissipation.

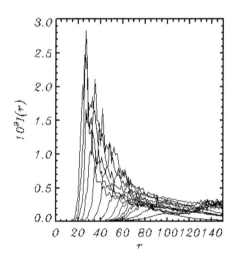

Fig. 6.17. The spectrum of energy input $I(r)$ integrated over angle ψ in the polar coordinates and averaged over approximately 100 units of non-dimensional time t.

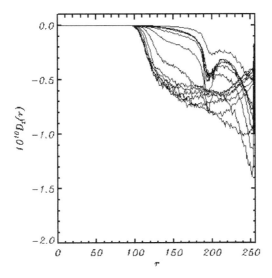

Fig. 6.18. The tail dissipation spectra $D_t(r)$ integrated over angle ψ in the polar coordinates and averaged over approximately 100 units of non-dimensional time t.

In the first case, the fictitious time step was made taking into account the terms described by Eqs. (6.41) and (6.42), while in the second case, the time step was made using the terms described by Eqs. (6.43)–(6.46).

The spectra of tail dissipation calculated similar to the spectra $I(r)$ are shown in Fig. 6.18. Dissipation occurs at the periphery of the spectrum, outside an ellipse with semi-axes $d_m M_x$ and $d_m M_y$. This is why this dissipation averaged over angles seems to affect a middle part of a 1D spectrum. The tail dissipation effectively stabilizes the solution. The breaking dissipation averaged over angles is presented in Fig. 6.19.

As seen in Fig. 6.19, the breaking dissipation has a maximum at the spectral peak. It does not mean that in the vicinity of wave peak the probability of large curvilinearity is quite high. A high rate of breaking dissipation can be explained by high wave energy in the vicinity of wave peak. The energy lost through breaking, described by the diffusion mechanism, correlates with the energy of breaking waves. Opposite to the high wave number dissipation that regulates

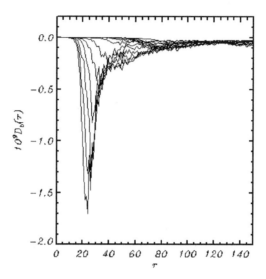

Fig. 6.19. Breaking dissipation spectra $D_b(r)$ integrated over angle ψ in the polar coordinates and averaged approximately 100 units of non-dimensional time t.

the shape of spectral tail, the breaking dissipation forms the main energy-containing part of spectrum.

The diffusion mechanism suggested in (6.43) and (6.46) modifies the elevation and surface stream function in the close vicinity of the breaking point. The amplitudes of the side perturbations are small and decrease very quickly over distance from a breaking point.

An example of a profile of the energy input due to breaking $D_b(x)$ is given in Fig. 6.20. As can be seen, the energy input fluctuates around the breaking point. The diffusion operator chosen for breaking parameterization not only decreases the total energy but also redistributes the energy between Fourier modes in Fourier space.

In general, for the specific conditions considered in the paper, the breaking is an occasional process taking place in a small part of domain. The kurtosis of the input energy due to breaking $D_b(\xi, \vartheta)$, i.e., the value

$$Ku = \overline{\overline{D_b^4}}(\overline{\overline{D_b^2}})^{-2} - 3, \tag{6.55}$$

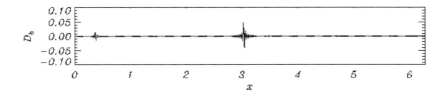

Fig. 6.20. An example of energy input due to breaking $D_b(x)$.

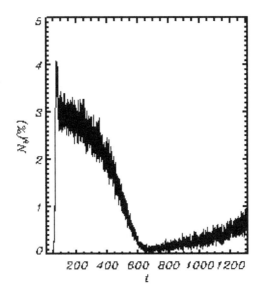

Fig. 6.21. Evolution of the number of wave breaking events N_b expressed as the percentage of the number of grid points $N_x \times N_y$.

is of the order of 10^3, which corresponds to a plain function with occasional separated peaks.

The number of breaking points in terms of percentage of the total number of points is given in Fig. 6.21. As can be seen, the number of breaking events decreases to $t = 600$ and then increases. The number of breaking events is not directly connected with the intensity of breaking, which is seen when comparing Fig. 6.21 and curve 2 in Fig. 6.13.

An integral term describing the nonlinear interaction $\overline{\overline{N}}$ in Eq. (6.26) is small, but the magnitude of the spectrum $N(r)$ is comparable with input $I(r)$ and dissipation $D_t(r)$ and $D_b(r)$ terms. The presentation of the term $N(r)$ in a form shown in the previous figures is not clear. This is why the spectra $10^8 N(r)$ averaged over an interval $\Delta t = 100$ are plotted separately in Fig. 6.22 for the last eight intervals (thick curves) together with the wave spectrum $10^6 S_h(r)$. In general, the shapes of the spectrum $N(r)$ agree with the conclusions of the quasi-linear Hasselmann [1962] theory. At low wave numbers, the spectrum of the nonlinear influx of energy is positive, while at higher values, it becomes negative. This process produces a shift of the spectrum to lower wave numbers (downshifting). Note that nonlinear interactions also produce broadening of the spectrum. In contrast to Hasselmann's theory, these results are obtained by solution of full 3D equations. It would be interesting to compare our results with the calculations of Hasselmann's integral. Unfortunately, neither of the available programs of such type permits doing calculations with such a high resolution that was used in the current model.

Obviously, nonlinearity is quite an important property of surface waves. The contribution of nonlinearity can be estimated, for example, by comparison of the kinetic energy of a linear component $E_l = 0.5\overline{(\bar{\varphi}_x^2 + \bar{\varphi}_y^2 + \bar{\varphi}_z^2)}$ and the total kinetic energy E_k (Fig. 6.23). The ratio E_l/E_k as a function of time remains very close to 1, which proves that the nonlinear part of energy makes up just a small percentage of the total energy. It does not mean that the role of nonlinearity is small; its influence can manifest itself over large timescales.

The time evolution of integral spectral characteristics is presented in Fig. 6.24.

Curve 1 corresponds to the weighted frequency ω_w

$$\omega_w = \left(\frac{\int kS dk dl}{\int S dk dl}\right)^{1/2}, \qquad (6.56)$$

where integrals are taken over the entire Fourier domain. The value ω_w is not sensitive to the details of spectrum, hence, it well characterizes the position of spectrum and its shifting. Curve 2 describes

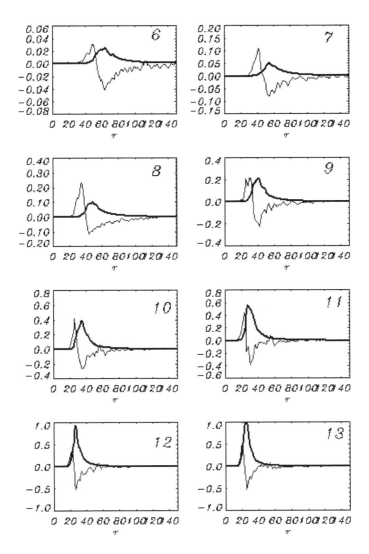

Fig. 6.22. A sequence of wave spectra $S_h(r)$ (thick curves) and nonlinear input term $N(r)$ (thin curves) averaged over the consequent eight periods of length $\Delta t = 100$ starting from sixth period.

the evolution of a spectral maximum. The step shape of curve corresponds to the fundamental property of downshifting. Opposite to the common views, development of the spectrum occurs not monotonically, but by appearance of a new maximum at the lower wave

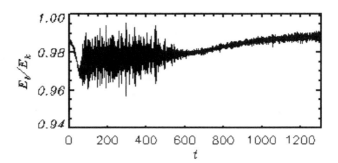

Fig. 6.23. Time evolution of ratio E_l/E_k.

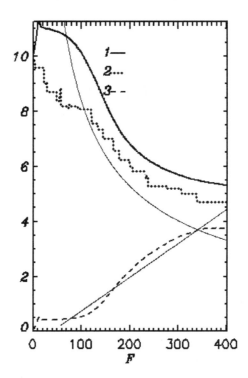

Fig. 6.24. Time evolution of: the weighted frequency ω_w (1) Eq. (6.34); spectral peak frequency ω_p (2); full energy E (3) Eq. (6.25). Thin curves are the empirical dependence for peak wave number and energy. F is the distance passed by the spectral peak.

number, as well as by attenuation of the previous maximum. Curve 3 describes the change of total energy $E = E_p + E_k$. As can be seen, all three curves have a tendency for saturation (decrease of the evolution rate).

The numerical experiment reproduces the case when development of the wave field occurs under the action of a permanent and uniform wind. This case corresponds to the JONSWAP experiment. Despite large scatter, the data allow us to construct empirical approximations of wave spectrum, as well as to investigate the spectral evolution as a function of fetch F. In particular, it is suggested that the frequency of the spectral peak changes as $F^{-1/3}$, while full energy grows linearly with F. Neither of the dependences can be exact, since they do not take into account the approach to a stationary regime. Besides, the dependence of frequency on fetch is singular at $F = 0$.

The value of fetch in a periodic problem can be calculated by integration of the peak phase velocity $c_p = |k|^{-1/2}$ over time:

$$F = \int_{t_0}^{t} c_p dt. \tag{6.57}$$

The JONSWAP dependencies for the wave number of spectral peak k_p and full energy E are shown in Fig. 6.24 by thin curves. Dependence $\omega_p \sim F^{1/3}$ is qualitatively valid. The dependence of the total energy on fetch does not look like a linear one, but it is worth noting that the JONSWAP dependence is evidently inapplicable at very small and large fetch.

In the current work, a 3D model was used for simulation of the development of the wave field under the action of wind and dissipation. The input energy is described by a single term, i.e., surface pressure p in Eq. (6.4). It is assumed that the complex pressure amplitude in Fourier space is linearly connected with the complex elevation amplitude with a complex coefficient β. Such simple formulations may be imperfect. Firstly, it is assumed that the wave field is represented by the superposition of linear modes with slowly changing amplitudes and a phase velocity which obeys the linear dispersive relation. This assumption is valid only for the low-frequency part of spectrum. In reality, the amplitudes of medium- and high-frequency modes undergo fluctuations created by reversible interactions. A solid dispersion relation does not connect their phase velocities with wave number. In addition, it is also possible that a suggestion of linearity of the connection between pressure and elevation amplitudes is not precise, i.e., β-function can depend on the amplitudes of the modes.

There are no observation data that can be used for formulation of a more sophisticated scheme for calculation of the input energy to the waves. The only method that can give more or less reliable results is the mathematical modeling of the statistical structure of the turbulent boundary layer above a curvilinear moving surface, the characteristics of which satisfy kinematic conditions. As a whole, the problem of the boundary layer structure seems even more complicated than the wave problem itself. Coupled models for boundary layer and potential waves, both written in conformal coordinates were developed in Chalikov and Rainchik [2014]. The calculations showed that the pressure field consists mostly of random fluctuations not directly connected with waves. A small part of these fluctuations is in phase with surface disturbances. The calculated values of β in Eq. (6.39) have large dispersion. However, since the volume of data was very large, the shape of the β-function was found with reasonable accuracy. Probably, the approximation of β used in the current work can be considered as adequate. The next step in investigations of the wave boundary layer (WBL) should use a 3D LES approach. Note that even availability of a large volume of data on the WBL structure does not make the problem of parameterization of wind input in a spectral wave models easily solvable, since the pressure is characterized by a broad continuous spectrum created by nonlinearity.

The wave breaking is obviously even more complicated than the input energy. Nevertheless, this problem can be simplified, if the common ideas used in numerical fluid mechanics are accepted. For example, in LES modeling, a more or less artificial viscosity is introduced to prevent too large local velocity gradients. The fact is that the numerical instability terminating computation precedes wave breaking. Hence, the scheme should prevent the breaking approach to preserve stability of the numerical scheme. Hence, a wave model should contain some algorithms preventing the appearance of the intervals with high curvilinearity. A criterion of breaking is introduced not for the recognition of the breaking itself, but for choosing the places where it might happen (or, unfortunately, might not happen). Finally, the algorithm should produce local smoothing of elevation (and surface potential). The algorithm should be highly selective so that "breaking" would occur within narrow intervals and not affect the entire area. The exact criteria of breaking events (most evident of them is the breaking itself) cannot be used for parameterization of breaking since in a coordinate system defined by (6.23) the numerical instability occurs long before breaking. In our opinion, the most sensitive parameter indicating potential instability is the curvilinearity (second derivative) of elevation.

In the current work, the breaking is parameterized by a diffusion algorithm with the nonlinear coefficient of diffusion providing high selectivity of smoothing. We admit that such an approach can be realized in various forms. The same situation is observed in a problem of turbulence modeling for parameterization of subgrid scales.

We finally conclude that the physics included in wave models is still based on a shaky ground. Nevertheless, the results of the calculations looks quite realistic, which convinces us that the approach deserves further development.

6.9. General Approach to the Phase-Resolving Wave Modeling

A numerical model of waves similar to that considered in this chapter has a lot of important applications. Such a model is most applicable for development of physical parameterization schemes in the spectral wave models. The main limitation of the current model as well as the HOS model [Ducroset et al., 2016; Chalikov et al., 2014] is an assumption of periodicity and constant depth. This limitation allows us to use a very convenient and exact Fourier transform method. Both models can be rewritten in a finite difference form which allows us to apply the model to basins with realistic shape and topography. In addition, a finite difference model can be effectively parallelized. Such a model could be used for simulation of waves within small harbors. Any model used for a long-term simulation of the wave field evolution should include algorithms describing transformation of energy, similar to those considered in this chapter.

Since wave motion is very conservative, high accuracy numerical schemes should be used for a good description of nonlinearity and spectral transformation. Such a universal approach is being developed at the Technical University of Denmark [see Engsig-Karup, 2009]. Actually, the models *ModelWave3D* developed at TUD are targeted at solutions of a variety of problems including such problems as modeling of wave interaction with submerged objects as well as the simulation of wave regimes in basins with real shapes and topography.

The model is based on the equations of potential flow with a free surface (6.19)–(6.21). An effect of variable bathymetry is taken into account by using a σ-coordinate:

$$\sigma \equiv \frac{z + h(x)}{\eta(x,t) + h(x)}. \tag{6.58}$$

The Laplace equation for velocity potential turns into an elliptical equation

$$\nabla^2(\Phi + \sigma\Phi_\sigma) + 2\nabla\sigma \cdot \nabla\Phi_\sigma + ((\nabla\sigma)^2 + \sigma_z^2)\Phi_{\sigma\sigma} = 0. \tag{6.59}$$

On the surface ($\sigma = 1$), Φ is equal to a surface value ϕ and on the bottom ($\sigma = 0$), the slip condition is assumed:

$$(\sigma_z + \nabla h \cdot \nabla\sigma)\Phi_\sigma + \nabla h \nabla\Phi = 0. \tag{6.60}$$

On the bottom and at vertical surfaces, a normal derivative of the velocity potential is equal to zero. A flexible-order approximation for spatial derivatives is used. The most time-consuming part of this model is the 3D equation for the velocity potential. The strategy of the model development is directed at exploiting architectural features of modern GPUs for the mixed precision computations. This approach is tested using a recently developed generic library for fast prototyping of partial differential equations (PDFs) solvers. The new wave tool is applicable for solving and analyzing of a variety of large-scale wave problems in coastal and offshore engineering. A description of the project and references can be found at site: http://www2.compute.dtu.dk/~apek/OceanWave3D/.

6.10. Conclusions

It seems that nowadays phase-resolving models are able to reproduce quasi-stationary wave motion dynamics in a large domain of the size of dozens of square kilometers, in a periodic domain as part of an infinite ocean [Ducroset *et al.*, 2016] or in real basins of complicated shape and bathymetry [Engsig-Karup *et al.*, 2012]. Those approaches are well-suited for simulation of wave dynamics in wave channels. The problem which cannot be considered as being resolved is the mechanics of generation, growth and decay of waves. The input of energy from wind is still described in terms of the theories suggested more than half a century ago. Most schemes for the energy input to waves are chosen "*ad hoc*" and they are not supported by reliable experimental data. Meanwhile, the mechanics of wave interaction with turbulent wind is hardly simpler than the surface wave processes. The wind flows over an essentially nonlinear wave field composed of many modes. In addition, the flux of energy to waves is formed in a thin layer above a wavy surface. This process is poorly investigated, the experimental data being scattered and occasional. As a good tool for investigations of such process, the coupled modeling of waves and turbulent wind can be used. The 2D prototype of such a model is a model by Chalikov and Rainchik [2014]. The ideal model should use LES techniques and it should be coupled with a 3D wave model similar to the HOS model or the model developed by Chalikov *et al.* [2014]. The parameterization of wind input should be obtained by processing of a vast volume of data on surface pressure and elevation.

Dissipation of wave energy through the breaking and interaction with currents and turbulence is still a poorly investigated phenomenon. Fortunately, the wave breaking is evidently a dominant process of dissipation. The unresolved breaking in the presence of input energy always terminates the computations. There are many views on the possible ways as to how to parameterize this phenomenon. It was shown in this chapter that the instability can be prevented by introduction of a high-selective operator, which produces local smoothing of the surface, decreases local energy and partly redistributes the energy between the wave components.

Despite the fact that the approaches demonstrated in this chapter allow the reproduction of reasonable statistical properties, we have to admit that currently we are just at the initial stage of understanding and describing the many processes responsible for the evolution of wave field.

Acknowledgments

The author thanks Mrs. O. Chalikova for her assistance in preparation of the chapter. This research was performed in the framework of the state assignment (theme No. 0149-2019-0015), supported in part (Section 6.7) by RFBR (project No. 18-05-01122).

References

Babanin, A.V. (2011). *Breaking and Dissipation of Ocean Surface Waves*, Cambridge University Press.

Beale, J.T. (2001). A convergent boundary integral method for three-dimensional water waves, *Math. Comput.*, **70**, 977–1029.

Benjamin, T.B. and Feir J.E. (1967). The disintegration of wave trains in deep water, *J. Fluid Mech.*, **27**, 417–430.

Bonnefoy, F., Ducrozet, G., Le Touzé, D. and Ferrant, P. (2010). Time-domain simulation of nonlinear water waves using spectral methods, in *Advances in Numerical Simulation of Nonlinear Water Waves*, Advances in Coastal and Ocean Engineering, Vol. 11, World Scientific, pp. 129–164, doi:10.1142/9789812836502_0004.

Causon, D.M., Mingham, C.G. and Qian, L. (2010). Developments in multi-fluid finite volume free surface capturing methods, in *Advances in Numerical Simulation of Nonlinear Water Waves*, World Scientific, pp. 397–427.

Chalikov, D.V. (1978). Numerical simulation of wind-wave interaction, *J. Fluid Mech.*, **87**, 561–582.

Chalikov, D.V. (1986). Numerical simulation of the boundary layer above waves, *Bound.-Layer Meteorol.*, **34**, 63–98.

Chalikov, D. (1995). The parameterization of the wave boundary layer, *J. Phys. Oceanogr.*, **25**, 1335–1349.

Chalikov, D. (2005). Statistical properties of nonlinear one-dimensional wave fields, *Nonlinear Process. Geophys.*, **12**, 1–19.
Chalikov, D. (2007). Numerical simulation of Benjamin–Feir instability and its consequences, *Phys. Fluid.*, **19**, 016602.
Chalikov, D. (2009). Freak waves: their occurrence and probability, *Phys. Fluid.*, **21**, 076602, doi:10.1063/1.3175713.
Chalikov, D. (2012). On the nonlinear energy transfer in the unidirected adiabatic surface waves, *Phys. Lett.*, **376**(44), 2755–2816.
Chalikov, D. (2016). *Numerical Modeling of Sea Waves*, Springer.
Chalikov, D. and Sheinin D. (1998). Direct modeling of one-dimensional nonlinear potential waves, in *Nonlinear Ocean Waves*, Perrie, W. (ed.), Advances in Fluid Mechanics, Vol. 17, pp. 207–258.
Chalikov, D. and Rainchik, S. (2010). Coupled numerical modelling of wind and waves and the theory of the wave boundary layer, *Bound.-Layer Meteorol.*, **138**, 1–41.
Chalikov, D. and Babanin, A.V. (2012). Simulation of wave breaking in one-dimensional spectral environment, *J. Phys. Oceanogr.*, **42**(11), 1745–1761.
Chalikov, D., Babanin, A.V. and Sanina, E. (2014). Numerical modeling of three-dimensional fully nonlinear potential periodic waves, *Ocean Dynam.*, **64**(10), 1469–1486.
Chalikov D. and Babanin A.V. (2016). Nonlinear sharpening during superposition of surface waves, *Ocean Dynam.*, **66**(8), 931–937.
Chalikov, D. and Belevich, M. (1992). One-dimensional theory of the wave boundary layer, *Bound.-Layer Meteorol.*, **63**, 65–96.
Chalikov, D. and Bulgakov, K. (2017). Numerical modeling of development of waves under action of wind, *Phys. Wave Phenom.*, **25**(4), 315–323.
Clamond, D. and Grue J. (2001). A fast method for fully nonlinear water wave dynamics, *J. Fluid Mech.*, **447**, 337–355.
Clamond, D., Fructus, D., Grue, J. and Krisitiansen O. (2005). An efficient method for three-dimensional surface wave simulations. Part II: Generation and absorption, *J. Comput. Phys.*, **205**, 686–705.
Clamond, D., Francius, M., Grue, J. and Kharif, C. (2006). Long time interaction of envelope solitons and freak wave formations, *Eur. J. Mech. B/Fluids*, **25**, 536–553.
Craig, W. and Sulem C. (1993). Numerical simulation of gravity waves, *J. Comput. Phys.*, **108**, 73–83.
Dalrymple, R.A., Gómez-Gesteira, M., Rogers, B.D., Panizzo, A., Zou, S., Crespo, A.J., Cuomo, G. and Narayanaswamy, M. (2010). Smoothed particle hydrodynamics for water waves, in *Advances in Numerical Simulation of Nonlinear Water Waves*, World Scientific, pp. 465–495.
Dommermuth, D. and Yue, D. (1987). A high-order spectral method for the study of nonlinear gravity waves, *J. Fluid Mech.*, **184**, 267–288.
Donelan, M.A., Babanin, A.V., Young, I.R., Banner, M.L. and McCormick, C. (2005). Wave follower field measurements of the wind input spectral func-

tion. Part I. Measurements and calibrations, *J. Atmos. Ocean. Technol.*, **22**, 799–813.

Donelan, M.A., Babanin, A.V. Young, I.R. and Banner, M.L. (2006). Wave follower field measurements of the wind input spectral function. Part II. Parameterization of the wind input, *J. Phys. Oceanogr.*, **36**, 1672–1688.

Ducrozet, G., Bingham, H.B., Engsig-Karup, A.P., Bonnefoy, F. and Ferrant, P. (2012). A comparative study of two fast nonlinear free-surface water wave models, *Int. J. Numer. Methods Fluids*, **69**, 1818–1834.

Ducrozet, G., Bonnefoy, F., Le Touzé, D. and Ferrant, P. (2007). 3-D HOS simulations of extreme waves in open seas, *Nat. Hazards Earth Syst. Sci.*, **7**, 109–122, doi:10.5194/nhess-7-109-2007.

Ducrozet, G., Bonnefoy, F., Le Touzé, D. and Ferrant, P. (2016). HOS-ocean: Open-source solver for nonlinear waves in open ocean based on high-order spectral method, *Comput. Phys. Commun.*, **203**, 245–254, doi:10.1016/j.cpc.2016.02.017.

Dysthe, K.B. (1979). Note on a modification to the nonlinear Schrödinger equation for application to deep water waves, *Proc. R. Soc. Lond. A*, **369**, 105–114.

Dysthe, K., Krogstad, H. and Muller, P. (2008). Oceanic rogue waves, *Annu. Rev. Fluid Mech.*, **40**, 287–310.

Engsig-Karup, A., Bingham, H. and Lindberg, O. (2009). An efficient flexible-order model for 3D nonlinear water waves, *J. Comput. Phys.*, **228**, 2100–2118.

Engsig-Karup, A., Madsen, M. and Glimberg, S. (2012). A massively parallel GPU-accelerated mode for analysis of fully nonlinear free surface waves, *Int. J. Numer. Methods Fluids*, **70**(1), 20–36, doi:10.1002/fld. 2675.

Fochesato, C., Dias, F. and Grilli, S. (2006). Wave energy focusing in a three-dimensional numerical wave tank, *Proc. R. Soc. A*, **462**, 2715–2735.

Fructus, D., Clamond, D., Grue, J. and Kristiansen, Ø. (2005). An efficient model for three-dimensional surface wave simulations. Part I: Free space problems, *J. Comput. Phys.*, **205**, 665–685.

Greaves, D. (2010). Application of the finite volume method to the simulation of nonlinear water waves, in *Advances in Numerical Simulation of Nonlinear Water Waves*, World Scientific, pp. 357–396.

Grilli, S., Guyenne, P. and Dias, F. (2001). A fully nonlinear model for three-dimensional overturning waves over arbitrary bottom, *Int. J. Numer Methods Fluids*, **35**, 829–867.

Gou, Y., Teng, B. and Yoshida, S. (2016). An extremely efficient boundary element method for wave interaction with long cylindrical structures based on free-surface green's function, *Computation*, **4**, 36.

Grue, J. and Fructus D. (2010). Model for fully nonlinear ocean wave simulations derived using Fourier inversion of integral equations in 3D, in *Advances in Numerical Simulation of Nonlinear Water Waves*, World Scientific, pp. 1–42.

Guyenne, P. and Grilli S.T. (2006). Numerical study of three-dimensional overturning waves in shallow water, *J. Fluid Mech.*, **547**, 361–388.

Hasselmann, K. (1962). On the non-linear energy transfer in a gravity wave spectrum, Part 1, *J. Fluid Mech.*, **12**, 481–500.

Hasselmann, K. et al. (1973). Measurements of wind-wave growth and swell decay during the Joint Sea Wave Project (JONSWAP), *Dtsch. Hydrogr. Z. Suppl.*, **A8**(12), 1–95.

Hasselmann, D. and Bösenberg, J. (1991). Field measurements of wave-induced pressure over wind-sea and swell, *J. Fluid Mech.*, **230**, 391–428.

Hsiao, S.V. and Shemdin, O.H. (1983). Measurements of wind velocity and pressure with a wave follower during Marsen, *J. Geophys. Res.: Oceans*, **88**, 9841–9849.

Iafrati, A. (2009). Numerical study of the effects of the breaking intensity on wave breaking flows, *J Fluid Mech.*, **622**, 371–411.

Issa, R., Violeau, D., Lee, E.-S. and Flament, H. (2010). Modelling nonlinear water waves with RANS and LES SPH models, in *Advances in Numerical Simulation of Nonlinear Water Waves*, World Scientific, pp. 497–537.

Janssen, P. (2003). Nonlinear four-wave interaction and freak waves, *J. Phys. Oceanogr*, **33**, 2001–2018.

Kharif, C. and Pelinovsky, E. (2003). Physical mechanisms of the rogue wave phenomenon, *Eur. J. Mech. B/Fluids*, **22**, 603–634.

Kharif, C., Pelinovsky, E. and Slunyaev, A. (2009). *Rogue Waves in the Ocean*, Springer.

Kim, K.S., Kim, M.H. and Park, J.C. (2014). Development of moving particle simulation method for multi-liquid-layer sloshing, *Math. Prob. Eng.*, **2014**, 350163, doi:10.1155/2014/350165.

Lake, B.M. and Yuen, H.C. (1978). A new model for nonlinear wind waves. Part 1. Physical model and experimental results, *J. Fluid. Mech.*, **88**, 33–62.

Liu, Y., Gou, Y., Teng, B. and Yoshida, S. (2016). An extremely efficient boundary element method for wave interaction with long cylindrical structures based on free-surface Green's function, *Computation*, **4**(3), 36, doi:10.3390/computation4030036.

Lubin, P. and Caltagirone, J.-P. (2010). Large eddy simulation of the hydrodynamics generated by breaking waves, in *Advances in Numerical Simulation of Nonlinear Water Waves*, World Scientific, pp. 575–604.

Ma, Q.W. and Yan, S. (2010). Qale-FEM method and its application to the simulation of free responses of floating bodies and overturning waves, in *Advances in Numerical Simulation of Nonlinear Water Waves*, World Scientific, pp. 165–202.

Madsen, P.A. and Fuhrman D.R. (2010). High-order Boussinesq-type modelling of nonlinear wave phenomena in deep and shallow water, in *Advances in Numerical Simulation of Nonlinear Water Waves*, World Scientific, pp. 245–285.

McLean, J.W. (1982). Instability of finite amplitude water waves, *J. Fluid Mech.*, **14**, 315–330.

Ma, Q.W. (ed.). (2010). *Advances in Numerical Simulation of Nonlinear Water Waves*. Advances in Coastal and Ocean Engineering, Vol. 11, World Scientific.

Pelinovsky, E., Shurgalina, E. and Chaikovskaya, N. (2011). The scenario of a single freak wave appearance in deep water: Dispersive focusing mechanism framework, *Nat. Hazards Earth Syst. Sci.*, **11**(1), 127–134.

Sanina, E.V., Suslov, S.A., Chalikov, D. and Babanin, A.V. (2015). Detection and analysis of coherent groups in three-dimensional fully-nonlinear potential wave fields, *Ocean Model.*, **103**, 73–86.

Shemer, L., Goulitski, K. and Kit, E. (2007). Evolution of wide-spectrum wave groups in a tank: An experimental and numerical study, *Eur. J. Mech. B/Fluids*, **26**, 193–219.

Snyder, R.L., Dobson, F.W., Elliott, J.A. and Long, R.B. (1981). Array measurements of atmospheric pressure fluctuations above surface gravity waves, *J. Fluid Mech.*, **102**, 1–59.

Su, M.Y., Bergin, M., Marler, P. and Myrick, R. (1982). Experiments on nonlinear instabilities and evolution of steep gravity wave trains, *J. Fluid Mech.*, **124**, 45–72.

Tanaka, M. (2001). Verification of Hasselmann's energy transfer among surface gravity waves by direct numerical simulations of primitive equations, *J. Fluid Mech.*, **444**, 199–221.

Toffoli, A., Onorato, M., Bitner-Gregersen, E. and Monbaliu, J. (2010). Development of a bimodal structure in ocean wave spectra, *J. Geophys. Res.: Oceans*, **115**, C03006, doi:10.1029/2009JC005495.

Tolman H. and Chalikov, D. (1996). On the source terms in a third-generation wind wave model, *J. Phys. Oceanogr.*, **26**, 2497–2518.

Tolman H. and the WAVEWATCH III® Development Group (2014). User manual and system documentation of WAVEWATCH III® version 4.18, Environmental Modeling Center Marine Modeling and Analysis Branch, Contribution No. 316.

Touboul, J. and Kharif, C. (2010). Two-dimensional direct numerical simulations of the dynamics of rogue waves under wind action, in *Advances in Numerical Simulation of Nonlinear Water Waves*, World Scientific, pp. 43–74.

West, B., Brueckner, K., Janda, R., Milder, M. and Milton, R. (1987). A new numerical method for surface hydrodynamics, *J. Geophys. Res.: Oceans*, **92**, 11803–11824.

Young, D.-L., Wu, N.-J. and Tsay, T.-K. (2010). Method of fundamental solutions for fully nonlinear water waves, in *Advances in Numerical Simulation of Nonlinear Water Waves*, World Scientific, pp. 325–355.

Xue, M., Xu, H., Liu, Y. and Yue, D.K.P. (2001). Computations of fully nonlinear three-dimensional wave–wave and wave–body interactions. Part 1. Dynamics of steep three-dimensional waves, *J. Fluid Mech.*, **438**, 11–39.

Zakharov, V.E. (1968). Stability of periodic waves of finite amplitude on the surface of deep fluid, *J. Appl. Mech. Tech. Phys. JETF* (English translation), **9**(2), 190–194.

Zakharov, V.E., Dyachenko, A.I. and Vasilyev, O.A. (2002). New method for numerical simulation of a nonstationary potential flow of incompressible fluid with a free surface, *Eur. J. Mech. B/Fluids*, **21**, 283–291.

Zhao, X., Liu, B.-J., Liang, S.-X. and Sun, Z.-C. (2016). Constrained interpolation profile (CIP) method and its application, *Chuan Bo Li Xue/J. Ship Mech.*, **20**, 393–402.

Zou, Z.L., Fang, K.Z. and Liu, Z.B. (2010). Inter-comparisons of different forms of higher-order Boussinesq equations, in *Advances in Numerical Simulation of Nonlinear Water Waves*, World Scientific, pp. 287–323.

Chapter 7

Extreme Conditions

Kevin Ewans[*,†] and Philip Jonathan[‡,§]

[*]*MetOcean Research Ltd., New Zealand*
[†]*Department of Ocean Engineering, University of Melbourne, Australia*
[‡]*Shell Research Ltd., UK*
[§]*Department of Mathematics and Statistics, Lancaster University, UK*

7.1. Introduction

This chapter addresses extreme wave conditions occurring in extreme events such as hurricanes and intense mid-latitude storms. The sea states associated with such events typically involve very large and dangerous waves and pose risk to humanity and infrastructure. Coastal locations in the path of hurricanes and tropical cyclones are subjected to heavy damage, sometimes catastrophic, as large waves riding on storm surge penetrate beyond breakwaters and sea walls and batter infrastructure. In the United States alone, at least 650,000 houses were damaged or destroyed by hurricane Sandy, with the vast majority of the damage caused by storm surge and waves, and there were 72 direct fatalities, most due to drowning [Blake *et al.*, 2013]. Every year ships are lost at sea due to storms; 598 ships were reported to have foundered between 2007 and 2016, most due to bad weather [Dobie *et al.*, 2017]. Offshore oil and gas production facilities have also suffered damage and destruction during intense storms. During the 2005 Atlantic hurricane season, the most active hurricane seasons on record, Category 5 hurricanes Katrina and Rita destroyed 112 platforms and severely damaged 53 others between them [Cruz and Krausmann, 2008].

Forecasting provides warning for impending extreme events, and while this enables opportunity to prepare for such events and reduce the risk to human life, infrastructure for the large part must weather whatever nature throws at them. Ships at sea can adjust their heading to avoid storms to some extent, but they too must be designed to survive extreme conditions. It is important therefore that the occurrence of extreme and very rare events, such as those with reoccurrence interval of 100 years or longer, and consequently the risk of damage to property is understood and quantified. Quantifying this risk involves extreme value analysis of one form or another.

Accurate and reliable estimates of probabilities of rare, extreme metocean events are critical for optimal design of offshore facilities. They ensure facilities are neither over- nor under-designed, allowing target reliability levels to be achieved without undue conservatism. Engineering design requires metocean parameters to be specified with a return period of 100 years, but often specification to a return period of 10,000 years is required.

Modern hindcast databases, typically used to derive extremal criteria, can be of limited extent, consisting of data for a few decades; hindcasts with periods of more than 50 years remain unusual. In addition, metocean criteria are often stratified by covariate — seasonality or directionality are common examples.

Specifying extreme conditions is generally not related to specifying the extremes of one parameter alone or a marginal distribution. Generally, interest lies in specifying an extreme or design sea state, which involves in its most general form the specification of a design wave spectrum or more usually a combination of wave parameters to be specified jointly. Also, specification of the joint occurrence of parameters at long return periods is necessary to avoid excessive conservatism, and such criteria may also need to be specified as functions of one or more covariates. Methods used by practitioners to meet these requirements are often somewhat *ad hoc*, based on experience and intuition.

As one might expect, estimating extreme events with low probability of occurrence, such as events with a return period of 10,000 years, is associated with large uncertainties. These must be understood and quantified.

For these reasons, extreme value analysis is often the cornerstone in understanding and quantifying extreme sea conditions. The majority of this chapter is devoted accordingly to outlining methods of extreme value analysis. The basics are given in Section 7.2. Section 7.3 addresses short-term variability or, for example, how to estimate extreme individual wave heights with a given return period. Consideration of covariates, such as wave direction and season, in estimating extremes is described in Section 7.4. Subsequently, Section 7.5 describes methods for estimating joint extremes, which might include the estimation of the associated wave period for a given extreme significant wave height.

A description of extreme seas around the globe is given in Section 7.6, and the chapter concludes with Section 7.7, which is devoted to the discussion of uncertainties in extreme conditions, both on the practicalities of estimating them and our fundamental lack of knowledge of how a changing climate will impact extremes.

7.2. Extreme Value Analysis

7.2.1. *General*

Extreme value analysis involves the estimation or inference of properties of rare events from historical information. Usually, this involves estimation of events with occurrence probabilities much lower than any that have occurred within the historical data, requiring extrapolation to recurrence intervals many times the length of the historical data. Typically, this also involves a purely statistical analysis without physical basis. In the realm of ocean waves, the most common problem is the estimation of the parameters of models for rare large events, such as the significant wave height of the sea state with a return period of 100 years. Such problems are addressed with the application of extreme value theory, which involves use of a distribution, motivated by asymptotic statistical considerations, to model the magnitude of occurrence of extreme events. It is important to recognize that the focus of extremal analysis is low probability events in a database, or those events that dominate the tail of the probability density function (pdf). The body of the probability density function describes less extreme more "normal" conditions, and these will generally have different stochastic properties to those of the extremes.

In this section, the basic principles of the extremal analysis are outlined, with application to the estimating of extreme significant wave heights.

7.2.2. Data requirements

Extreme value analysis requires a set of historical data for analysis. Proceeding with significant wave height as our parameter of interest, this dataset might consist of a sample of the long-term population of sea state significant wave heights. By virtue of it being a sample, rather than the whole population, the dataset has inherent sample variability, but the implicit assumption is that the stochastic properties of the historical data are representative of what can be expected in the future. Accordingly, the effects of climate change or other long-term temporal variations that are not well represented in the historical data will not be reflected well in the estimation of the rare events. To minimize these deficiencies, it is preferred that the historical dataset covers many years, so that inter-annual variability is well-characterized, and long-term climate cycles are included to some degree; but it is unlikely that a climate change signature will be sufficiently strong for that effect to be significant in the dataset. In general, uncertainties associated with estimates from the sample are a function of the length of data available for analysis and increase with decreasing probability of the event being estimated.

The main sources of wave data are *in-situ* measurements, remote sensing data, and numerical model data. Measurement databases are usually too short to enable low-probability events to be estimated to an acceptable reliability, and they often contain gaps due to instrument malfunction. However, measured wave data, either locally or remotely, remain important for the development, calibration, and validation of numerical models, and the baselining and specification of more detailed wave descriptions such as spectra and individual wave heights and crest elevations. This is particularly important in coastal areas where the prediction of waves is further complicated by shallow-water, bottom topography and coastal boundary effects. Altimetry data from satellites have been used to provide global extremes [Vinoth and Young, 2011], but limitations of spatial coverage and repeat cycle remain an issue. Wave data from hindcast studies are the preferred datasets for developing design criteria. For most ocean basins in the world, adequate historical meteorological records exist to allow numerical simulation of hindcast data, some for periods as long as 100 years. For example, the proprietary GOMOS hindcast dataset covers the period 1900–2017 [Oceanweather, 2018]. Such hindcasts include the relevant parameters of the wind and sea state and provide a good basis upon which analysis can be undertaken, including joint extremal analysis.

A suitable database will provide sea state parameters estimated or measured at regular intervals, often 3-hourly. The tacit assumption is that the sea state is stationary during this interval, an assumption that may not be valid when conditions are changing rapidly, such as during a tropical cyclone or hurricane. For this reason, hindcast databases that have tropical cyclones that produce the extreme conditions will often be sampled at 1-hourly intervals and sometimes even shorter intervals.

Attention must be given to the extreme events, such as the more intense cyclones or tropical cyclones in the database; generally, there will be few of these, but they will dominate the tail of the probability density function, the focus of the extreme value analysis.

7.2.3. *Fundamentals*

This section introduces the fundamentals of extreme value analysis. The usual statistics nomenclature is adopted, in which a random variable or variate is denoted with upper-case letters, and when that takes a specific value, that value is denoted with a lower-case letter. The cumulative distribution function is denoted by an upper-case letter, and its derivative, the probability density function, is denoted with a lower-case letter. Thus, for example, $\Pr[X \leq x]$ is the probability that the variate X is less than a specific value x, and is given by the cumulative distribution function, or distribution $P_X(x)$,

$$P_X(x) = \Pr[X \leq x],$$

and the probability density function $p_X(x)$ is given by

$$p_X(x) = \frac{dP_X(x)}{dx}.$$

Now, suppose we have observations of the significant wave height, H_s, at hourly intervals, obtaining values $h_{s_1}, h_{s_2}, h_{s_3}, \ldots$. Each h_{s_i} can be considered a realization of the corresponding random variable H_{s_i} from the probability distribution $P_{H_{s_i}}(h_{s_i})$. The maximum of n such observations, $H_{s_{\max,n}}$, is

$$H_{s_{\max,n}} = \max\{H_{s_1}, H_{s_2}, H_{s_3}, \ldots, H_{s_n}\}. \tag{7.1}$$

If the set of observations corresponds to an interval of one year, and $n = 8766$ observations (on average), then $H_{\max,n}$ is the annual maxima $H_{s_{\mathrm{AM}}}$.

If the $\{H_{s_i}\}$ are independent and identically distributed, an estimate of the probability distribution $P_{H_s}(h_s)$ of H_s could be obtained from a sample of observations $\{h_{s_i}\}$, perhaps made over a number of years. Accordingly, an estimate of the probability distribution of the annual maximum $H_{s_{\mathrm{AM}}}$ could be obtained from

$$P_{H_{s_{\mathrm{AM}}}}(h_s) = [P_{H_s}(h_s)]^n \tag{7.2}$$

and for the probability distribution of the r-year maximum $H_{s_{\mathrm{AM}r}}$ from

$$P_{H_{s_{\mathrm{AM}r}}}(h_s) = [P_{H_s}(h_s)]^m \tag{7.3}$$

where $m = nr$ (e.g., $m = 876,600$ for the probability distribution of the maximum 100-year H_s).

In reality, Eqs. (7.2) and (7.3) provide poor estimates of the annual and r-year maxima for a number of reasons, primarily because measurements of hourly significant wave height would not meet the independent and identically distributed requirement. Consecutively occurring sea states are unlikely to be independent for typical sampling intervals, and in most cases, they are not identically distributed. The latter point is particularly relevant here. The climate at any location will consist of a range of weather conditions; a large proportion of the time a location may experience relatively mild conditions, often referred to as "normal" conditions, but occasionally extremes occur — in low latitudes, the extreme conditions result from tropical cyclones, at mid-latitudes, the extreme conditions result from severe extra-tropical or winter storms. As a result, the probability distribution of H_s in the extreme conditions is different from that in normal conditions: the latter contributing to the body of the distribution of H_s and the former to the tail. The focus of extreme value analysis is the tail of the distribution. Nevertheless, the approach based on Eq. (7.3), sometimes referred to as the initial distribution method, has been used to estimate extremal criteria [e.g., Lopatoukhin *et al.*, 2000; Vinoth and Young, 2011], and it is useful to continue with this approach here as it enables key aspects of the statistics of extremes to be illustrated.

Accordingly, suppose that $h_{s_{AMr}}$ is the value of the annual maximum $H_{s_{AM}}$ with exceedance probability $1/r$, then

$$P_{H_{s_{AM}}}(h_{s_{AMr}}) = 1 - 1/r. \qquad (7.4)$$

Here, $h_{s_{AMr}}$ is referred to as the return level or value associated with the return period r. That is, the probability that the r-year return value is exceeded in any random year is $1/r$; so, for example, the 100-year return value has a probability of 0.01 of being exceeded in a random year. Equivalently, $h_{s_{AMr}}$ is the quantile of the distribution of the annual maximum with non-exceedance probability $1 - 1/r$.

Further, insight into the return period value is gained from its probability of being exceeded over a duration corresponding to the return period. For example, Fig. 7.1 demonstrates this approach applied to H_s data generated from a Weibull 2-parameter distribution. It is assumed that the data relate to 1-hour sea states. The thin blue continuous line is the empirical probability distribution of H_s estimated directly from the data. The dashed blue line is the Weibull 2-parameter fit to the empirical distribution, representing our inference of the probability distribution of the population 1-hour H_s. It gives the probability that H_s over a randomly selected one-hour period will not be exceeded. The blue dash-dotted line is the probability density function of H_s. The red and black dashed lines

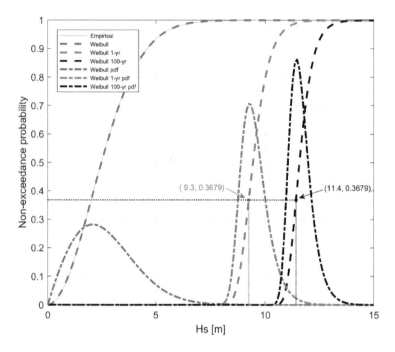

Fig. 7.1. Probability distributions and density functions of H_s values derived from a Weibull 2-parameter distribution. Continuous blue line is the empirical distribution estimated from H_s values, the remaining dash and dash-dotted lines are estimates of the cumulative distribution and probability density functions of the hourly wind speeds (blue), the annual maximum hourly H_s values (red), and the 100-year maximum hourly H_s values (black).

Note: The ordinate axis label refers to the distributions, but in the case of the probability density functions, the ordinate axis is probability density.

are the 1-year and 100-year maximum 1-hour H_s cumulative distribution functions derived from the probability distribution of H_s with application of Eqs. (7.2) and (7.3), respectively. The red and black dash-dotted lines are the probability density functions of the 1-year and 100-year maximum 1-hour H_s. The horizontal black dotted line denotes the non-exceedance probability of 0.3679 (strictly, $\exp(-1)$), and the corresponding values of the 1-year and 100-year maximum 1-hour H_s, where this line intersects each, are given next to the respective intersection points. These values of H_s also correspond to the mode of the corresponding probability densities.

Extreme Conditions

A question that often arises is what is the probability that the r-year return period value is exceeded in r years? With the definitions thus far, this is easily answered, again assuming that annual maxima are independently and identically distributed. With the r-year return period value $h_{s_{AMr}}$ defined by Eq. (7.4), the probability that it is not exceeded in a random year is

$$\Pr[H_{s_{AM}} \leq h_{s_{AMr}} \text{ in one year}] = 1 - \frac{1}{r},$$

and the probability that it is not exceeded in r-years is

$$\Pr[H_{s_{AM}} \leq h_{s_{AMr}} \text{ in } r \text{ years}] = \left(1 - \frac{1}{r}\right)^r$$

$$\approx \frac{1}{e}$$

$$\approx 0.3679$$

for large r.

Hence, the probability that the r-year return period value is exceeded in r-years is given by

$$\Pr[H_{s_{AM}} > h_{s_{AMr}}] = 1 - \frac{1}{e}$$

$$\approx 0.6321.$$

The same logic allows the expression for the distribution of the annual maximum $H_{s_{AM}}$ in terms of the distribution of the 1-hour sea state H_s. The annual maximum is the return value of the 1-hour sea state H_s with non-exceedance probability $1 - 1/n$, where n is the number of sea states in one year. Hence

$$\Pr[H_{s_{AM}} \leq h_{s_{AM}}] = \Pr[H_s \leq h_{s_{AM}} \text{ in } n \text{ 1-hour sea states}]$$

$$= (\Pr[H_s \leq h_{s_{AM}}])^n$$

$$= \left(1 - \frac{1}{n}\right)^n$$

$$\approx \frac{1}{e}.$$

This probability is noted on the curves in Fig. 7.1, indicating the 1-year return period value to be 9.8 m and the 100-year return period value to be 11.4 m. The values with a non-exceedance probability of 0.3679 converge (with increasing return period) to the most probable values (the mode or maximum of the probability density function) for the given return period.

In addition, the number of occurrences that the r-year return period value is exceeded in a given number of years can be estimated with the binomial or Poisson distributions, and we would therefore expect on average one such occurrence in r years. Thus, the r-year return period value can be expected to be exceeded on average once in r years; for example, the probability that the 50-year return period value is exceeded in 50 years is 0.6321.

Another frequent question is what is the probability that a random 1-hour sea state H_s value exceeds the r-year return period value — i.e., $\Pr[H_s > h_{s_{\text{AMr}}}]$. We know that the r-year return period value has a probability of 0.3679 of not being exceeded in a random r-year period, when r is large. Therefore

$$\Pr[H_{s_{\text{AM}}} \leq h_{s_{\text{AMr}}} \text{ in } r \text{ years}] \approx 0.3679.$$

Assuming our 1-hour sea state H_s values are independent and identically distributed, we seek the value h_s for which

$$P_{H_s}(h_s)^{(365.25 \times 24 \times r)} \approx 0.3679$$

therefore

$$P_{H_s}(h_s) \approx 0.99989^{1/r},$$

and hence

$$\Pr[H_s > h_{s_{\text{AMr}}}] \approx (1 - 0.99989^{1/r})$$

$$\approx 1/(365.25 \times 24 \times r)$$

since $1 - (1-\epsilon)^{1/r} \approx 1 - (1-\epsilon/r) \approx \epsilon/r$ when ϵ is small and positive. Thus, $\Pr[H_s > h_{s_{\text{AMr}}}]$ is approximately simply the inverse of the number of 1-hour sea states in r years, as might be expected, and setting $r = 1$ in the equation above gives the probability that a random 1-hour sea state exceeds the 1-year return period value. Of course, consecutive measurements of H_s are typically not iid, because storm events evolve over time periods corresponding to multiple sea states, causing temporal dependence of H_s.

Statistical estimates of extreme wave environments focus on analysis of the largest events in the data using methods of extreme value analysis. There are generally two approaches: one motivated by analysis of the maxima from blocks of observations (typically in time), and the other the analysis of exceedances of a high threshold. There are close analogies between the approaches, as will be demonstrated in what follows. The first method considered involves block maxima.

Suppose we have many years of hourly H_s values. We could take the maximum value of blocks of these data, giving us block maxima. For example, yearly blocks give us a sample from $\{H_{s_{\text{AM}_i}}\}$, but we could block over any period of interest. If the blocks are big enough, then by the extreme value theorem [e.g., Coles, 2001], the probability distribution of $H_{s_{\text{AM}}}$ is given by the generalized extreme value (GEV) distribution

$$P_{H_{s_{\text{AM}}}}(h_s) = \exp\left[-\left[1 + \frac{\xi}{\sigma}(h_s - \mu)\right]^{-1/\xi}\right]. \quad (7.5)$$

Here, μ is the location parameter, σ is the scale parameter and $\sigma > 0$, and ξ is the shape parameter.

Estimates of the three parameters are obtained from fitting Eq. (7.5) to a sample $\{h_{s_{\text{AM}_i}}\}$ from $\{H_{s_{\text{AM}_i}}\}$, and an estimate of the r-year return value can then be obtained from Eq. (7.5),

$$h_{s_{\text{AM}r}} = \begin{cases} \mu - \dfrac{\sigma}{\xi}\left[1 - \left\{-\log\left(1 - \dfrac{1}{r}\right)\right\}^{-\xi}\right] & \text{for } \xi \neq 0, \\ \mu - \sigma \log\left[-\log\left(1 - \dfrac{1}{r}\right)\right] & \text{for } \xi = 0, \end{cases}$$

where r is the return period in years.

Historically, and formally, the GEV distribution can be described as the combination of three types of extreme value distribution, depending on the value of the shape parameter [Embrechts et al., 2003]. The Type I distribution occurs when $\xi = 0$, which can be found from Eq. (7.5) by letting $\xi \to 0$. This is the Gumbel family of distributions. Return values for this family of distribution are unbounded. The Type II distribution occurs when $\xi > 0$. This is the Fréchet family of distributions. Return values for this family of distribution are unbounded. The Type III distribution occurs when $\xi < 0$. This is the (reversed) Weibull family of distributions. Return values for this family of distributions type are bounded above by $\mu - \sigma/\xi$. Thus, the GEV provides a description of all the three types of distribution naturally, the specific distribution depending on the data, without the need for any *a priori* assumptions about model type.

A practical problem with extremal analysis of block maxima is that it requires many blocks of data for reliable extremes to be estimated; in the case of annual maxima, only one datum is available per year. The annual maximum will have occurred during the most intense storm during the year, but there are likely others that are of interest and these could even produce sea states more severe than the annual maximum in other years. Since our interest is in the probability distribution of the population of extremes, we would like to include all events belonging to this population. The second method considered allows all events that exceed some threshold, say $H_s = \psi$, to be included in the extreme value analysis, allowing for the possibility of more storms to be considered in the analysis. In this sense, analysis of threshold exceedances can be more efficient than that of block maxima. If the threshold is set high enough, and if the GEV distribution would be applicable for block maxima of our data, then the probability distribution of H_s, conditional on $H_s > \psi$, is given by the generalized Pareto distribution (GPD)

$$P_{H_s}(h_s|H_s > \psi) = 1 - \left(1 + \frac{\xi(h_s - \psi)}{\tilde{\sigma}}\right)^{-1/\xi} \qquad (7.6)$$

for $(1+\xi(H_s - \psi)/\tilde{\sigma}) > 0$, and

$$\tilde{\sigma} = \sigma + \xi(\psi - \mu).$$

The parameters μ, σ, and ξ are those of the GEV in Eq. (7.5).

The GPD has three parameters, ψ, $\tilde{\sigma}$, and ξ. The threshold ψ is chosen sufficiently high so as not to undermine the basis of the

GPD applicability and not too high so as not to limit the number of data exceeding it, and the remaining two parameters are determined by fitting Eq. (7.6) to sample $\{h_{s_i}\}$ of exceedances. Estimates of the return value can then be obtained from Eq. (7.6),

$$h_{s_r} = \begin{cases} \psi + \dfrac{\tilde{\sigma}}{\xi}[(r\lambda)^\xi - 1] & \text{for } \xi \neq 0, \\ \psi + \tilde{\sigma}\log(r\lambda) & \text{for } \xi = 0, \end{cases}$$

where λ is the mean number of threshold exceedances per year, and r is the return period in years.

As for the GEV, if $\xi \geq 0$, the generalized Pareto distribution is unbounded above, but if $\xi < 0$, the distribution has an upper bound, in this case of $\psi - \tilde{\sigma}/\xi$.

The assumption of independence of the data is usually invalid for consecutive observations, including those exceeding a threshold. This can be overcome if the data are sufficiently separated in time. Typically, in time series data, such as a historical record of consecutive 1-hourly or 3-hour H_s values, the values will exceed a given threshold for a period of time before once again falling below the level. If the threshold is set sufficiently high, the values will remain below the threshold for longer periods than they exceed it, and the times at which the threshold is first exceeded will denote the time of the onset of a storm, and it can be safely concluded that single H_s values selected from different storms will be independent. This is the basis of the peaks over threshold (POT) method. In this case, only the H_s values corresponding to the storm peaks H_s^{sp} of the separate threshold exceedances or storms are selected for extreme value analysis. Since each datum exceeding a sufficiently high threshold is deemed to have a probability given by the generalized Pareto distribution, the maximum of each threshold exceedance is also distributed according to the generalized Pareto distribution, and if consecutive threshold exceedances or storms are sufficiently separated in time, the independence assumption is justified.

It is usual to model the probability of the number of occurrences of threshold exceedances or storms over a period with the Poisson distribution, and the size of the peaks with a generalized Pareto distribution. Accordingly, the probability of k storms occurring in a given period, say a particular year, is given by

$$p(k;\lambda) = \frac{\lambda^k}{k!}\exp(-\lambda),$$

where λ is the number of storms per year.

Recognizing that the occurrence of zero storms is mutually exclusive of the occurrence of exactly one storm, which is mutually exclusive of the occurrence of exactly two storms, and so on, the probability that the storm peak H_s value h_s^{sp} is not exceeded in a random year is given by

$$\begin{aligned}\Pr[H_s^{sp} < h_s^{sp} \text{ in 1 year}] &= \sum_{k=0}^{\infty}\frac{\lambda^k}{k!}\exp(-\lambda)P_{H_s^{sp}}^k(h_s^{sp})\\ &= \exp\left[-\lambda(1-P_{H_s^{sp}}(h_s^{sp}))\right]\\ &= \exp\left[-\lambda\left(1+\frac{\xi h_s^{sp}}{\tilde{\sigma}}\right)^{-1/\xi}\right].\end{aligned}$$

A common extreme value analysis approach in developing met-ocean design criteria to is to adopt a specific distribution (other than GPD or GEV), and then to estimate the relevant parameters of that distribution. This involves a somewhat arbitrary *a priori* choice of model form that then predetermines the behavior of the tail. Use of the GPD, or the GEV in the case of block maxima, avoids this shortcoming by allowing the data to determine the tail form.

7.3. Including Short-Term Variability

Until now, the discussion has been focused on extremes of the significant wave height H_s. The significant wave height is a statistical measure of the severity of the sea states, and its value is a function of a number of physical phenomena, but most importantly the wind fields. The wind field climate is the most important contributor to the long-term variability in H_s. Except under extreme forcing as expected in a severe cyclone, for example, a sea state is usually assumed to be stationary for periods up to several hours. This implies that the sea state is a random process with a specific parent or population value of significant wave height H_{s_i}, say; but a specific observation h_{s_i} will have variability about H_{s_i} depending on how the sample was obtained. For example, Donelan and Pierson [1983] demonstrate that this variability is approximately 12% at the 90% confidence levels for surface elevation measurements that are 17 minutes long. Other sea state parameters, such as the peak frequency, spectral peakedness, and the various mean wave periods, will also have sampling variability [e.g., Young, 1986, 1995; Saulnier, 2013; Ewans and McConochie, 2018].

Of course, the variability of the sea state parameters arises due to the fundamental random nature of the sea surface elevation, which leads to randomness in the wave heights and crest elevations. The randomness within a sea state, and the subsequent variability in the estimate of any statistical parameter from a sample of the sea state is referred to as short-term variability. This variability is captured in the statistical distributions for the sea surface elevation, which is usually assumed to be near Gaussian, and the distributions of the wave height and crest elevations, for which there exists a vast literature base.

Many statistical distributions for wave height and crest elevation have been proposed, largely distinguished by the degree to which various nonlinear aspects have been incorporated. For example, the Rayleigh distribution of wave heights and crest elevations assumes that all processes are linear (and also narrow-banded), while the crest model of Forristall [2000] is based on consideration of second-order effects, and that of Tayfun [2011] considers higher-order effects. In addition, there are numerous papers on these distributions for finite water depths (e.g. Katsardi *et al.*, 2018). Thus, there is a perhaps bewildering choice of distributions for both wave height and crest elevation, but, in particular, crest elevation for which the nonlinear effects are most manifest.

Once chosen, the distribution can be used to estimate the probability distribution of the maximum within a sea state. For example, if $P_H(h)$ is the probability distribution of the individual wave heights within a sea state, then in a similar way to the estimation of block maxima, the probability distribution of the maximum wave height in the sea state containing N waves, assuming the individual heights are independent, is given by $[P_H(h)]^N$. In design, the maximum wave height or crest elevation that can be expected with a given return period is an important quantity for determining the extreme loading on an offshore structure and, for example, the air gap needed between the mean still sea level and the deck of steel jacket structures standing on the seabed. Thus, in addition to significant wave height, estimates of the maximum wave height and crest elevation with long return periods are needed. This requires incorporating short-term variability with the long-term variability in estimates of extremes. Various approaches have been proposed, and a number of these are described in Tucker and Pitt [2001].

The simplest but flawed approach is to undertake the extreme value analysis of the values of H_s to estimate the r-year return period value $h_{s_{\text{AM}r}}$ of the significant wave height, and then to estimate the maximum wave height or crest based on this value, using the appropriate short-term distribution. However, this method does not account for the possibility that the highest wave height or crest could come from a sea state with a significant wave height less than $h_{s_{\text{AM}r}}$. An elegant approach to overcome this shortcoming is given by Tromans and Vanderschuren [1995]. The method is referred to as the storm-based approach, as it is based on considering sea states about the peaks of storms that have been selected by the POT method, with extremal analysis performed on the most probable maximum wave height or crest within a storm. The analysis is based on the theory outlined by Leadbetter that the time history of sea states in extreme storms have similar behavior near the peak and the distribution of, for example, the maximum individual wave height $P_{H_{\text{mxs}}|H_{\text{mp}}}(h|h_{\text{mp}})$ in a storm can be described by

$$P_{H_{\text{mxs}}|H_{\text{mp}}}(h|h_{\text{mp}}) = \exp\left[-\exp\left(-[\log_e N]\left(\left(\frac{h}{h_{\text{mp}}}\right)^2 - 1\right)\right)\right], \tag{7.7}$$

where H_{mp} is the most probable maximum *in the whole storm* and N is the number of waves occurring in the storm. The theory suggests that H_{mp} is controlled by sea states within 80% of the storm peak significant wave height, H_s^{sp}. We note from Eq. (7.7), setting $h = h_{\mathrm{mp}}$, that $P_{H_{\mathrm{mxs}}|H_{\mathrm{mp}}}(h|h_{\mathrm{mp}}) = 1/e$.

We proceed as follows. Peaks over threshold analysis is first used to identify storms, with appropriate filtering, for example, to maintain a minimum time separation ensuring that consecutive storms are decorrelated. An appropriate distribution for individual wave height or crest elevation is then adopted.

For example, consider a typical storm consisting of a set of sea states. We assume that the characteristics of individual wave height H for the jth sea state are conditional on the value H_{s_j} of sea state significant wave height such that the characteristics of H for the whole storm are conditional on the set $\{H_{s_j}\}$. Given H_{s_j}, the cumulative distribution function for individual wave height in the jth sea state is $P_{H|H_{s_j}}(h|h_{s_j})$. If the sea state contains N_j waves, then the cumulative distribution function for the maximum individual height within the sea state is $[P_{H|H_{s_j}}(h|h_{s_j})]^{N_j}$, assuming individual wave heights to be independent. Further, the corresponding cumulative distribution function $P_{H_{\mathrm{mxs}}|\{H_{s_j}\}}(h|\{h_{s_j}\})$ of the largest wave in the whole storm conditional on the set of values $\{h_{s_j}\}$ for all sea state characteristics is given by

$$P_{H_{\mathrm{mxs}}|\{H_{s_j}\}}(h|\{h_{s_j}\}) = \prod_j [P_{H|H_{s_j}}(h|h_{s_j})]^{N_j}.$$

According to the theory, this distribution converges to the asymptotic form $P_{H_{\mathrm{mxs}}|H_{\mathrm{mp}}}(h|h_{\mathrm{mp}})$ in Eq. (7.7) which is conditional on H_{mp} only, rather than on the whole set of parameters $\{H_{s_j}\}$. Since $P_{H_{\mathrm{mxs}}|H_{\mathrm{mp}}}(h_{\mathrm{mp}}|h_{\mathrm{mp}}) = 1/e$, we can use the equation above to estimate h_{mp} for any storm of interest, obtaining set $\{h_{\mathrm{mp}_i}\}$ for a set of storms indexed by i. Hence, since Eq. (7.7) is conditional only on H_{mp} per storm, we can now use it to evaluate $P_{H_{\mathrm{mxs}}|H_{\mathrm{mp}}}(h|h_{\mathrm{mp}})$ for any storm of interest, given a value h_{mp} for H_{mp}.

The long-term statistical distribution of H_{mp} can be assumed to follow a GPD, with probability density $p_{H_{\mathrm{mp}}}(h_{\mathrm{mp}})$ say. Therefore, the cumulative distribution function of the maximum individual wave height in a random storm $P_{H_{\mathrm{mxrs}}}(h)$ is obtained by combining the short-term distribution $P_{H_{\mathrm{mxs}}|H_{\mathrm{mp}}}(h|h_{\mathrm{mp}})$ with the best-fit GPD to the $\{h_{\mathrm{mp}_i}\}$, giving

$$P_{H_{\mathrm{mxrs}}}(h) = \int_{h_{\mathrm{mp}}} P_{H_{\mathrm{mxs}}|H_{\mathrm{mp}}}(h|h_{\mathrm{mp}})\, p_{H_{\mathrm{mp}}}(h_{\mathrm{mp}})\, dh_{\mathrm{mp}}.$$

As before, and assuming the occurrence of storms in time is Poisson distributed, with a mean of λ storms per year, the probability distribution of the maximum individual wave height in a random year $P_{H_{\mathrm{AM}}}(h)$ is given by

$$P_{H_{\mathrm{AM}}}(h) = \exp[-\lambda(1 - P_{H_{\mathrm{mxrs}}}(h))],$$

and the distribution of the maximum individual wave height in r years, assuming independent and identically distributed years, is $[P_{H_{\mathrm{AM}}}(h)]^r$. If r is large, this reduces to $[P_{H_{\mathrm{mxrs}}}(h)]^{\lambda r}$.

Distributions of the maximum individual crest height are obtained by using the appropriate short-term distributions of individual crest heights in a sea state.

Forristall [2008] concluded that the storm-based method of Tromans and Vanderschuren was the most appropriate method for deriving extremes of individual wave height for engineering design, and the method produced results that matched those from long-time series simulations.

7.4. Including Covariates

A fundamental assumption in extreme value analysis is that the data are independent and identically distributed. The latter point will be violated if data from different statistical processes are combined, such as would occur if H_s data were combined from different storm types, for example hurricanes and winter storms, or more commonly as would occur if H_s data were combined from sea states with different directions, where for example one direction may be fetch limited and another not. Combining H_s data from different seasons of the year can also be expected to result in a dataset that is not statistically homogeneous. Combining data in this way is likely to lead to inconsistencies in the design criteria, as demonstrated by Anderson et al., [2001] in the case of seasonal H_s criteria, and by Forristall [2004] in the case of directional H_s criteria. Jonathan et al. [2008] show that incorporating the effect of directionality into the extreme value model generally explains the observed variation of extremes significantly better, and therefore directional and omni-directional criteria derived from a directional model are more consistent and should be preferred in general.

A simple approach and one often practiced in engineering, to deal with covariate effects, is to partition the data by covariate and perform independent analysis on the data in each partition. For example, directional modeling may be performed for a set $\{I_i\}$ of directional sectors or months of the year in which the data are assumed homogeneous with respect to directional sector or month, respectively. Extremal analyses are then performed on the H_s values in each sector or each month of the year. Accordingly, if $P_{H_s|I_i}(h)$ is the probability distribution of H_s for partition I_i, and assuming that the partitions are independent, the probability distribution for all partitions combined, being referred to as the omni-directional or all-year distribution, is given by

$$P_{H_s}(h) = \prod_i P_{H_s|I_i}(h). \qquad (7.8)$$

Design values with a given return period can then be determined for each directional sector or month from the $P_{H_s|I_i}(h)$ and for all directions or all-year from the $P_{H_s}(h)$. A common practice is to estimate the omni-directional or all-year distributions from the complete set of H_s values ignoring direction or month, but this approach will often lead to inconsistencies in design criteria, resulting in the probability of non-exceedance of a given H_s when calculated from the directional criteria (Eq. (7.8)) being different from that obtained from omni-directional criteria [see, e.g., Forristall, 2004].

It is now becoming common practice to incorporate the effect of covariates by allowing the parameters of the model to vary smoothly with covariate. This is achieved by defining the parameters of the extremal model as a function of an appropriate basis, such as a Fourier series or a spline basis. Jonathan and Ewans [2007] demonstrate the Fourier series approach for estimating directional extremes associated with hurricanes in the Gulf of Mexico and how this model might be used to provide directional extremal criteria for the design of offshore structures. Randell et al. [2013] model extreme value parameters with a spline basis and show how the spline basis can be extended to include additional covariates, resulting in an extremal model involving direction, season, and space as covariates. Such approaches offer flexibility in terms of the choice of directional sectors or season in setting design criteria and also consistency between directional (seasonal) and omni-directional (all-year) criteria.

Following the POT method by way of example, Eq. (7.6) would be adjusted to allow for a smooth variation of the distribution parameters with covariate θ as

$$P_{H_s|\theta}(h_s|H_s > \psi(\theta); \theta) = 1 - \left(1 + \frac{\xi(\theta)(h_s - \psi(\theta))}{\tilde{\sigma}(\theta)}\right)^{-1/\xi(\theta)}. \quad (7.9)$$

If a Fourier series expansion is chosen for the basis, the parameters would be expressed, e.g., using $\xi(\theta)$ as

$$\xi(\theta) = \sum_{i=0}^{n} (A_i \cos i\theta + B_i \sin i\theta),$$

where n is the order of the Fourier model, with $n = 0, 1, 2, 3, \ldots$ corresponding to a constant, first-order, second-order, third-order models, etc.

The $\{A_i\}$ and $\{B_i\}$ may be estimated using maximum likelihood, and the order chosen is dependent on the data; a high-order Fourier form ensures the directional extreme value model is sufficiently flexible to characterize variation in extremal behavior with the covariate. A roughness penalty, which can be estimated by cross-validation, may be imposed to ensure that extreme value estimates have an appropriate smoothness, consistent with the data [see, e.g., Jonathan and Ewans, 2007a].

A similar formalism is followed in the case of a spline basis, referred to as B-splines, where, e.g.,

$$\xi(\theta) = \sum_{i=1}^{n} \alpha_i B_{i,p}(\theta),$$

where $\{B_{i,p}(\theta)\}$ are polynomial basis functions of order p, n is the number of basis functions, and the coefficients $\{\alpha_i\}$ are determined by fitting to the data. If a penalty function is applied to impose smoothness and avoid overfitting, the spline is referred to as a P-spline for "penalized B-spline".

Accordingly, the domain of the covariate is sub-divided into adjacent intervals, with a B-spline basis function defined for each interval. In the case of the covariate being direction, 45° intervals might be chosen, resulting in eight intervals in total.

B-splines can be extended to higher dimensions by forming the tensor-product of the B-splines over each dimension. For example, if covariates direction θ and season ϕ were to be considered, the combined B-spline basis B would be given by

$$B = B_\theta \otimes B_\phi.$$

Background to the B-splines and the estimation of the spline coefficients can be found in Eilers and Marx [1996], and examples of the application of B-splines to estimate extreme significant wave heights, including covariates are given by Randell et al. [2013], Jonathan et al. [2014], Feld et al. [2015], and Randell et al. [2015].

The incorporation of season as a temporal covariate is relatively straightforward. Extending the description to include longer term temporal effects associated with cyclical variations and climate trends is also possible. Méndez et al. [2006] describe a statistical model to estimate long-term trends in the frequency and intensity of severe storm waves based on a time-dependent version of the POT model applied to NOAA buoy significant wave height data. The model allows consideration of seasonality, trends, and atmosphere–ocean-related index relationships. The distribution parameters in the paper were modeled with a Fourier series to describe the seasonality, an exponential function for the trend, and a linear function of the Southern Oscillation Index (SOI) and the Pacific–North American Index (PNA) to account for the El Nino-Southern Oscillation (ENSO) and Pacific–North American teleconnection patterns. The dataset was too short to consider climate effects with longer periodicity, such as the Pacific Decadal Variability, but with a sufficiently long dataset, these could be included. The goodness of fit of the model was improved by including the time-dependencies.

7.5. Joint Criteria

Significant wave height provides a good first-order specification of a sea state, but wave periods and directions within a sea state are also important for most offshore applications. For example, personnel transfer from a floating vessel to the landing on a fixed platform may be conducted safely in a sea state with a significant wave height of one meter, when the wave periods are generally longer than 10 s, but the operation may not be safe in a sea state with the same significant wave height but with shorter wave periods or if the wave directions are unfavorable. Generally, the specification of a design sea state necessitates that other parameters in addition to the significant wave height are specified, such as the spectral peak wave period, the mean wave period, the mean wave direction, and the directional spreading; and many applications require a spectral description, some even needing the frequency-direction spectrum to be specified.

The design and structural assessment of offshore structures require extreme sea states to be specified to a much greater detail than is possible with the significant wave height alone. For example, the spectrum for an extreme sea state, such as the 100-year return period sea state, is often required for assessing dynamic loads, and it is common practice to derive this spectrum from a spectral form expressed in terms of sea state parameters, such as the significant wave height and the spectral peak period. This requires estimating the parameters, so they can be specified jointly to the given return period.

More generally, physical systems respond to all the environmental conditions acting — the pitch of a vessel is as much a function of the wave period or wavelength as it is of the wave height, and this will depend on the orientation of the vessel relative to the waves, where the orientation is strongly dependent on the wind and current conditions. Thus, the problem of estimating extreme conditions is generally a multivariate problem.

The approaches used by the offshore industry to calculate joint extreme environmental conditions essentially fall into two camps: response-based and response-independent. The response-based approach relies on the specification of a response model, for example, giving load as a function of environment and permitting a back calculation of the environmental variables once an extreme load has been established. The response-independent or environmental approach involves developing directly joint criteria for the environmental variables, associated with rare return periods. An outline of both approaches follows.

7.5.1. *Response-based methods*

Response-based methods involve calculating a key response or several key responses via a response function in which the variables are environmental variables. For example, Tromans and Vanderschuren [1995] describe generic response functions for the mud-line base shear and over-turning moment of steel jacket structures. Each response function is given in terms of a sum of terms involving variables of the winds, waves, and currents. The coefficients of the terms are determined by calibration of a large number of conditions with a given wave kinematics model and current profile on a 1-meter diameter vertical column. The response model of Ewans [2003] describes a method in which the response variable is the pipe weight per unit pipe length required for a pipe to be stable to the action of waves and currents, thus involving terms pertaining to near-bottom wave-induced and current water particle velocities. The coefficients of the terms in this response model were obtained from calibration against simulations with on-bottom pipe dynamic response software.

With a given response function, a long-term dataset of environmental variables can be converted into an equivalently long-term dataset of responses, allowing an extreme value analysis of the response variable to be undertaken. Estimates of the extremes of the response variable can be made to a given annual probability of exceedance or return period, and this value can be used in the original response function to back-calculate the environmental variables, to establish an appropriate design set of environmental variables for detailed engineering design. It should be noted that the response variable calculated from the response function need not be an actual engineering response or load, but it must have the same statistical behavior with the environmental variables as an actual engineering response or load.

It should also be noted that there is no unique combination of values of the parameters that define for example the 100-year return period sea state, in terms of response, as the combination of values is dependent on what specific response function is involved. For example, the values of a particular combination of parameters that generate the 100-year return period surge of a floating vessel will likely differ from the those that would generate the 100-year return period roll.

The back-calculation of the environmental variables from the response function is not trivial. In its simplest form, the back-calculation involves establishing an optimum combination of environmental variables, based on relationships established from the data and assumptions that these relationships will also apply in the extreme. Usually, one of the variables, such as the wave height in the case of the steel jacket response functions, is assumed to be dominant and the value of this variable is set at the return period of interest. The other variables are then determined from their respective relationships for this value of the dominant variable. The optimum set of variables when substituted into the response function would give the desired extreme value of the response variable.

The optimum choice of variables can also be determined by extending a response-independent analysis, typically of environmental variables, to include one or more response variables. Distributions of the values of the environmental variables conditional on an extreme response variable can then be established, and an appropriate choice, such as the most probable value of each variable, can be made.

A consequence of the response-based approach, and in particular of having access to the probability distribution for the response or load variable, is that it is possible, in principle, to calculate the reliability of a structure against failure due to the environmental loading and in turn to estimate the joint conditions responsible for the failure.

This is achieved as follows. The response Y of a structure can be considered as a deterministic function of the sea state variables X. This assumes that the variability in the environment (randomness in X) dominates, so that the variability in Y given X is negligible. The condition of the structure for every x can be described by a safety margin $g(x)$, such that the structure fails if $g(x) \leq 0$ and is safe if $g(x) > 0$. In the case of a steel jacket structure, the response considered might be the resistance R to environmental loading E, such that $g(X) = R - E$; that is, the structure will fail when the environmental load is greater than the resistance. The usually multidimensional function $g(x)$ is also referred to as the failure surface. The probability of failure p_F of the structure is then defined by

$$p_F = \int_{g(x) \leq 0} p_X(x) dx,$$

where $p_X(x)$ is now the joint probability density function of the set of environmental variables.

In practice, the integral is difficult to evaluate since both $p_X(x)$ and the integration boundary $g(x) = 0$, the failure surface, are multidimensional, usually nonlinear, and often not known sufficiently well. More fundamentally, $g(x)$ is both structure-dependent and response-dependent, and characterization of extreme environments leading to structural failure must incorporate knowledge of the interaction between environment and structure. It is therefore of interest to define the extreme environmental conditions independently of any specific structure response, and application of the inverse FORM or I-FORM method [see, e.g., Winterstein et al. 1993] has proved popular with practitioners, for determining joint environmental contours of a combination of metocean parameters alone. The deterministic response function with a return period corresponding to the contours can be estimated along the contours, and then checked whether or not the response would correspond to a failure condition.

Accordingly, the problem is simplified by transforming the random vector $X = (X_1, X_2, \ldots, X_n)$ of n variables, suitably ordered, to a vector of *independent*, conditional random variables $(\tilde{X}_1, \tilde{X}_2, \ldots, \tilde{X}_n)$, with $\tilde{X}_1 = X_1, \tilde{X}_2 = X_2|X_1, \tilde{X}_3 = X_3|X_2, X_1$ etc., and cumulative distribution functions $P_{\tilde{X}_1}, P_{\tilde{X}_2}, \ldots, P_{\tilde{X}_n}$. These independent random variables can now be transformed in turn to standard normal random variables $U_1, U_2, \ldots$ via the probability integral transform

$$P_{\tilde{X}_i}(x) = \Phi(u_i) \quad \text{for } i = 1, 2, \ldots,$$

$$u_i = \Phi^{-1}(P_{\tilde{X}_i}(x_i)),$$

where Φ is the cumulative distribution function of the standard normal distribution. The probability of failure is then evaluated using

$$p_F = \int_{g_u(u) \leq 0} \phi(u) du$$

for the transformed failure surface g_u, where ϕ is the probability density function of a set of independent standard normal random variables. Thus, the contours of constant probability density of the integrand are now concentric circles (bivariate case) or hyper-spheres in higher dimensions.

To facilitate solution, the integration boundary $g_u(u)$ is simplified by truncating its Taylor expansion about an as yet unknown point u^*, to first-order (FORM) or second-order (SORM). u^* is the point that has the highest probability density on $g_u(u) = 0$ to minimize accuracy loss (the integrand function quickly diminishes away from the expansion point) and is referred to as the most probable point (MPP).

The MPP is found by minimizing $\|u\|$ for $g_u(u) = 0$, the minimum distance from the origin to the failure surface. The minimum distance $\beta = \|u^*\|$ is called the reliability index, and the probability of failure is now simply $p_F = 1 - \phi(\beta)$, where ϕ is now the probability density function of a single normal random variable. The value of u^* can be transformed back to a corresponding x^* (in terms of the original variables) to establish the failure design set (the values of the set of environmental parameters causing the failure).

7.5.2. Response-independent methods

Response-independent methods for establishing combinations of environmental variables for design, involve joint distributions that describe the behavior of the variables when one or more is extreme. The joint distributions are established directly from the environmental variables themselves. A particular combination of variables with a given low probability of occurrence can then be specified. Reference to a response variable is not required but is possible. In this sense, response-based methods are different only in that they involve finding the most likely combination of environmental variables to produce a target response value. Response-independent methods usually result in the specification of multi-dimensional contours of environmental parameters and are therefore sometimes referred to as environmental contour methods.

In the case of FORM or SORM, the failure surface is the target and a failure probability is calculated, but conversely if the target is a failure probability, a design point can be calculated on an associated failure surface. Winterstein *et al.* [1993] demonstrate this approach, which they refer to as inverse FORM, to calculate probability contours of joint occurrences of environmental variables. The design point, u^*, is found by minimizing $g_u(u)$ for $\|u\| = \beta$. The FORM and SORM failure surfaces are tangential to the contour at u^*, but for design, the behavior of the system can be checked to ensure the actual failure surface is outside the contour for that probability. An example of the use of inverse FORM to derive joint metocean conditions is given by Nerzic *et al.* [2007], for a West Africa location, and Fig. 7.2 shows the contours of equal probability density for significant wave height H_s and spectral peak period T_p following the joint probability model proposed by Haver and Nyhus [1986] and using inverse FORM.

The specific joint model used to produce Fig. 7.2 consists of a combined lognormal-Weibull distribution for the marginal distribution of H_s. That is, the logarithm of H_s is assumed normally-distributed, denoted $\log_e(H_s) \sim \mathcal{N}(\mu, \sigma^2)$, for $H_s \leq \psi$, and Weibull-distributed, $H_s \sim \mathcal{W}(\sigma, \xi)$, for $H_s > \psi$. The distribution for T_p conditional on H_s

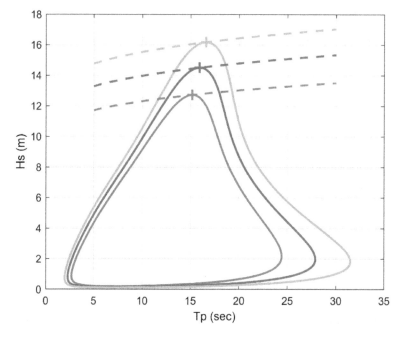

Fig. 7.2. 10- (blue), 100- (red), and 1000-year (orange) $H_s - T_p$ contours. Crosses mark respective design point sea states, corresponding to the median maximum crest elevation for the contours. Dashed lines are lines of constant median maximum crest elevation corresponding to the respective design point sea states.

is modeled as a lognormal distribution. Thus, for $H_s \leq \psi = 3.27$,

$$\Pr[H_s \leq h \text{ for } H_s \leq \psi] = P_{H_s}(h; \mu, \sigma) = \Phi\left(\frac{\log_e(h) - \mu}{\sigma}\right),$$

where Φ is the cumulative distribution function of the standard normal distribution, and μ and σ are, respectively, the mean (0.836) and standard deviation of $\log_e H_s$ (0.914). For $H_s > \psi = 3.27$,

$$\Pr[H_s \leq h \text{ for } H_s > 3.27] = P_{H_s}(h; \alpha, \xi) = 1 - \exp\left(-\left(\frac{h}{\alpha}\right)^\xi\right),$$

where α is the scale and ξ is the shape parameter, set to 2.822 and 1.547 in this example. The distribution of T_p conditional on H_s is given by

$$\Pr[T_p \leq t_p | H_s = h_s]$$
$$= P_{T_p|H_s}(t_p; \mu(h_s), \sigma(h_s)) = \Phi\left(\frac{\log_e(t_p) - \mu(h_s)}{\sigma(h_s)}\right),$$

where

$$\mu(h_s) = 1.59 + 0.42 \log_e(h_s + 2),$$
$$\sigma^2(h_s) = 0.005 + 0.085 \exp(-0.13 h_s^{1.34}).$$

The probabilities p_r associated with the return periods r in Fig. 7.2 are in this case obtained (assuming sea states have 3-hour durations) from

$$p_r = \frac{3}{365.25 \times 24 \times r}.$$

In bivariate normal probability space (U_1, U_2), a set of points (u_1, u_2) on the circle centered at the origin with radius $\beta = \sqrt{u_1^2 + u_2^2}$, where $\beta = \Phi^{-1}(1 - p_r)$ can be defined, and drawn as the corresponding set (h_s, t_p) in (H_s, T_p) space using the probability integral transform

$$h_s = P_{H_s}^{-1}(\Phi(u_1)), \quad t_p = P_{T_p|H_s}^{-1}(\Phi(u_2|h_s)).$$

To demonstrate how these contours might be used to determine a specific design set of conditions, we continue to follow Winterstein et al. [1993], and determine, the 10-, 100-, and 1000-year sea state that would produce the median maximum crest elevation in the sea state. For demonstration purposes, we will use the Rayleigh distribution for the crest elevations E_c where for this specific case E_c denotes a random variable

$$\Pr(E_c \leq \eta | H_s = h_s) = P_{E_c|H_s}(\eta|h_s) = 1 - \exp\left[-8\left(\frac{\eta}{h_s}\right)^2\right].$$

The distribution of the maximum crest in the sea state (assuming the individual crests are independent) is given by

$$P_{E_{\text{cmx}}|H_s}(\eta|h_s) = [P_{E_c|H_s}(\eta|h_s)]^N,$$

where N is the number of waves in the sea state, often determined by D/T_{02}, where D is the duration of the sea state and T_{02} is the period derived from the ratio of the zeroth and second moments of the wave spectrum. For the purposes of the example, we might assume our return period sea state can be specified by a JONSWAP spectrum, for which we can make use of the relationship $T_p/T_{02} = 1.28$.

The design point sea state for a given r-year return period is the pair (h_{sr}, t_{pr}) that maximizes median maximum crest, $\eta_{\text{cmx}|H_s,0.5}$ as determined from $\eta_{\text{cmx}|H_s,0.5} = P^{-1}_{E_{\text{cmx}}|H_s}(0.5)$, given by

$$\eta_{\text{cmx}|H_s,0.5} = \sqrt{-\frac{1}{8}\frac{h_{sr}^2}{\log_e(1 - 0.5^{(1/N)})}}.$$

These points are marked on the respective contours in Fig. 7.2 and show that the median extreme wave crest for each return period corresponds to the largest H_s on each contour, which corresponds to the marginal return period value of H_s. Contours of $\eta_{\text{cmx}|H_s,0.5}(h_s, t_p)$ are also plotted (dashed lines) for each of the return periods in Fig. 7.2. In a similar way, contours could be drawn for other response functions that were functions of H_s and T_p, and if the function corresponded to values such that the structure would fail is exceeded, the contours would represent failure surfaces.

This model was based on fitting the distributions to a joint frequency table of (H_s, T_p) from 3-hourly wave measurements made in the northern North Sea and is therefore strongly influenced by the body of the sample of wave data rather than the extremes. The form of lognormal parameter estimates for $T_p|H_s$ fitted within the body of the sample may not be relevant for extrapolation to extremes. Accordingly, the validity of the results of such models for extremes is not known. In addition, inverse FORM is difficult to model beyond two variables, requiring a model for the probability density of a variable conditional on the occurrence of the others. For example, if we were also interested in the joint occurrence of H_s and current speed C with the wave direction of the sea state, for which we could use the direction, Θ_p, associated with the spectral peak frequency as a measure, the objective would be to estimate $P_{C,H_s,\Theta_p}(c, h_s, \theta_p)$. Since we are interested in how our conditions vary with direction, our model might be based on conditioning on direction,

$$P_{C,H_s,\Theta_p}(c, h_s, \theta_p) = P_{C|H_s,\Theta_p}(c|h_s, \theta_p)P_{H_s|\Theta_p}(h_s|\theta_p)P_{\Theta_p}(\theta_p),$$

but the estimation of $P_{C|H_s,\Theta_p}(c|h_s, \theta_p)$ is not straightforward, requiring an expression for the probability density of C in terms of H_s and Θ_p that must be estimated from C values that have been stratified by H_s and Θ_p involving a relatively small number of values compared with the original dataset. The difficulty of this is increased with each additional parameter added to the mix. A useful discussion on the environmental contours method is given by Ross et al. (2019).

The conditional extremes model of Heffernan and Tawn [2004] provides a more general framework based on an asymptotic limit assumption. It involves modeling the conditional distribution of one or more variables when the value of a conditioning variate is large, but a distinct advantage over a typical FORM analysis is that no prior knowledge of the forms of the distributions is required. Instead, asymptotic distributional forms are used.

The method is most clearly and most easily described in the case of two variables $(\dot{X}, \dot{Y})$ but is easily extended to higher dimensions; the dot notation indicates original physical scale, for convenience, here. The steps involved are as follows:

(a) perform marginal peaks over threshold analysis for variates $\dot{X}$ and $\dot{Y}$ separately;
(b) fit threshold exceedances with a GPD;
(c) transform $\dot{X}$ and $\dot{Y}$ in turn to Gumbel (or Laplace) scale (to X and Y) using the probability integral transform;
(d) estimate conditional extremes models for $Y|X$ of the form

$$(Y|X = x) = \alpha x + x^\beta Z \quad \text{for } x > \psi$$

for an appropriate high threshold ψ, where $\alpha \in (0, 1]$ (Gumbel; or $\alpha \in [-1, 1]$ for Laplace) is a slope parameter, and $\beta \in (-\infty, 1]$ is a scale parameter, and Z is a random variable, independent of X, converging with increasing ψ to a non-degenerate limiting distribution, G (which is assumed Gaussian for model fitting purposes only);

(e) retain the estimates for model parameters α and β, and residuals $\{z_i\}$ where $\{x_i, y_i\}$ is the sample on Gumbel (or Laplace) scale, and

$$z_i = \frac{y_i - \alpha x_i}{x_i^\beta} \quad \text{for } i = 1, 2, \ldots;$$

(f) simulate joint extremes (X, Y) on the standard Gumbel scale under the model; and finally
(g) transform realizations to the original scale using the probability integral transform.

The threshold ψ is selected by inspecting a number of model fit diagnostics [see, e.g., Jonathan et al., 2010]; the smallest value of threshold for which acceptable model fit diagnostics are observed, admitting the largest possible sample for model fitting, is generally used.

By way of example, an application of the Heffernan and Tawn method to wave data from several locations is given in Jonathan et al. [2010]. Figure 7.3 is a plot of measured storm peak significant wave height and associated spectral peak period from measurements in the northern North Sea, together with estimates from simulations for $H_s^{sp} > 15$ m from conditional extremes modeling. The plot shows a generally increasing trend in associated T_p with H_s^{sp}. The most probable value of associated T_p, which appears to be between 16 s and 17 s for $H_s^{sp} > 15$ m, is significantly less than the longest in the measured data. Jonathan et al. [2010] demonstrates the improved performance of the conditional extremes model with respect to the model of Haver and Nyhus [1986] for simulated samples from known multivariate distributions. The article also suggests how the model of Haver and Nyhus [1986] might be improved to provide estimates with improved statistical characteristics.

The conditional extremes model is extended to incorporate covariates, Θ, in a similar way to the marginal distributions described

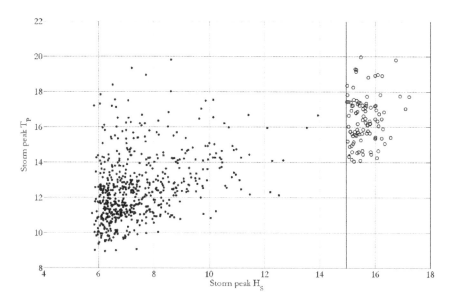

Fig. 7.3. Storm peak value of associated T_p versus H_s^{sp} for measured NNS storm peak data [Jonathan et al., 2010].

above. Accordingly, the covariate dependence of the threshold and marginal GPD is defined in terms of a basis matrix consisting of the Kronecker product of the respective covariate B-splines, and the GPDs are transformed into Gumbel or Laplace scales using the probability integral transform, retaining the covariate dependence.

For positively dependent random variables $X(\theta)$, $Y(\theta)$ with standard Gumbel or Laplace marginal distributions for any θ, the equation for the Heffernan and Tawn [2004] model becomes

$$(Y(\theta)|X(\theta) = x) = \alpha(\theta)x + x^{\beta(\theta)}Z(\theta).$$

When Gumbel scale is used, extended forms are available for potentially negatively dependent variables; these are not necessary when modeling on Laplace scale. Details of the model and the estimation procedure can be found in Jonathan et al. [2013], which gives an example of the application of the method for hindcast storm peak significant wave height, H_s^{sp}, and associated T_P in the northern North Sea. The objective is to model the distribution of T_P for large H_s^{sp}

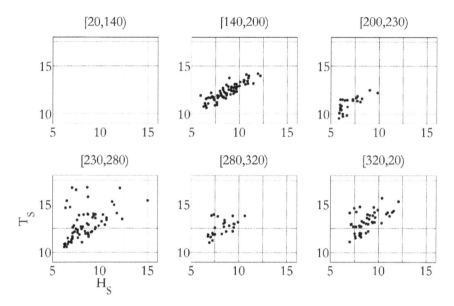

Fig. 7.4. Plots of storm peak H_s^{sp} (in meters, horizontal) versus associated T_P (in seconds, vertical) for the most severe (20%) storms emanating from six directional sectors (ordered, clockwise from 20°). Note that there are no data for sector [20,140). [After Jonathan et al., 2013]

as a function of storm direction. The location is particularly useful for application of the model as the wave field has identifiable characteristics for various directional sectors, as can be seen in Figs. 7.4 and 7.5. Storms with the largest sea states are those occurring in the north, south, and southwest to west sectors; less severe sea states are associated with storms from the northwest sector; and virtually no storms occur that cause waves from the easterly sector. Further, it can be seen in Fig. 7.4 that storm peak sea states from the northwest and southwest–west sectors are associated with the longest associated T_P values. These characteristics should be evident in the conditional extremes modeling and can serve as an indicator of the success of the modeling. The joint distribution of H_s^{sp} and associated T_P below the threshold is modeled using quantile regression.

Conditional associated T_P values corresponding to H_s^{sp} values with exceedance probability of 0.01 are illustrated in Fig. 7.5 (right).

Fig. 7.5. Left: Northern North Sea location and directional sectors with distinctive wave characteristics. Right: Return values of H_s^{sp} (inner circle) and conditional values of associated T_P (outer circle). Inner dashed lines (on common scale): storm peak H_s^{sp} with non-exceedance probability 0.99 (in 34 years), with (white) and without (black) directional effects. Outer solid lines (on common scale): median associated T_P with (white) and without (black) directional effects; outer dashed lines give corresponding 2.5% and 97.5% values for associated T_P. [After Ewans and Jonathan, 2013]

The inner (black and white) dotted curves, drawn on the same scale, illustrate estimates of the H_s^{sp} return value. The inner white dotted curve is an estimate for the directional variation of the H_s^{sp} return value. For comparison, the inner black dotted curve is an estimate for the same return value ignoring directional effects. The influence of longer fetches from south (in particular), the Atlantic and Norwegian Sea are visible.

The outer (black and white) curves, drawn on the same scale, illustrate estimates for return values of associated T_P conditional on exceedances of the corresponding H_s^{sp} value. Solid lines represent median values, and dashed lines 95% uncertainty bands, incorporating (white) or ignoring (black) directional effects. The results clearly show increased associated periods from the Atlantic and Norwegian sectors, as expected. When directionality is ignored, associated T_P values are underestimated for some sectors and overestimated for others. The importance of this difference for design can be seen in the response of a simple system with a transfer function characteristic of the roll or heave of a floating system with a natural period of around 17 s. Response is overestimated by more than 30% in directional sectors with short fetches, but underestimated by as much as 20% in sectors with long fetches, particularly the Atlantic sector [see Jonathan et al., 2013].

7.6. Regional Considerations

7.6.1. *General*

The wave climate varies substantially around the globe, both in terms of the normal conditions and extremes. In the tropics, the predominant wind forcing is from the trade winds, which are persistent but moderate, resulting in a corresponding persistent but mild wave climate. In the mid-latitudes, the prevailing winds are often very strong; extratropical cyclones dominate the more extreme wind forcing, but occasionally tropical cyclones intrude from lower latitudes, where they have formed. The wave climate in the mid-latitudes is substantially more extreme than in the tropical regions. The wave climate in the polar regions is limited by the land mass of Antarctica in the south and the polar ice cap in the north.

The more extreme wave conditions occur at high latitudes. Young [1999] presented global distributions of mean monthly values of the significant wave height, which showed that on average the Southern Ocean is the roughest ocean on Earth. The Southern Ocean is also the source of persistent swell that propagates into the Indian, South Pacific, and South Atlantic, and even into the North Pacific during the Southern Hemisphere winter. The mean monthly value distributions of Young [1999] also showed that the northern North Atlantic and North Pacific Oceans experienced high sea states on average. These results were consistent with earlier findings by Young [1994], when GEOSAT altimeter data were used to estimate statistics of global significant wave height, and he concluded that the most extreme sea conditions are in the North Atlantic Ocean and the Southern Ocean between Africa and Australia, based on 90-percentile occurrence estimates. Similar results were reported by Cox and Swail [2001], following analysis of data from a 40-year global hindcast.

However, average conditions are not necessarily indications of extreme conditions; the body of the probability distribution does not provide details on the tail of the distribution that describe the extremes. As we have seen already, even the 99-percentile sea is not an extreme sea state, being less even that the one-year return period sea state. Estimates of the global 100-year significant wave height have been made by Alves and Young [2003], using satellite data, and by Caires and Sterl [2005], based on the ECMWF 40-year hindcast dataset. Their results are qualitatively similar, showing that the most extreme wave conditions occur in the North Atlantic, with 100-year significant wave height values more than 20 m. Both studies acknowledge that tropical cyclones are not well resolved in the respective databases and that their return values in the region of tropical storms are therefore likely underestimated. A number of studies have largely confirmed these global patterns of the extremes [e.g., Izaguirre et al., 2011; Breivik et al., 2014].

Cardone et al. [2011] report a study of very extreme sea states (VESS) for which they adopted a threshold value of 14 m for significant wave height. They searched the GlobWave database satellite altimeter significant wave height (denoted HS in the paper) estimates for occurrences of VESS and ultimately ended with a population of 120 storm events in which there was at least one altimeter estimate of HS > 16 m. They found about five storms per year that met this criterion. The frequency of storms was found to be greatest in the North Atlantic Ocean, followed in order by the North Pacific Ocean, the Southern Indian Ocean, the South Pacific Ocean, and the South Atlantic Ocean. Only four of the 120 storm events were associated with tropical cyclones, but again this is believed to be a consequence of insufficient resolution. The highest HS value of 20.6 m was observed in the North Pacific Ocean. By comparison, the highest *in situ* measurement of significant wave height of 19.0 m was recorded by the UK Met Office K-5 buoy in the eastern North Atlantic (Northwest of the UK) on 4 February 2013 [Cardone et al., 2014].

Clearly in many regions of the World's oceans, extratropical storms are responsible for the extreme sea states — particularly so in the North Atlantic, the North Pacific, and the Southern Oceans; but at lower latitudes, tropical cyclones are responsible for driving the extreme sea states. Tropical cyclones form over the warm ocean waters in tropical regions, deriving their energy through the evaporation of water from the ocean surface, but they rarely form within 5° of the equator, as the Coriolis force is insufficient to produce the strong rotating winds. Cold currents on the west coasts of South America and Africa that flow northward into tropical regions prevent tropical cyclones from forming in those regions.

Tropical cyclone paths tend to be westward initially, with a slight poleward drift, but may eventually move far enough from the equator to move into areas dominated by westerly winds, which may cause the path to change eastward. While tropical cyclones have strong winds and sometimes very extreme wind speeds and wave forcing, they are smaller (typically around 500 km radius to the outermost closed isobar) than extratropical storms (typically around 1000 km radius to the outermost closed isobar) and therefore have smaller fetches than extratropical cyclones. Thus, the waves forced by tropical cyclones are usually steep and extreme, but due to the smaller fetches, they may not produce as large sea states as an extratropical storm with lower wind speeds.

At high latitudes, small-scale ($\sim$1000 km), short-lived (typically a few days), but intense cyclones, called polar lows, can occur when cold polar air is advected over warmer water, which often happens at sea ice margins. Because their features resemble tropical cyclones, they are sometimes referred to as "Arctic hurricanes". Due to their small scale and short life, their wave-generating potential is relatively small by comparison with extratropical and tropical storms. They can, however, generate more substantial wave-fields in the region of the wind field that has the same direction as the direction of movement of the low [Dysthe and Harbitz, 1987], which has also been observed to occur in tropical cyclones. Even so, the wave heights associated with Polar Lows are not expected to reach the 100-year significant wave height values in the region that they exist [Rasmussen and Turner, 2003].

As a tropical cyclone moves poleward, it may leave the tropics, losing tropical cyclone characteristics and gaining extratropical storm characteristics, resulting in a new weather system that is a cross between a tropical cyclone and an extratropical cyclone.

This process is called extratropical transition (ET) and meteorologists often simply call the resulting storms ETs. With time, ETs will end up being mostly extratropical, but they retain some tropical traits, with the potential to generate large sea states. Some of Western Europe's worst extratropical storms were former ETs, which had traveled across the entire Atlantic Ocean.

As a result of the regional nature of the various storm types, extreme sea states are driven by different meteorological systems for different basins. In mid-latitude regions, such as the North Atlantic, the extreme sea states are driven by extratropical storms, which in some cases had transitioned from tropical cyclones. In tropical regions at latitudes greater than 5°, such as the Gulf of Mexico, tropical cyclones dominate the extreme conditions. In tropical regions at latitudes less than 5°, sea states are relatively benign, with extreme sea states being generated by monsoons, such as occurs in the south of the South China Sea, which tend to be a mix of windsea and swell from further up fetch in the Monsoon where the winds are stronger, or predominantly swell, such as occurs offshore the Niger delta, where the extreme sea states are due to swell from extratropical storms in the South Atlantic.

7.7. Uncertainties

7.7.1. *General*

Uncertainties can be classified as natural (inherent, or aleatory) or modeling (sampling, or epistemic) uncertainties. Aleatory uncertainty refers to the fundamental natural randomness of the phenomenon being considered and cannot be avoided or reduced by more measurements. An example of this is the variability of the sea state significant wave height over time and consequently the extreme quantile corresponding to the r-year return period value. Epistemic uncertainty refers to uncertainties associated with lack of knowledge, due to limitations in the number and quality of the data and inadequacies in the modeling of the process. Epistemic uncertainty can be reduced by increasing the sample size, improving the accuracy of the measurements, and improving the accuracy of the model.

An informative example of aleatory uncertainty is the distribution of the significant wave height r-year return period value. The r-year maximum significant wave height is a random variable, and even if the epistemic uncertainties could be reduced to zero, it would have a distribution given by an extreme value distribution, such as the GEV or GPD, which is skewed with a long tail towards large significant wave height values. The extent of the skewness and the rate at which the tail decays are dependent on the parameters of the distribution, for example, the ratio of the 95th percentile of the distribution of the r-year maximum to its 37th percentile varies with the shape parameter ξ of the GP used. This ratio is greater in Gulf of Mexico conditions, where $\xi \approx -0.1$, than in the northern North Sea conditions, where $\xi \approx -0.2$. Thus, for example, even if the 100-year return period value of the significant wave height were the same for the Gulf of Mexico and the northern North Sea, exceptionally large values would be expected more often in the Gulf than in the northern North Sea [Jonathan and Ewans, 2007b].

In practice, epistemic uncertainty cannot be reduced to zero, and it contributes also to the shape of the r-year return period value distribution. For example, a location for which large high-quality datasets are available can be expected to have a lower ratio of the 95th percentile of the distribution of the r-year maximum to its 37th percentile [e.g., Randel et al., 2015].

Approaches for estimating the epistemic uncertainties of extreme values are given in Section 7.7.2. The effect of climate change on wave climate extremes is an interesting example of a potential aleatory uncertainty that is difficult to resolve due to large epistemic uncertainties. Discussion on this follows in Section 7.7.3.

7.7.2. *Estimating uncertainties*

Asymptotic Methods: If the parameters of extreme value models are estimated using the maximum likelihood method, approximate confidence limits in the estimates can be obtained from asymptotic considerations (i.e., for large numbers of data). Asymptotically, parameter estimates are normally distributed. That is, $\hat{\theta}_i \sim N(\theta_i, \phi_i^2)$, where θ_i is the true parameter, $\hat{\theta}_i$ its estimate, and ϕ_i^2 a term bounded below by the (Fisher) information which can be estimated from the log-likelihood function. The $(1 - \alpha)$ confidence interval for θ_i is given by

$$\hat{\theta}_i \pm z_{\alpha/2}\sqrt{\hat{\phi}_i^2},$$

where z is standard normal and $\hat{\phi}_i^2$ is an estimate of ϕ_i^2 [Coles, 2001].

In a similar way, maximum likelihood estimate of the return value is asymptotically normally distributed, allowing confidence limits to be established using the so-called delta method [Coles, 2001; Jonathan and Ewans, 2007].

Bootstrapping Methods: Bootstrapping is a process of randomly resampling the original sample with replacement and using inferences on the resamples to quantify uncertainty. Large numbers of resamples can be taken, allowing a bootstrap distribution of an extreme value estimate to be established, from which confidence limits can be determined. The method is simple and asymptotically more accurate than the standard intervals based on the assumptions of normality [DiCiccio and Efron, 1996]. In addition, it can be used to estimate uncertainties for more complicated cases where for example data are not independent; for example, when site-pooling of data is used to improve the reliability of extreme wave height estimates in hurricane-dominated regions, since data are sparse for any single site. In this case, a block-bootstrapping approach can be used to estimate the uncertainties in the extremal estimate derived from site-pooling in which each draw consists of taking on individual from each of all the (spatial dependent) sites [Jonathan and Ewans, 2007].

Bayesian Methods: Bayesian inference is motivated by the premise that our knowledge or beliefs about the parameters of the probability distribution, for example, a GPD for peaks over threshold of storm peak significant wave heights say, can be expressed as a probability distribution, referred to as the prior distribution before we undertake any measurement. If $p_\Theta(\theta)$ is the density of the prior distribution of one or more parameters Θ, the likelihood of observing a sample of data $= \{x_1, x_2, x_3, \ldots, x_n\}$ is given by the density for the whole the sample given parameter value, $p_{X|\Theta}(x|\theta)$. For the case of a single parameter, Bayes theorem then states

$$p_{\Theta|X}(\theta|x) = \frac{p_{X|\Theta}(x|\theta)p_{\Theta}(\theta)}{\int_{\theta}' p_{X|\Theta}(x|\theta')p_{\Theta}(\theta')d\theta'}.$$

Now, $p_{\Theta|X}(\theta|x)$ is referred to as the posterior density and provides a description of our updated knowledge or beliefs about the parameter and its uncertainty. The denominator, referred to as the (Bayesian or model) evidence, quantifies the strength of our belief in the form of model we are fitting, and is useful in comparing different model forms. The posterior also enables prediction and modeling uncertainty to be estimated directly. For example, if q is an extreme quantile of interest, which can be expressed as a *deterministic* function $q(\theta)$ of parameter θ, and $p_{\Theta|X}(\theta|x)$ is the posterior distribution of the parameter given inference using observations x, then the posterior mean is given by

$$q_x = \int_{\theta} q(\theta)p(\theta|x)d\theta.$$

Here, q_x "integrates out" the uncertainty in the value of q due to the parameter, and can be viewed as our best estimate of q incorporating this uncertainty.

If M is (say) the maximum value observed in some period of interest, then M is a random variable (i.e., it exhibits aleatory uncertainty regardless of parameter Θ). It has probability density function $p_{M|\Theta}(m|\theta)$ which depends on the parameter. In this case, we can estimate the posterior predictive density of M given x using

$$p_{M|X}(m|x) = \int_{\theta} p_{M|\Theta}(m|\theta)p(\theta|x)d\theta.$$

Here, $p_{M|X}(m|x)$ represents our best estimate of the distribution of M incorporating the uncertainty in the parameter.

A recent example of the application of Bayesian inference, to quantify the joint directional and seasonal variation of storm peak significant wave height at a northern North Sea location and estimate predictive directional–seasonal return value distributions, is given by

Randell et al. [2016]. Hansen et al. (2019) present a model to estimate joint extremal characteristics of the ocean environment incorporating non-stationary marginal and conditional extreme value analysis, and thorough uncertainty quantification, within a fully Bayesian framework.

7.7.3. *Climate change*

The effect of climate change on the occurrence and intensity of extreme sea states is a hot topic, but the uncertainties associated with attempts to predict the effect are large. The shortness and variable quality of appropriate historical databases from which predictions might be made is one source of uncertainty, and this will continue to hamper attempts to identify long-term trends in historical wave climate extremes. In addition, even if a trend in the historical can be detected unequivocally, it cannot be assumed that this trend will continue. The use of future climate prediction data provides the opportunity to estimate the effects of different emission scenarios, but large uncertainties are currently associated with sea state extreme predictions from these data.

General circulation models (GCMs) used to make global climate predictions are typically run at relatively coarse spatial grid sizes of the order of 100 km and do not include estimates of sea state parameters. Thus, predictions of wave fields from GCMs involve downscaling from the coarse GCM grid parameter fields, usually sea level pressure or wind fields. There are two principal methods for downscaling: dynamical and statistical. Dynamical downscaling involves running numerical wave prediction models using the GCM data as boundary conditions. This approach is computationally expensive, and there are few examples where attempts have been made to estimate wave extremes. Grabemann and Weisse [2008] simulated sea states in the North Sea for the 30-year period 2071–2100 using the wave model WAM driven by wind fields from two GCMs. They considered 99th percentiles of significant wave height and concluded a moderate increase of the most severe wave conditions in the North Sea towards the end of the 21st century but noted a rather large uncertainty related to differences among the climate models they considered. Similarly, Hemer et al. [2013a] made global sea state predictions for the period 2070–2099 using the wave model WAVEWATCH III® from two GCMs, downscaled to around 50 km spatial resolution. Although they considered the 99th percentile of significant wave height, conclusions were made only about mean values. Mori et al. [2010] produced predictions with the SWAN wave model from global climate projections from the 20 km resolution Japanese Meteorological Research Institute and Japan Meteorological Agency (MRI/JMA) GCM. They observed increases in significant wave height percentiles greater than the 99th percentile for the Southern Indian Ocean and the Pacific south of Japan but essentially no change east of Japan and for the Antarctic Ocean. Focusing on the northeast Atlantic, Aarnes et al. [2017] found differences in the tendency of the upper tail of the distribution of significant wave height, depending on which of the two emission scenarios (RCP4.5 and RCP8.5) is employed in CMIP5 wind models used to drive the WAM wave model. The annual maxima from the runs with the RCP4.5 emission scenario showed little to no significant change by comparison with the historical period, but the 10- and 20-year return period values showed an increase west of the British Isles. The runs from the RCP8.5 emission scenario indicated a significant decrease in the annual maximum significant wave heights but essentially no change in the 10- and 20-year return period values. Similar results were found by Gleeson et al. [2017], who found projections using the EC-Earth GCM of extreme significant wave height values were location-dependent amplified by comparison with historical values in some regions but lower in others. They also found a dependence of the extremes on the North Atlantic Oscillation and differences between the RCP4.5 and RCP8.5 emission scenarios but noted uncertainty in the projections of extremes.

Examples in which statistical downscaling has been used are Wang et al. [2014] and Towe et al. [2017]. Wang et al. [2014] made statistical projections of changes in ocean wave heights using sea-level pressure from 20 CMIP5 GCMs for the 21st century. They predicted, for the RCP8.5, emission scenario that projected 2070–2099 10-year extreme wave heights would likely double or triple in several coastal regions around the world, compared with values estimated for the period 1986–2005. Towe et al. [2017] developed a multi-step statistical downscaling algorithm, employing the conditional extremes model discussed in Section 7.5.2, in using data from a low resolution GCM and local-scale hindcast data to make predictions of the extreme wave climate in the next 50-year period at locations in the North Sea. Their projections of the extreme wave climate of the North Sea indicated that the sizes of the largest waves as well as their directions are changing within the lifetime of current offshore designs, but the uncertainties were too large to draw conclusions about return levels of 100 years or more.

The uncertainties in these projections are difficult to quantify, but they can be expected to be large. Hemer et al. [2013b] considered the uncertainties in the estimates of sea state parameter in five studies results contributed to the Coordinated Ocean Wave Climate Project (COWCLIP) — four involved dynamical downscaling and one statistical downscaling. They concluded that a "broad range of uncertainty" surrounds the projections, dominated by uncertainties in downscaling methodologies. Hemer et al. [2016] showed that the skill of the wave predictions made with the GCMs varied with region and that they have greater skill in representing mean conditions than extremes. They also found little to no improvement in the wave predictions using a higher resolution CMIP5 GCM (with resolution of ~150 km) compared with a lower resolution GCM (~300 km), leading to their recommendation of the need to evaluate the performance of a GCM for the marine meteorological climate independently of the performance of the GCM for the "standard" climate variables such as surface temperature and precipitation.

Wang et al. [2015] used analysis of variance approaches to quantify the climate change signal and uncertainty in multimodel ensembles of statistical simulations of significant wave height based on the CMIP5 historical, RCP4.5 and RCP8.5 forcing scenario simulations of sea level pressure. They found increases that were statistically significant from zero in the annual maximum significant wave height over increasing areas of the ocean over the period

2005–2099, with the strongest signal in the eastern tropical Pacific. They also noted model uncertainty (differences between GCMs) to be about 10 times as large as the scenario uncertainty between RCP4.5 and RCP8.5 scenarios. The scenario uncertainty was found to be largest in the eastern tropical Pacific, suggesting that wave heights in the eastern tropical Pacific are most sensitive to climate change.

A good overview of the uncertainties associated with wave climate projections is given by Martínez-Asensio et al. [2016]. They also compared a dynamical wave climate simulation for the entire 21st century for the North Atlantic Ocean with a set of statistical projections using atmospheric variables or large-scale climate indices as predictors. They found that statistical models that use wind speed as the independent variable predictor can capture a larger fraction of the long-term significant wave height changes projected by the dynamical simulation. Conversely, regression models using climate indices, sea level pressure and/or pressure gradient as predictors, do not reproduce the dynamically projected long-term trends in significant wave height over the North Atlantic.

The influence of climate change on extreme wave conditions is an active area of research, but despite this attention, more understanding is needed to characterize behavior more quantitatively. Studies indicate that climate change projections based of various GCMs and different emission scenarios suggest increases extreme sea states in some parts of the World oceans and not in others, but uncertainties associated with both GCMs and scenarios are presently large, hindering progress.

References

Alves, J. and Young, I. (2003). On estimating extreme wave heights using 297 combined Geosat, Topex/Poseidon and ERS-1 altimeter data, *Appl. Ocean Res.*, **25**, 167–186.

Anderson, C.W., Carter, D.J.T. and Cotton, P.D. (2001). Wave climate variability and impact on offshore design extremes, Report Commissioned from the University of Shefield and Satellite Observing Systems for Shell International.

DiCiccio, T.J. and Efron, B. (1996). Bootstrap confidence intervals, *Stat. Sci.*, **11**, 189–228.

Breivik, Ø., Aarnes, O., Abdalla, S., Bidlot, J. and Janssen, P. (2014). Wind and wave extremes over the world oceans from very large ensembles, *Geophys. Res. Lett.*, **41**, 5122–5131.

Blake, E., Kimberlain, T., Berg, R., Cangialosi, J. and Beven, J. (2013). Tropical cyclone report hurricane Sandy (AL182012), 22–29 October 2012, National Hurricane Center, NOAA Report.

Caires, S. and Sterl, A. (2005). 100-year return value estimates for ocean wind speed and significant wave height from the ERA-40 data, *J. Climate* **18**, 1032–1048.

Cardone, V., Cox, A., Morrone, M. and Swail, V. (2011). Global distribution and associated synoptic climatology of Very Extreme Sea States (VESS), in *12th Int. Workshop on Wave Hindcasting and Forecasting*, Hawaii, October 31–November 4, 2011.

Cardone, V., Callahan, B., Chen, H., Cox, A., Morrone, M. and Swail, V. (2014). Global distribution and risk to shipping of very extreme sea states (VESS), *Int. J. Climatol.*, **35**, 69–84.

Coles, S. (2001). *An Introduction to Statistical Modelling of Extreme Values*, Springer.

Cox, A. and Swail, V. (2001). A global wave hindcast over the period 1958–1997: Validation and climate assessment, *J. Geophys. Res.: Oceans*, **106**(C2), 2313–2329.

Cruz, A. and Krausmann, E. (2008). Damage to offshore oil and gas facilities following hurricanes Katrina and Rita: An overview, *J. Loss Prevent Proc.*, **21**, 620–626.

Dobie, G., Whitehead, J., Kidston, H. and Collins, S. (2016). Safety and shipping review 2017, Allianz Global Corporate & Specialty Report.

Dysthe, K. and Harbitz, A. (1987). Big waves from polar lows? *Tellus A*, **39**(5), 500–508.

Eilers, P. and Marx, B. (1996). Flexible smoothing using B-splines and penalized likelihood (with Comments and Rejoinder), *Stat. Sci.*, **11**, 89–121.

Embrechts, P., Klueppelberg, C. and Mikosch, T. (2003). *Modelling Extremal Events for Insurance and Finance*, Springer-Verlag.

Ewans, K.C. (2003). A response-based method for developing joint metocean criteria for on-bottom pipeline stability, *J. Offshore Mech. Arct. Eng.*, **125**, 119–125.

Ewans, K.C. and Jonathan, P. (2013). Evaluating environmental joint extremes for the offshore industry using the conditional extremes model. *Journal of Marine Systems*, **130**, 134–130.

Ewans, K.C. and McConochie, J. (2018). On the uncertainty of estimating JONSWAP spectrum peak parameters. OMAE2018-78386, *Proc. ASME 2018 37th Int. Conf. Ocean, Offshore and Arctic Engineering*, June 17–22, 2018, Madrid, Spain.

Feld, G., Randell, D., Wu, Y., Ewans, K. and Jonathan, P. (2015). Estimation of storm peak and intra-storm directional-seasonal design conditions in the North Sea, *J. Offshore Mech. Arct. Eng.*, **137**, 021102-1–021102-15.

Forristall, G.Z. (2000). Wave crest distributions: Observations and second order theory, *J. Phys. Oceanogr.*, **30**, 1931–1943.

Forristall, G.Z. (2004). On the use of directional wave criteria, *J. Waterw. Port C-ASCE*, **130**, 272–275.

Forristall, G.Z. (2008). How should we combine long and short term wave height distributions? OMAE2008-58012. in *Proc. ASME 2008 27th Int. Conf. Ocean, Offshore and Arctic Engineering*, June 15–20, 2008, Estoril, Portugal.

Gleeson, E., Gallagher, S., Clancy, C. and Dias, F. (2017). NAO and extreme ocean states in the Northeast Atlantic Ocean, *Adv. Sci. Res.*, **14**, 23–33.

Grabemann, I. and Weisse, R. (2008). Climate change impact on extreme wave conditions in the North Sea: An ensemble study, *Ocean Dynam.*, **58**, 199–212.

Hansen, H., Randell, D., Zeeberg, A., and Jonathan, P. (2019). Directional-seasonal extreme value analysis of North Sea storm conditions. *Ocean Engineering* (In Press.)

Haver, S. and Nyhus, K. (1986). A wave climate description for long term response calculations, *Proc. 5th OMAE Symp.*, Vol. IV, pp. 27–34.

Heffernan, J. and Tawn, J. (2004). A conditional approach for multivariate extreme values, *J. R. Stat. Soc. B*, **66**, 497.

Hemer, M.A., Katzfey, J. and Trenham, C. (2013a). Global dynamical projections of surface ocean wave climate for a future high greenhouse gas emission scenario. *Ocean Model.*, (Special Issue on Surface Ocean Waves), pp. 221–245.

Hemer, M.A., Fan, Y., Mori, N., Semedo, A. and Wang, X.L. (2013b). Projected changes in wind–wave climate in a multi-model ensemble, *Nat. Clim. Change*, **3**, 471–476.

Hemer, M.A. and Trenham, C.E. (2016). Evaluation of a CMIP5 derived dynamical global wind-wave climate model ensemble, *Ocean Model.*, **103**, 190–203.

Izaguirre, C., Méndez, F.J., Menéndez, M. and Losada, I.J. (2011). Global extreme wave height variability based on satellite data, *Geophys. Res. Lett.*, **38**, L10607.

Jonathan, P. and Ewans, K.C. (2007a). The effect of directionality on extreme wave design criteria, *Ocean Eng.*, **34**, 1977–1994.

Jonathan, P. and Ewans, K.C. (2007b). Uncertainties in extreme wave height estimates for hurricane dominated regions, *J. Offshore Mech. Arct. Eng.*, **129**, 300–305.

Jonathan, P., Ewans, K.C. and Forristall, G.Z. (2008). Statistical estimation of extreme ocean environments: The requirement for modelling directionality and other covariate effects, *Ocean Eng.*, **35**, 1211–1225.

Jonathan, P., Flynn, J. and Ewans, K. (2010). Joint modelling of wave spectral parameters for extreme sea states. *Ocean Eng.*, **37**, 1070–1080.

Jonathan, P., Ewans, K. and Randell, D. (2013). Joint modelling of extreme ocean environments incorporating covariate effects. *Coast. Eng.*, **79**, 22–31.

Jonathan, P., Randell, D., Wu, Y. and Ewans, K. (2014). Return level estimation from non-stationary spatial data exhibiting multidimensional covariate effects. *Ocean Eng.*, **88**, 520–532.

Katsardi, V., de Luto, L. and Swan, C. (2013). An experimental study of large waves in intermediate and shallow water depths. Part I: Wave height and crest height statistics. *Coastal Engineering*, **73**, 43–57.

Lopatoukhin, L.J., Rozhkov, V.A., Ryabinin, V.E., Swail, V.R., Boukhanovsky, A.V. and Degtyarev, A.B. (2000). Estimation of extreme wind-wave heights, WMO/TD-No. 1041, JCOMM Technical Report No. 9.

Marx, B.D. and Eilers, P.H.C. (1998). Direct generalised additive modelling with penalised likelihood, *Comput. Stat. Data Anal.*, **28**, 193–209.

Méndez, F., Menéndez, M., Luceño, A. and Losada, I. (2006). Estimation of the long-term variability of extreme significant wave height using a time-dependent POT model, *J. Geophys. Res.: Oceans*, **111**, C07024.

Mori, N., Yasuda, T., Mase, H., Tom, T. and Oku, Y. (2010). Projection of extreme wave climate change under the global warming, *Hydrological Res. Lett.*, **4**, 15–19.

Nerzic, R., Frelin, C., Prevosto, M. and Quiniou-Ramus, V. (2007). Joint distributions of wind, waves and current in West Africa and derivation of multivariate extreme i-form contours, in *Proc. 17th Int. Offshore and Polar Engineering Conference*, Lisbon, Portugal, July 1–6.

Oceanweather Inc. (2018). GOMOS2017: Gulf of Mexico Oceanographic Study Project Description, Oceanweather Inc. May 2018.

Randell, D., Wu, Y., Jonathan, P. and Ewans, K. (2013). Modelling covariate effects in extremes of storm severity on the Australian north west shelf, in *Proc. ASME 2013 32nd Int. Conf. Ocean, Offshore and Arctic Engineering*, OMAE2013-10187, June 9–14, 2013, Nantes, France.

Randell, D., Feld, G., Ewans, K. and Jonathan, P. (2015). Distributions of return values for ocean wave characteristics in the South China Sea using directional–seasonal extreme value analysis, *Environmetrics*, **26**, 442–450.

Randell, D., Turnbull, K., Ewans, K. and Jonathan, P. (2016). Bayesian inference for nonstationary marginal extremes. *Environmetrics*, **27**, 439–450.

Rasmussen, E. and Turner, J. (2003). *Polar Lows Mesoscale Weather Systems in the Polar Regions*. Cambridge University Press, Cambridge.

Ross, E., Astrup, O., Bitner-Gregersen, E., Bunn, N., Feld, G., Gouldby, B., Huseby, A., Liu, Y., Randell, D., Vanem, E., and Jonathan, P. (2019). On environmental contours for marine and coastal design. *Ocean Engineering*, doi.org/10.1016/j.oceaneng.2019.106194.

Saulnier, J. (2013). Uncertainty in peakedness factor estimation by JONSWAP spectral fitting from measurements. OMAE2013-10004, in *Proc. ASME 2013 32nd Int. Conf. Ocean, Offshore and Arctic Engineering*, June 9–14, 2013, Nantes, France.

Tayfun, M.A. (2011). On the distribution of large wave heights: Nonlinear effects, in *Marine Technology and Engineering*, eds. Guedes Soares *et al.*, Taylor & Francis Group, London.

Tromans, P. and Vanderschuren, L. (1995). Risk based design conditions in the North Sea: Application of a new method, in *Offshore Technology Conf.*, Houston, OTC-7683.

Tucker, M.J. and Pitt, E.G. (2001). *Waves in Ocean Engineering*, Elsevier.

Vinoth, J. and Young, I.R. (2011). Investigation of trends in extreme wave height and wind speed, *J. Climate*, **24**, 1647–1665.

Wang, X. and Swail, V. (2006). Historical and possible future changes of wave heights in northern hemisphere oceans, *WIT Trans. State Art Sci. Eng.*, **23**, 1755–8336.

Wang, X.L., Feng, Y. and Swail, V.R. (2014). Changes in global ocean wave heights as projected using multimodel CMIP5 simulations, *Geophys. Res. Lett.*, **41**, 1026–1034.

Wang, X.L., Feng, Y. and Swail, V.R. (2015). Climate change signal and uncertainty in CMIP5-based projections of global ocean surface wave heights, *J. Geophys. Res.: Oceans*, **120**, 3859–3871.

Winterstein, S., Ude, T., Cornell, C., Bjerager, P. and Haver, S. (1993). Environmental parameters for extreme response: inverse FORM with omission factors, in *Proc. 6th Int. Conf. Structural Safety and Reliability*, Innsbruck, Austria.

Young, I.R. (1986). Probability distributions of spectral integrals. *J. Waterw. Port. C-ASCE*, **112**, 338–341.

Young, I.R. (1994). Global ocean wave statistics obtained from satellite observations, *Appl. Ocean Res.*, **16**, 235–248.

Young, I.R. (1995). The determination of confidence limits associated with estimates of the spectral peak frequency, *Ocean Eng.*, **22**, 669–686.

Young, I.R. (1999). Seasonal variability of the global ocean wind and wave climate, *Int. J. Climatol.*, **19**, 931–950.

Chapter 8
Satellite Observations and Climate

Ian Young

University of Melbourne, Parkville VIC 3010, Australia

8.1. Introduction

An understanding of global ocean wind and wave climate is critical for many reasons, including: offshore and coastal design of structures, the stability of beaches, prediction of coastal inundation, ship routing, air–sea interaction and climate change. A comprehensive understanding of the global wind and wave climate can only be obtained from two sources: numerical modeling and satellite remote sensing. Global phase-averaging spectral models such as WAM [Komen *et al.*, 1994] and WAVEWATCH [Tolman, 2011] (see Chapter 3) have a remarkable degree of skill and have consistently shown their ability to generate long-duration reanalyzes such as ERA-Interim [Dee *et al.*, 2011], ERA-5 [C3S, 2017] and GOW2 [Perez *et al.*, 2017]. Despite the remarkable performance of such models, direct measurements are still critically important and here Satellite Remote Sensing plays a critical role.

The satellite record of observations of wind speed and wave height commenced in 1985 and now spans more than 30 years [Young and Ribal, 2019; Ribal and Young, 2019]. Over this period, three instrument types have provided global coverage: altimeter, radiometer and scatterometer. Altimeter measures both wind speed and wave height, radiometer measures wind speed and scatterometer measures wind speed and direction. In addition to these three systems, Synthetic Aperture Radar (SAR) has also been operational, providing the potential to measure the full directional wave spectrum [Hasselmann et al., 1996]. However, as SAR does not provide global coverage, it is not considered here.

8.2. Satellite Measurement Basics

Each of the instruments considered here (altimeters, radiometers, scatterometers) measure wind speed and wave height as a result of the way electromagnetic radiation interacts with the wavy or wind-ruffled water surface. Essentially, these are indirect measurements of the quantities under consideration (wave height, wind speed, wind direction). As such, theoretic considerations are important, but in each case, validation or calibration against buoy data becomes an important element of algorithm development.

8.2.1. *Altimeter*

The altimeter (ALT) is an active instrument, transmitting radar pulses which are averaged to provide measurements of significant wave height, H_s, and wind speed (at 10 m height and averaged over 10 min), U_{10}. The footprint of the ALT varies between 8 km and 10 km and data are provided approximately every 1 s (approx. 10 km) along the satellite track. ALT missions have been placed in a variety of near-polar orbits. Depending on the details of the orbit, the satellite will re-trace its ground tracks after a period between 5 and 20 days. This duration is termed an exact repeat mission (ERM). The ERM defines the ground track separation, with a long ERM corresponding to a relatively small ground track separation. The ground track separation decreases with increasing latitude. At the Equator, values range between 100 km and 400 km. Thus, the ALT has relatively high resolution along track (10 km) and low resolution across track (100–400 km). This low cross-track resolution, together with the long ERM, means that, although the ALT provides global coverage, it is possible that storms may be under-sampled or completely missed.

Based on the shape and intensity of the returned radar signal from the footprint, the ALT can estimate the significant wave height, H_s, and wind speed, U_{10} [Cheney et al., 1987; Walker, 1995; Chelton

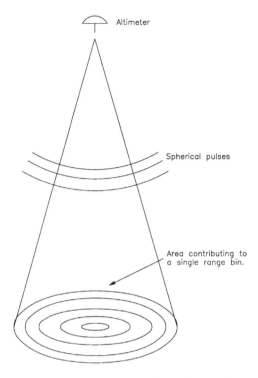

Fig. 8.1. Spherical altimeter pulses radiating from the satellite antenna and areas illuminated in successive range gates [After Young, 1999].

et al., 2001; Queffeulou, 2004; Young, 1994, 1999; Zieger et al., 2009]. Figure 8.1 shows the ground area imaged by the ALT radar pulse.

The significant wave height H_s can be determined from the slope of the leading edge of the radar return [Chelton et al., 2001; Holthuijsen, 2007]. This is based on the assumption that a calm sea would act like a mirror and the return pulse would approximate a "square wave" (i.e., near vertical leading edge to the return signal), whereas a "wavy" surface will result in a leading edge, the slope of which decreases with increasing H_s. Figure 8.2 shows typical examples of the return pulse for both calm and rough seas.

The wind speed, U_{10}, is determined from the ratio of the intensity of the incident to the reflected radar energy (radar cross-section, σ_0) [Chelton et al., 2001]. The relationship between U_{10} and σ_0 is nonlinear, with σ_0 decreasing as U_{10} increases. The assumption is that the

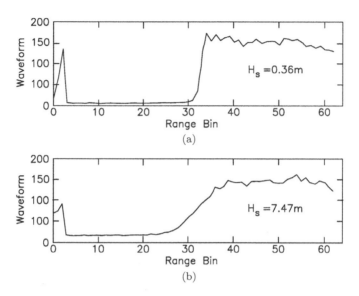

Fig. 8.2. Examples of the one second average return pulse for the GEOSAT altimeter for (a) $H_s = 0.36$ m and (b) $H_s = 7.47$ m. [After Young, 1999]

high wave number components of the surface wave spectrum respond almost immediately to the local wind, and that these relatively steep, short wavelength components act as scatters, increasingly scattering radar energy as it increases. Figure 8.3 shows a typical relationship between U_{10} and σ_0.

The exact quantity related to the high wave number spectrum upon which U_{10} depends is not known, although it is reasonable to assume that the slope of the high wave number components plays an important role. Hwang et al. [1998] and Plant [2002] provided a detailed analysis showing that the ALT responds to surface tilting slopes approximately 3–5 times longer than the electromagnetic radiation.

8.2.2. Radiometer

In contrast to the altimeter, the radiometer (RAD) is a passive instrument, measuring the emissivity of the ocean surface at a number of frequencies, this quantity being characterized by the brightness temperature, T_B. While the ALT is a nadir-looking instrument,

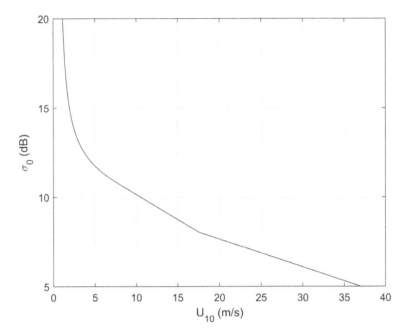

Fig. 8.3. Relationship between radar cross-section, σ_0 and wind speed, U_{10} for Ku band altimeters. For $U_{10} < 18\,\mathrm{ms}^{-1}$ the relationship is given by Abdalla [2007] and for $U_{10} \geq 18\,\mathrm{ms}^{-1}$ by Young [1993].

measuring along a line directly below the satellite, the RAD measures over a broad swath of approximately 1400 km width, as shown in Fig. 8.4. The resolution of the measurements within the swath is 25 km (both across and along the track). As such, a single RAD mission will image almost the full globe twice per day.

Wentz [1983, 1992, 1997] showed that there is a relationship between the brightness temperature, T_B, and the three geophysical quantities: near surface wind speed, U (ms^{-1}), columnar water vapor, V (mm), and columnar cloud liquid water, L (mm). The relationship between these quantities is defined by a radiative transfer equation, which has been expressed by Wentz [1983] in a closed form. As the radiometer measures at a number of frequencies, there is sufficient data to solve the radiative transfer equation for each of these quantities. Although the primary dependence of T_B is on the three parameters above, there is also a secondary dependence on: sea-surface

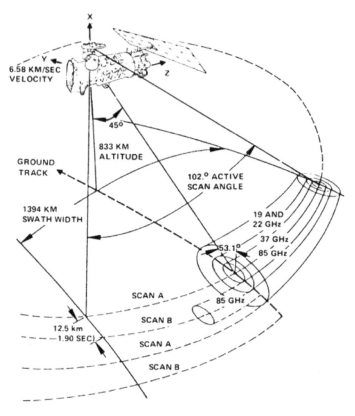

Fig. 8.4. The scan geometry of a radiometer instrument, showing the approximately 1400 km wide swath and the fact that the instrument senses emissions at multiple frequencies. [After Hollinger et al., 1990]

temperature, T_S (°K), effective atmospheric temperature, T_E (°K), effective atmospheric pressure, P (hPa) of the water column, and the wind direction, ϕ. The first three of these quantities are determined from climatological values and ϕ is included as a fourth quantity to be obtained in the solution of the radiative transfer equation. As with ALT data, the model is calibrated against buoy data [Wentz, 1997]. It should be noted that the signal is degraded during rain events and hence the above analysis is strictly applicable in non-rain conditions.

8.2.3. Scatterometer

Like the ALT, the scatterometer (SCAT) is an active instrument, transmitting radar energy which is reflected from the ocean surface by high-frequency Bragg scattering ocean waves. These high-frequency waves are assumed to be in equilibrium with the local wind and hence contain information on both the wind speed and direction. The radar cross-section, σ_0, can be related through a geophysical model function (GMF) as [Stoffelen, 1998; Donneley et al., 1999] follows:

$$\sigma_0 = \text{GMF}(U_{10}, \phi, \theta_S, p_S, \lambda_S), \qquad (8.1)$$

where U_{10} is the wind speed, ϕ the wind direction, θ_P the incidence angle, p_S the polarization, and λ_P the wavelength of the microwave radiation. Scatterometers often have a circular rotating antenna such that they can image a point from multiple angles, using multiple beams and with radiation of multiple polarizations. A typical example of the sea surface imaged by a rotating beam SCAT is shown in Fig. 8.5. As with the ALT, the SCAT images over a wide swath

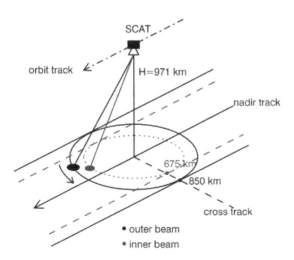

Fig. 8.5. The scan geometry of a rotating bean scatterometer, showing the swath width and a location imaged by multiple beams. As the scatterometer flies past the location, it will again image the location by the circular scan on the rear-side of the circle.

and with resolution of approximately 25 km. Of the three instruments considered here, the SCAT is the only platform which can measure wind direction. In contrast to the RAD, the SCAT can measure through rain. As a result, scatterometers have often been used to measure wind vectors during tropical cyclones.

8.3. Global Satellite Dataset and Calibration

The combined dataset from these three platforms (ALT, RAD and SCAT) now spans 34 years (1985–2019), as shown in Fig. 8.6. A number of studies have attempted to combine some of these missions and calibrate the systems in a consistent manner [Zieger *et al.*, 2009; Young *et al.*, 2017; Ribal and Young, 2019]. Such calibration approaches use a large database of ocean buoy measurements of wind speed and wave height for the calibration process. The desire is to form a co-location database of coincident observations from the buoys and the satellites under consideration. Zieger *et al.* [2009] assumed that if the satellite passed within 50 km of a buoy and 30 min of the buoy sampling then a calibration matchup occurred. The extensive buoy database used by Young *et al.* [2017] is shown in Fig. 8.7 and an example of the resulting calibration shown in Fig. 8.8.

Based on these buoy co-locations, Young *et al.* [2017] and Ribal and Young [2019] have developed linear least squares calibrations, which can be applied to the raw data distributed by the respective satellite data suppliers. As the desire is to obtain a combined dataset which is consistent over time, these multi-platform calibrations also examine the consistency of these calibrations as a function of time. This is intended to address issues where there is a change in calibration or drift in the calibration. This has been achieved by examining the difference between buoy and satellite measurements of U_{10} and H_s as a function of time. A typical example of such a comparison, showing a discontinuity in the calibration is shown in Fig. 8.9. When such changes in calibration were identified, they were removed by undertaking a piece-wise calibration with different calibration relations either side of the observed discontinuity.

As one of the common applications of satellite data is for the investigations of extreme wind speeds and wave heights, it is desirable

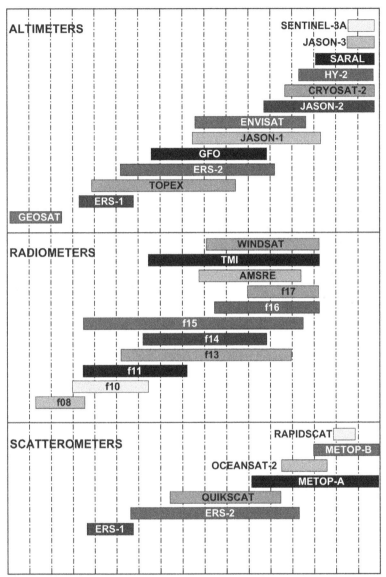

Fig. 8.6. Satellite missions from altimeter, radiometer and scatterometer which have been operational since 1985.

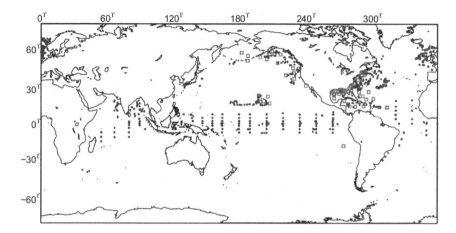

Fig. 8.7. Buoy co-location points used for satellite calibration. [After Young *et al.*, 2017]

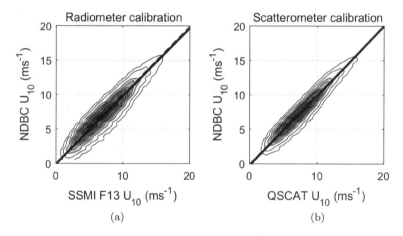

Fig. 8.8. An example of a match-up calibration between buoy and (a) radiometer and (b) scatterometer. [After Young *et al.*, 2017]

that the calibrations include information about the performance of the various satellite systems at high wind speeds and wave heights. Zieger *et al.* [2009], Young *et al.* [2017] and Ribal and Young [2019] all examined the calibrations at extreme values using Q–Q plots and, where necessary, applied corrections for these values. A typical such plot for RAD data is shown in Fig. 8.10.

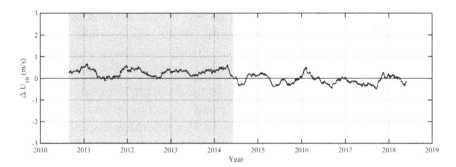

Fig. 8.9. Difference in wind speed ΔU_{10} between buoy data and CRYOSAT-2 altimeter. A discontinuity is clear around mid-2014. In this case, separate calibration relations were applied either side of the discontinuity. [After Ribal and Young, 2019]

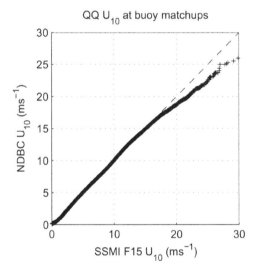

Fig. 8.10. Q–Q plot between the f15 radiometer and buoy data, showing the radiometer measures higher than the buoys above approximately $18\,\mathrm{ms}^{-1}$. [After Young et al., 2017]

Finally, to ensure the calibrations are consistent, two approaches have been applied. Young et al. [2017] used two independent buoy datasets. Calibration was undertaken against the extensive NDBC dataset [Evans et al., 2003]. An independent buoy dataset was then used for validation. The second approach to ensure consistency is to examine match-ups between satellites. Again, co-locations are considered to occur when satellite ground track cross-overs within 50 km occur within 30 min of the respective satellite passes. Such satellite–satellite match-up datasets tend to be much larger than the buoy calibration datasets and have much greater geographic coverage. Such satellite–satellite validations have been undertaken for altimeter–altimeter, altimeter–radiometer and altimeter–scatterometer combinations. A typical example of such satellite–satellite validations is shown in Fig. 8.11.

The resulting multi-platform datasets provide the basis for the investigations of a range of properties of global wind speed and wave height. The calibration studies have, however, demonstrated a number of aspects of the performance of these various systems. These issues include the following.

- The ALT datasets are much smaller than RAD of SCAT (factor of 20 smaller). This occurs because of the sampling pattern of the ALT (nadir pencil beam). As a result, ALT data may under-sample some meteorological systems.
- Both RAD and some SCAT systems underestimate U_{10} compared to buoys and platform data at high wind speeds. Corrections have been applied for such situations.
- RAD data cannot measure wind speeds during heavy rain. As a result, the data has a significant "fair-weather" bias.
- All three systems measure U_{10} indirectly, by inferring wind speed due to the impact on high-frequency components of the surface wave spectrum. Young and Donelan [2018] have shown that, as a result, they effectively measure wind speed as a height generally less than the reference height of 10 m. For neutral atmospheric boundary layers, this is not an issue, as they have all been calibrated against anemometers referenced to a height of 10 m. The marine boundary layer is, however, seldom neutral and due to differences in air and water temperatures unstable boundary layers (water warmer than air) are very common. In such situations, the shape of the boundary layer is different to the neutral (logarithmic)

Satellite Observations and Climate 333

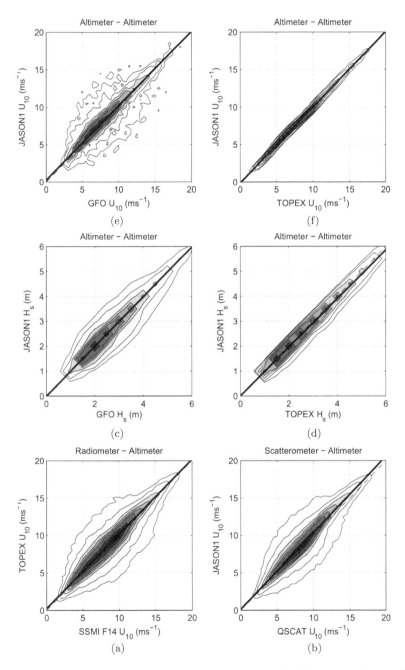

Fig. 8.11. Satellite–satellite validations for a number of different satellite combinations. [After Young et al., 2017]

form assumed for calibration purposes. These effects can result in errors up to 3% in measurements of wind speed obtained by such satellite systems.

8.4. Seasonal Variations in Ocean Wind and Wave Climate

The global datasets described above offer the potential to investigate seasonal variations in the global wind and wave climate. Such analyses have been undertaken by Young [1999] and Young and Donelan [2018]. Initially, the data from the various combined datasets is binned into $2° \times 2°$ bins. Mean monthly and 90th percentile monthly values can then be evaluated for each $2°$ bin. The results are shown below for the altimeter as color shaded global plots in Figs. 8.12–8.15.

Mean monthly U_{10} for January is shown in Fig. 8.12(a) and monthly 90th percentile U_{10} for January in Fig. 8.12(b). January corresponds to the northern hemisphere winter and mean monthly U_{10} in the North Pacific and North Atlantic reach values of $13\,\mathrm{ms}^{-1}$. Although in the southern hemisphere it is summer, the strong westerlies of the Southern Ocean are still clear, reaching values as large as $11\,\mathrm{ms}^{-1}$. Considerable zonal detail is visible in Fig. 8.12(a), corresponding to the trade wind belts in both hemispheres. The 90th percentile values (Fig. 8.12(b)) again show the strongest winds in the northern hemisphere, reaching values of $18\,\mathrm{ms}^{-1}$ in the North Atlantic. Values of the 90th percentile U_{10} in the Southern Ocean are again significant, reaching $12\,\mathrm{ms}^{-1}$. For both the mean and 90th percentile, there is a maximum in the Southern Ocean between South Africa and Australia. The wind climate across the Southern Ocean is, however, spatially quite uniform.

Mean and 90th percentile U_{10} for July are shown in Figs. 8.13(a) and 8.13(b), respectively. As July now corresponds to the southern hemisphere winter, the strongest winds are now in the Southern Ocean. Both the mean and 90th percentile maximum wind speeds in the southern hemisphere are comparable to the values seen in the northern hemisphere winter (Fig. 8.12(a)). In contrast to the southern hemisphere, however, where summer mean and 90th percentile

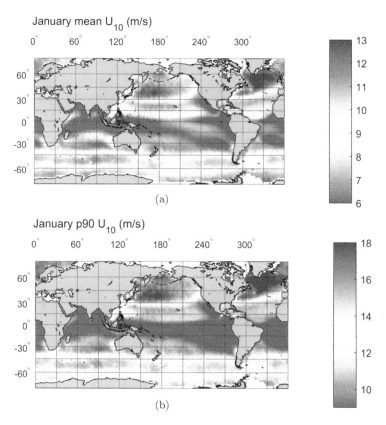

Fig. 8.12. Global wind speed, U_{10} during the month of January from ALT data [unit: ms^{-1}]. (a) Mean monthly wind speed (b) 90th percentile wind speed.

values were still large, the northern hemisphere is quite calm in summer. The variation in wind speed between summer and winter in the northern hemisphere is far larger than the southern hemisphere, where the Southern Ocean is windy year-round.

As noted for January (Fig. 8.12(a)), the July mean U_{10} plot (Fig. 8.12(b)) also shows clear trade wind belts across many of the major oceanic basins. Also clear in Figs. 8.12(a) and 8.12(b) is the region of strong winds east of the Horn of Africa. This is the well-known Somali Jet, associated with the onset of the Asian Monsoon.

The corresponding monthly values for significant wave height H_s are shown in Fig. 8.14 for January and Fig. 8.15 for July. The seasonal

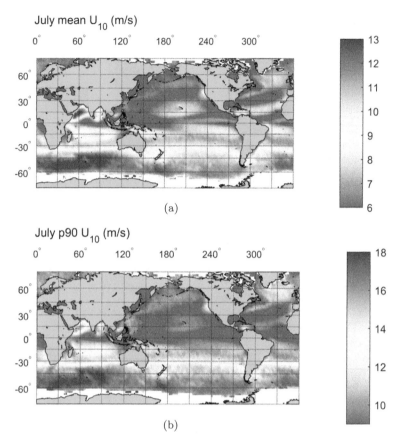

Fig. 8.13. Global wind speed, U_{10} during the month of July from ALT data [unit: ms^{-1}]. (a) Mean monthly wind speed (b) 90th percentile wind speed.

cycle at high latitudes largely follows the observations for U_{10}. The magnitudes of the monthly values are comparable between the hemispheres during their respective winters (5 m for mean monthly values and 7 m for the monthly 90th percentiles). Also, as for the wind speed, the Southern Ocean is relatively rough year-round. In contrast, the North Pacific and North Atlantic have a large variation in H_s from summer to winter. The Somali Jet is also clear for both mean and 90th percentile values during July (Fig. 8.15).

The striking difference between the U_{10} and H_s spatial distributions is the lack of trade wind belts in the wave height plots (compare

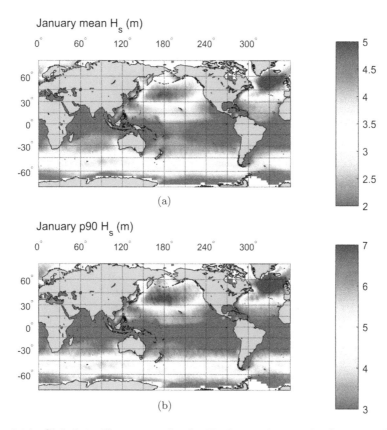

Fig. 8.14. Global significant wave height H_s during the month of January from ALT data [unit: m]. (a) Mean monthly significant wave height (b) 90th percentile significant wave height.

Fig. 8.13(a) with Fig. 8.15(a)). This occurs because, once generated, waves propagate across oceanic basins. As a result, the spatial distribution of wave height is much smoother than for wind speed and strong zonal banding does not exist due to the propagation of swell in the meridional direction. This feature indicates that wind speed has a shorter decorrelation distance than wind speed, as noted in the context of extreme value analysis by Meucci et al. [2018].

The results shown in Figs. 8.12–8.15 are for the ALT. Very similar results have been produced for RAD [Young and Donelan, 2018] and also exist for SCAT.

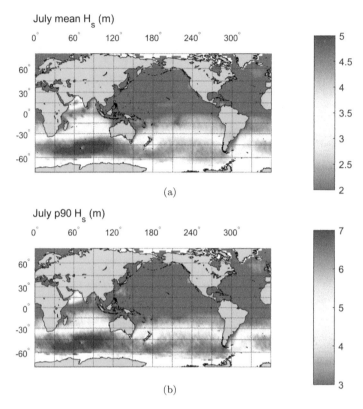

Fig. 8.15. Global significant wave height, H_s during the month of July from ALT data [unit: m]. (a) Mean monthly significant wave height (b) 90th percentile significant wave height.

8.5. Extreme Value Estimates

The satellite datasets described above now span almost 34 years, longer than many *in situ* buoy datasets. As such, they offer the potential to be used for extreme value analysis, that is, to estimate the 1 in 100-year return period values of significant wave height, H_s^{100}, and wind speed, U_{10}^{100}. Extreme value analysis (EVA) typically involves fitting a pre-defined probability distribution function to the measured data and extrapolating to the desired probability level (1 in 100 years or a probability of 0.01 in any year). Takbash *et al.* [2019] have investigated using altimeter data gridded at 2°, as above for this purpose.

They adopted a Peaks over Threshold (PoT) approach in which all independent peaks above a defined threshold are used to define the low probability tail of the probability distribution function (pdf). In such a case, probability theory indicates that the extreme values should follow a generalized Pareto distribution, defined by

$$F(x) = 1 - \left[1 + k\left(\frac{x-A}{B}\right)^{-1/k}\right], \tag{8.2}$$

where $F(x)$ is the probability distribution function (pdf), x is the parameter under consideration (H_s or U_{10}), k and B are shape and scale parameters, respectively, and A is the chosen threshold. Following Takbash et al. [2019], the 90th percentile has been chosen as the threshold value.

Figure 8.16 shows color filled contour plots of U_{10}^{100} (Fig. 8.16(a)) and H_s^{100} (Fig. 8.16(b)) from the ALT PoT analysis. The contours have been drawn on the 2° × 2° grid without further smoothing. In comparison to earlier findings of Vinoth and Young [2011], the results show a far smoother spatial distribution. The contours of U_{10}^{100} show much greater zonal structure than H_s^{100}. This is consistent with mean monthly climatology [Young, 1994, 1999; Young and Donelan, 2018] and is caused by the dispersive nature of waves. Once generated, waves propagate across oceanic basins as swell [Young et al., 2013], ensuring a smoother distribution of H_s than U_{10}. This is also the case for the extreme values, as shown in Fig. 8.16.

Figure 8.16(a) shows the maxima of U_{10}^{100} that are approximately 36 ms^{-1} occur in the North Atlantic and North Pacific. Although the Southern Ocean is consistently windy year-round and monthly means in winter are comparable to the northern hemisphere [Young, 1999; Young and Donelan, 2018], the extremes are not as great. The maximum values of U_{10}^{100} in the Southern Ocean are approximately 32 ms^{-1}. This maximum tends to occur south of the Indian Ocean (between Australia and South Africa). The maxima in the North Atlantic and North Pacific tend to be displaced towards the western boundaries of these basins. Takbash et al. [2019] showed this agrees with the storm tracks of tropical cyclones (and tropical low-pressure systems).

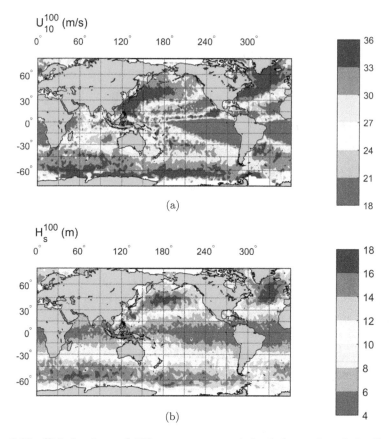

Fig. 8.16. Global values of 100-year return period wind speed and significant wave height obtained from ALT data. (a) 100-year return period wind speed, U_{10}^{100} [unit: ms^{-1}], (b) 100-year return period significant wave height, H_s^{100} [unit: m].

Takbash et al. [2019] showed that North Atlantic and Pacific tropical storms track east to west across the tropics of each ocean basin, respectively, before turning north along the western boundary of each basin. Because of the small spatial scale of tropical cyclones and the relatively large distance between altimeter tracks, it is likely that these systems are under-sampled in the present analysis. As such systems move north, they tend to increase in size, making it more likely that they are observed by the altimeter. This is clear in the region of the western North Atlantic, where extreme winds are predicted (Fig. 8.16(a)) north of 30°N, but there is no clear indication of tropical cyclones moving across the tropical regions of the Atlantic (east to west). In contrast, extreme winds along the western boundary of the Pacific are predicted as far south as 10°N. There is then a clear path of intense winds shown across the Pacific equatorial regions. North Pacific tropical cyclones (typhoons) tend to be larger in spatial extent than North Atlantic tropical cyclones (hurricanes) [Knaff et al., 2014]. They are also more frequent, making them less affected by under-sampling in the altimeter dataset. This explains why the east-west tropical track is clear in the western Pacific (10°N) but not the western Atlantic.

A number of other storm track features can also be seen in the values of U_{10}^{100} in Fig. 8.16(a). The region of high occurrence of tropical cyclones near the central American Pacific coast is reflected in a "hot spot" of extreme wind of approximately 36 ms^{-1} in that region. There is a region of reduced U_{10}^{100} in the central Indian Ocean. Takbash et al. [2019] showed that this corresponds to a region almost devoid of tropical cyclones, between the western Australian and eastern African basins. Less clearly, there is also a band of slightly elevated U_{10}^{100} from north-east of New Zealand to east of New Guinea. Again, this corresponds to the track regions for South Pacific storms. A further "hot spot" of elevated U_{10}^{100} can be seen in the Bay of Bengal, another region of high occurrence of tropical cyclones.

The eastern side of the South Atlantic (off Africa) shows relatively low values of U_{10}^{100} with the exception of a band of slightly increased values along the equator between South America and North Africa. It is probable that this is the signature of storm activity in the Intertropical Convergence Zone (ITCZ). A triangular region of low U_{10}^{100} bounded by the equator west of South America is also clear in Fig. 8.16(a).

Many of the same features described above are also apparent in model calculations of U_{10}^{100} [Breivik et al., 2014; Meucci et al., 2018]. Both the location and magnitudes of the maximum values in the North Atlantic and Pacific Oceans are comparable to Fig. 8.16(a). Also, the relatively low values in the triangular region west of South America and across the Atlantic west of Africa are found in both the model and altimeter data in Fig. 8.16(a). However, features which we have attributed to small-scale tropical cyclone activity are not clear in the model results. This includes tropical cyclone activity across the Pacific, north of the equator, or in the Pacific Ocean east of Australia or the low extremes area in the central Indian Ocean. It should be pointed out that neither the model results, nor the altimeter dataset are optimal for investigating tropical cyclone extremes. The spatial resolution of the models [e.g., of order 100 km] means that tropical cyclone winds will not be resolved. In contrast, the altimeter will measure tropical cyclone winds [Young, 1993], provided there is a ground track close to the tropical cyclone. However, as noted above, these storms will be under-sampled. Therefore, the differences between U_{10}^{100} from model data and altimeter are as one would expect.

Figure 8.16(b) shows color filled contours of H_s^{100} calculated using the PoT method and altimeter data. As noted previously, there is much less small-scale variability than for U_{10}^{100} (Fig. 8.16(a)). The largest values of H_s^{100} are once again in the North Atlantic and North Pacific, with values of approximately 18 m. Again, the regions with the largest extreme waves are displaced towards the western boundaries of these basins, but not to the same extent as the wind U_{10}^{100} (Fig. 8.16(a)). Similar to the extreme winds, the largest values of H_s^{100} in the Southern Hemisphere are found south of the Indian Ocean between Australia and South Africa, with values of approximately 17 m. Values of H_s^{100} gradually decrease from these maximum regions in each hemisphere towards the equator. In the equatorial regions H_s^{100} reaches only approximately 4 m. These results are much smoother (spatially) than the PoT results of Vinoth and Young [2011] and agree well with model results [Breivik et al., 2014; Meucci et al., 2018] both in magnitude and spatial distribution.

Takbash et al. [2019] compared ALT U_{10}^{100} and H_s^{100} against buoy data and obtained remarkably good agreement. Although such a validation is reassuring, the relatively sparse ground track pattern of the ALT raises concerns about whether such an instrument can adequately sample extremes and, as a result, whether the results in Fig. 8.16 are biased due to such. Although the ALT is the only satellite system which can be used to determine extreme value significant wave height, both RAD and SCAT can be applied to wind speed. It should be remembered, however, that both of these instruments tend to overestimate high wind speeds compared to buoys. Despite the fact that Ribal and Young [2019] proposed corrections for high winds, the present application is a demanding test of platform performance at high wind speeds. Takbash et al. [2019] applied RAD data to EVA and found that the "fair weather bias" in such data rendered it unusable for EVA. Storms are often associated with strong winds and, although the RAD has a broad observation swath, it effectively under samples many extremes due to rain and hence poorly reproduces the low probability tail of the pdf.

The SCAT has a broad swath like the RAD and hence images most locations on the Earth twice per day, however, it can measure wind speed during rain events. Therefore, it should not have the same limitations as the RAD. Figure 8.17 shows the global distribution of U_{10}^{100} obtained from the SCAT data. This figure can be directly compared with the ALT result Fig. 8.16(a).

The agreement between Fig. 8.16(a) (ALT) and Fig. 8.17 (SCAT) is remarkable. The overall spatial distributions and the magnitudes of the 100-year return period wind speeds are very similar. Both platforms show the highest values of U_{10}^{100} in the North Pacific and North Atlantic, with slightly lower values at comparable latitudes in

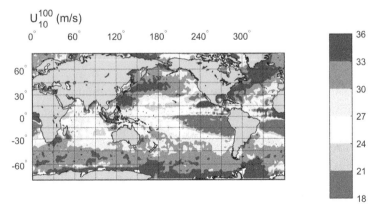

Fig. 8.17. Global values of 100-year return period wind speed, U_{10}^{100} obtained from SCAT data [unit: ms^{-1}].

the Southern Ocean. The SCAT dataset has approximately 20 times more data than the ALT. Therefore, it could be expected that it would be far more adept at identifying tropical cyclone extremes. Despite this, both datasets show the extremes in the North Atlantic and North Pacific are displaced towards the westerns side of these basins, conforming to the storm tracks. Both datasets show relatively low values of U_{10}^{100} in the triangular region south of the equator in the Pacific west of South America. This region is slightly more defined in the ALT data. Both datasets show a region of very high extreme winds in the Pacific west of Mexico. This is associated with tropical cyclones and shows slightly higher values in the SCAT dataset. Similarly, the region north of the equator in the Pacific east of Asia is also slightly stronger in the SCAT data than the ALT. Again, this is a region influenced by tropical cyclones. Interestingly, the band of elevated U_{10}^{100} across the Pacific north of the equator, identified in the ALT data assumed to be the result of tropical cyclone tracks is less distinct in the SCAT result. This indicates that the ALT result may be overestimated in this region. The apparently calm region in the Pacific identified in the ALT data and explained by Takbash *et al.* [2019] as a region devoid of tropical cyclones is also clear in the SCAT data but at slightly larger values.

There are some areas of small-scale tropical cyclone activity identified by the SCAT which are not apparent in the ALT data. The SCAT shows high values of U_{10}^{100} in the Gulf of Mexico, the Bay of Bengal, north-west Australia and east of Madagascar. These are all areas of known tropical cyclone activity, which are apparently undersampled in the ALT result.

Despite the differences, the ALT generally remarkably reliable in determining U_{10}^{100} compared to the SCAT. However, issues of undersampling in the ALT data are, not surprisingly, apparent in tropical cyclone areas.

8.6. Trends in Wind Speed and Wave Height

Determining whether global values of wind speed and wave height have been changing in recent decades is important for a number of reasons. Ocean waves are an important element of coastal flooding and beach erosion. Changes in wave climate have the potential to significantly change beach alignment and inundation of coastal areas. Understanding the historical record and the physical processes responsible for changes is also an important part of projecting future changes as a result of long-term changes in climate. Ocean winds are also important for determining the offshore wind resource and the magnitude of evaporation. Also, ocean winds and the waves they generate play an important part in defining the roughness of the water surface and hence fluxes of gas, energy and heat between the atmosphere and the ocean.

The most obvious method to determining such trends is from *in situ* buoy data. However, changes in buoy hulls, instrument type, anemometer height and processing methods over time have resulted in a data record which is far from homogeneous [Gemmrich *et al.*, 2011]. As a result, trend analysis from such buoy data tends to be unreliable [Young *et al.*, 2011].

Estimates of wind speed and wave height trends from model data [Aarnes *et al.*, 2015] have also proved unreliable. This is partly due to the long-term accuracy of such models but also due to the lack of homogeneity in reanalysis products which have including variable amounts of assimilated data over time.

Noting the above limitations, long-duration satellite observations become an attractive way to explore trends on a global scale. It is essential, however, that such datasets are carefully calibrated over the full duration of the record, so as to ensure the time series is homogeneous. Therefore, datasets such as the ones described earlier become a valuable resource to investigate such trends.

Young *et al.* [2011] investigated the period 1985–2008 using ALT data and found generally positive trends in mean wind speed and stronger trends for 90th percentile wind speeds. Trends in significant wave height have similar results, although the magnitudes of the trends were significantly reduced.

Young and Ribal [2019] considered the longer period 1985–2018 using the three datasets (ALT, RAD, SCAT) shown in Fig. 8.6. Figure 8.18 shows trends for both the mean wind speed, $\bar{U}_{10}$ and

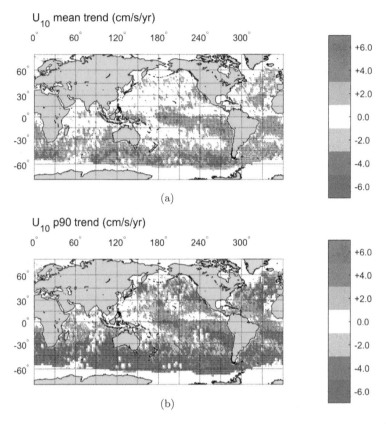

Fig. 8.18. Global trends in ALT wind speed. (a) Mean monthly wind speed, $\bar{U}_{10}$, (b) 90th percentile monthly wind speed, U_{10}^{90} [units cm/s/yr].

the 90th percentile wind speed, U_{10}^{90}. The corresponding trends for mean significant wave height, $\bar{H}_s$ and 90th percentile, H_s^{90} are shown in Fig. 8.19.

Consistent with the results of Young et al. [2011], Fig. 8.18(a) shows positive trends in $\bar{U}_{10}$ across the Southern Ocean, the equatorial Pacific and parts of the south Atlantic of approximately 3 cm/s/yr. The positive trend in these regions is statistically significant. At the 90th percentile (Fig. 8.18(b)), the U_{10}^{90} trends remain positive, increase in magnitude and the geographic extent of statistically significant values increases. Across the Southern Ocean, the trend becomes approximately 5 cm/s/yr, resulting in an increase of

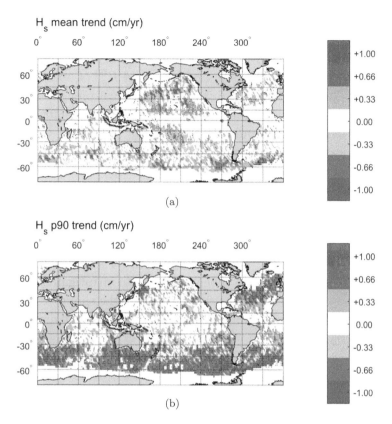

Fig. 8.19. Global trends in ALT significant wave height. (a) Mean monthly significant wave height, $\bar{H}_s$, (b) 90th percentile monthly significant wave height, H_s^{90} [units cm/yr].

approximately $1.5\,\mathrm{ms}^{-1}$ over the 33-year period of the data (approx. 8% increase). Large areas of the Indian, South Pacific and Atlantic Oceans have 90th percentile trends of approximately $3\,\mathrm{cm/s/yr}$. In all of these areas, the trend is statistically significant.

The trends in significant wave height (Fig. 8.19) are significantly less positive than for wind speed. The trends for mean significant wave height $\bar{H}_s$ are quite small with areas of statistically significant positive trends ($0.3\,\mathrm{cm/yr}$) in the Southern Ocean south of the Pacific. There are also regions of both the north and south Pacific with weak negative trends of approximately $-0.3\,\mathrm{cm/yr}$. Only some of these regions of negative trend are statistically significant. At the 90th percentile (Fig. 8.19(b)), there are large areas of the Southern Ocean and North Atlantic with positive values of H_s^{90} of approximately 1 cm/yr. This results in an increase in H_s^{90} of approximately 30 cm or 5% over the 33-year duration of the record for the Southern Ocean. There is no clear trend in the 90th percentile significant wave height in the Pacific or Indian oceans.

As outlined by Young and Ribal [2019], it is surprising that there is a stronger trend in wind speed than wave height. If the waves are in equilibrium with the local winds, then $H_s \approx U_{10}^2$. This would suggest that there would be a stronger trend in U_{10} than H_s. This suggest that the waves are not in equilibrium with the local wind. This may occur because the waves are not generated locally (i.e., remotely generated swell) or that the wind has not blown for a sufficient time for the waves to reach full development. In order to investigate these possibilities, Young and Ribal [2019] investigated the shape of the wind and wave pdfs by considering trends in the mode of the distribution for both U_{10} and H_s.

Figure 8.20(a) shows the global trend in the U_{10} mode and Fig. 8.20(b) shows the trend in the H_s mode. Note that mode is a noisier quantity than mean and hence greater statistical variability can be expected. The trend in U_{10} mode (Fig. 8.20(a)) has a similar spatial distribution to $\bar{U}_{10}$ (Fig. 8.18(a)) but the trend is more positive globally. The largest trends are still seen in the Southern Ocean and the equatorial Pacific, as for $\bar{U}_{10}$ but now there are significant regions of positive trend over most regions of the earth. In contrast to both the $\bar{H}_s$ trend (Fig. 8.19(a)) and the U_{10} mode trend (Fig. 8.20(a)), the H_s mode trend (Fig. 8.20(b)) is strongly negative over the whole earth.

In order to try and interpret what is happening with the mode trends in Fig. 8.20, Young and Ribal [2019] examined the details of how the pdf has changed over the 33-year measurement period. Figure 8.21 shows pdfs for both U_{10} and H_s averaged over a region west of the southern tip of South America (240°E to 270°E and –60°S to −40°S). Probability distribution functions are shown for two time windows (1998–2000 and 2015–2017). Consistent with Figs. 8.18(a)

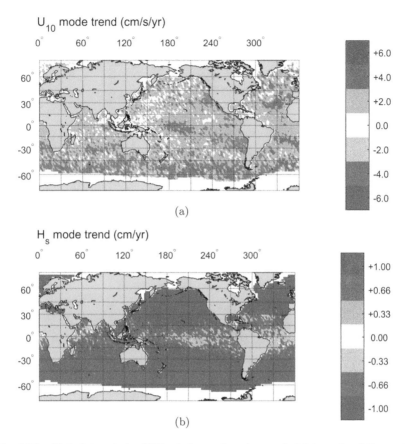

Fig. 8.20. Global trends in ALT wind speed and wave height mode of the probability distribution. (a) Wind speed trend [units cm/s/yr]. (b) Significant wave height trend [units cm/yr].

and 8.20(a), the wind speed pdf shows that over this period, the mode of the pdf has moved to higher values of wind speed, as has the mean. In addition, the pdf has broadened slightly (i.e., a more uniform distribution of wind speeds). The wave height pdf, however, behaves quite differently. In response to the broadening of the wind speed pdf, there is a rapid growth in the percentage of wave heights less than the peak of the pdf. The pdf also slightly broadens. As a result, there is little change in the mean of the pdf and a significant decrease in the mode (see Figs. 8.19(a) and 8.20(b)).

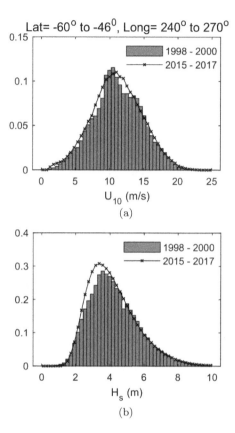

Fig. 8.21. Probability distributions functions for ALT wind speed and wave height for a region west of South America. (a) Wind speed pdf. (b) Significant wave height pdf.

Young and Ribal [2019] speculated that this behavior was caused by changes in the duration of winds. The broadening of the wind speed pdf results in a greater occurrence of wind speeds both above and below the peak of the pdf. The increase in the occurrence of lower wind speeds is reflected by an increase in the occurrence of smaller waves. However, the increase in the occurrence of winds above the peak does not cause a similar increase in higher waves, as to generate these higher waves required not only an increase in wind speed but that the winds must blow for a sufficient duration. It appears that the stronger winds may be associated with a shorter duration, explaining the shape of wave height pdf. This also explains why there is little change in the mean wave heights when there has been an increase in the mean winds.

The increase in the magnitude of both the wind speed and wave height trends at the 90th percentile needs to be treated with some caution. Determining accurately small changes in mean monthly wind speed and wave height over the 33-year period is a demanding undertaking. Accurately determining the monthly 90th percentiles values is even more challenging.

Young and Ribal [2019] went to considerable length to test if the 90th percentile ALT trends were robust. As the determination of 90th percentile values on a monthly basis may be influenced by under-sampling, they repeated the trend estimates using annual values. This approach resulted in very similar magnitude and spatial distribution of both mean and 90th percentile trends for both wind speed and wave height.

As is clear in Fig. 8.6, the number of ALT missions has increased in recent years. This may mean that, with more observations in recent years, the ALT may sample more storms. This would potentially result in an erroneous positive bias in the trend measurements, the impact being greater at higher percentiles. To investigate this possibility, Young and Ribal [2019] randomly decimated ALT measurements to test the accuracy of the 90th percentile measurements with differing numbers of observations per month. Finally, they selectively removed satellite missions in recent years, such that there was an approximately constant number of observations across the full period of the observations (30 years). All of these tests confirmed that both the mean and 90th percentile trend estimates were not greatly impacted by any of these sampling issues.

The ultimate test, however, would be to investigate the mean and 90th percentile values from the RAD or SCAT measurements. Both of these instruments measure over a broad swath. As a result, they image individual locations approximately 10 times more frequently than the ALT. As such, they should be much less affected by sampling issues. Young and Ribal [2019] examined both mean and 90th percentile wind speed trends (note that neither RAD nor SCAT measures wave height). As noted earlier, the RAD is impacted by a "fair-weather bias" which degrades statistics of extreme wind speeds. As a result, the 90th percentile trends were unreliable. The mean trends in $\bar{U}_{10}$ were, however, similar to the ALT. The spatial distribution was identical and the magnitude of the trend approximately 25% less in magnitude than for the ALT. Young and Ribal [2019] showed that this difference in magnitude of the trend was consistent with differing time of day at which the ALT and RAD sampled.

As the SCAT does not have issues with rain and hence a "fair-weather bias", it is a likely candidate to determine trends in both $\bar{U}_{10}$ and U_{10}^{90}. Both the ALT and RAD datasets have been more

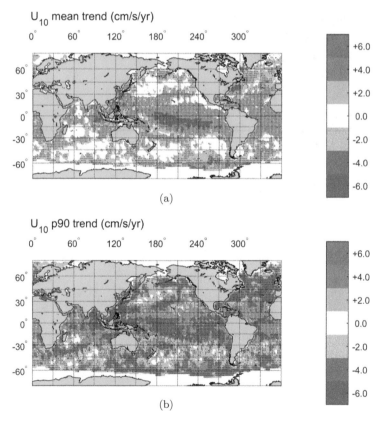

Fig. 8.22. Global trends in SCAT wind speed. (a) Mean monthly wind speed, $\bar{U}_{10}$, (b) 90th percentile monthly wind speed, U_{10}^{90} [units cm/s/yr].

extensively tested than the SCAT. Hence, the homogeneity of the SCAT record is yet to be extensively tested. Noting this, the trend was determined from the SCAT data for both $\bar{U}_{10}$ and U_{10}^{90}. The results appear in Fig. 8.22. The trends for $\bar{U}_{10}$ shown in Fig. 8.22(a) can be compared to Fig. 8.18(a) for the ALT. Similarly, the trends for U_{10}^{90} in Fig. 8.22(b) can be compared with Fig. 8.22(b) for the ALT. Although there are some differences in spatial distribution, the SCAT and ALT results are in good agreement. Importantly, the 90th percentile trends are stronger than the means, consistent with the ALT results.

8.7. Future Satellite Missions

The results above clearly show the potential for well-calibrated multi-mission oceanographic satellites. The application and accuracy of such data will grow as the number of satellites in orbit at any one time and the length of the data record both increase. In addition, there will be a number of new satellites which will fly in the coming years. Firstly, the new generations of SAR-mode altimeters have the potential to provide much higher resolution measurements of significant wave height, particularly near coastlines. The CFOSat satellite with its SWIM package is a broad swath instrument which will measure the full directional spectrum. Such instruments will continue to enhance the large dataset of global observations which now exists.

References

Aarnes, O.J., Abdalla, S., Bidlot, J.-R. and Breivik, O. (2015). Marine wind and wave height trends at different ERA-interim forecast ranges, *J. Climate*, **28**, 819–837, doi:10.1175/JCLI-D-14-00470.1.

Breivik, O., Aarnes, O.J., Abdalla, S., Bidlot, J.-R. and Janssen, P.A. (2014). Wind and wave extremes over the world oceans from very large ensembles, *Geophys. Res. Lett.*, **41**, 5122–5131, https://doi.org/10.1002/2014GL060997.

Chelton, D.B., Ries, J.C., Haines, B.J., Fu, L.-L. and Callahan, P.S. (2001). *Satellite Altimetry, Satellite Altimetry and Earth Sciences: A Handbook of Techniques and Applications*, Academic Press, Chap. 1, pp. 1–131.

Cheney, R.E., Douglas, R.W., Miller, L., Porter D.L. and Doyle, N. (1987). *GEOSAT Altimeter Geophysical Data Record: User Handbook*, NOAA, Rockville.

Copernicus Climate Change Service (C3S) (2017). *ERA5: Fifth Generation of ECMWF Atmospheric Reanalyses of the Global Climate*. Copernicus Climate Change Service Climate Data Store (CDS), https://cds.climate.copernicus.eu/cdsapp#!/home.

Dee, D.P. et al. (2011). The ERA-interim reanalysis: configuration and performance of the data assimilation system, *Q. J. Royal Meteorol. Soc.*, **137**, 553–597.

Donnelcy, W.J., Carswell, J.R., McIntosh, R.E., Chang, P.S., Wilkerson, J., Marks, F. and Black, P.G. (1999). Revised ocean backscatter models at C and Ku-band under high wind conditions, *J. Geophys. Res.: Oceans*, **104**(C5), 11485–11497.

Evans, D., Conrad, C.L. and Paul, F.M. (2003). Handbook of automated data quality control checks and procedures of the national data buoy center, Technical Document 03-02, NOAA National Data Buoy Center.

Gemmrich, J., Thomas, B. and Bouchard, R. (2011). Observational changes and trends in northeast Pacific wave records, *Geophys. Res. Lett.*, **38**, L22601, doi:10.1029/2011GL049518.

Hasselmann, S., Brüning, C., Hasselmann, K. and Heimbach, P. (1996). An improved algorithm for the retrieval of ocean wave spectra from synthetic aperture radar image spectra, *J. Geophys. Res.: Oceans*, **101**(C7), 16615–16629, doi:10.1029/96JC00798.

Hollinger, J.P., Peirce, J.L. and Poe, G.A. (1990). SSM/I instrument evaluation, *IEEE Trans. Geosci. Remote Sensing*, **28**, 781–790.

Holthuijsen, L.H. (2007). *Waves in Oceanic and Coastal Waters*, Cambridge University Press.

Hwang, P.A., Teague, W.J., Jacobs, G.A. and Wang, D.W. (1998) A statistical comparison of wind speed, wave height and wave period derived from satellite altimeters and ocean buoys in the Gulf of Mexico region, *J. Geophys. Res.: Oceans*, **103**, 10451–10468.

Knaff, J.A., Longmore, S.P. and Molenar, D.A. (2014). An objective satellite-based tropical cyclone size climatology, *J. Climate*, **27**, 455–476, https://doi.org/10.1175/JCLI-D-13-00096.1.

Komen, G.J., Cavaleri, L., Donelan, M., Hasselmann, K., Hasselmann, S. and Janssen, P.A.E.M. (1994). *Dynamics and Modelling of Ocean Waves*, Cambridge University Press.

Meucci, A., Young, I.R. and Breivik, O. (2018). Wind and wave extremes from atmosphere and wave model ensembles, *J. Climate*, **31**, 8819–8843, doi: 10.1175/JCLI-D-18-0217.1.

Perez, J., Menendez, M. and Losada, I.J. (2017). GOW2: A global wave hindcast for coastal applications, *Coast. Eng.*, **124**, 43405.

Plant, W.J. (2002). A stochastic, multiscale model of microwave backscatter from the ocean, *J. Geophys. Res.: Oceans*, **107**(C9), 3-1–3-21, doi:10.1029/2001JC000909.

Queffeulou, P. (2004). Long-term validation of wave height measurements from altimeters, *Mar. Geod.*, **27**, 495–510, doi:10.1080/01490410490883478.

Tolman, H.L. (2011). *WAVEWATCH III User Manual*, U. S. Department of Commerce, Washington, D.C.
Ribal, A. and Young, I.R. (2019). 33 years of globally calibrated wave height and wind speed data based on altimeter observations, *Sci. Data*, 6(1), 77.
Stoffelen, A.C.M. (1998). Toward the true surface wind speed: Error modeling and calibration using triple collocation, *J. Geophys. Res.: Oceans*, **103**(C4), 7755–7766.
Takbash, A., Young, I.R. and Breivik, O. (2019). Global wind speed and wave height extremes derived from long-duration satellite records, *J. Climate*, **32**, 109–126, doi:10.1175/JCLI-D-18-0520.1.
Vinoth, J. and Young, I.R. (2011). Global estimates of extreme wind speed and wave height, *J. Climate*, **24**(6), 1647–1665, doi:10.1175/2010JCLI3680.1.
Walker, D.M. (1995). Measurement techniques and capabilities of the Geosat Follow-On (GFO) radar altimeter, in *2nd Topical Symp. Combined Optical — Microwave Earth and Atmosphere Sensing*, pp. 226–228, doi:10.1109/COMEAS.1995.472314.
Wentz, F.W. (1983). A model function for ocean microwave brightness temperatures, *J. Geophys. Res.: Oceans*, **88**(C3), 1892–1908.
Wentz, F.J. (1992). Measurement of oceanic wind vector using satellite microwave radiometers, *IEEE Trans. Geosci. Remote Sensing*, **30**, 960–972.
Wentz, F.J. (1997). A well-calibrated ocean algorithm for special sensor microwave/imager, *J. Geophys. Res.: Oceans*, **102**(C4), 8703–8718.
Young, I.R. (1993). An estimate of the geosat altimeter wind speed algorithm at high wind speeds, *J. Geophys. Res.: Oceans*, **98**(C11), 20275–20285.
Young, I.R. (1994). Global ocean wave statistics obtained from satellite observations, *Appl. Ocean Res.*, **16**, 235–248.
Young, I.R. (1999). *Wind Generated Ocean Waves*, Elsevier Sciences Ltd.
Young, I.R., Zieger, S. and Babanin, A.V. (2011). Global trends in wind speed and wave height, *Science*, **332**, 451–455, doi:10.1126/science.1197219.
Young, I.R., Babanin, A.V. and Zieger, S. (2013). The decay rate of ocean swell observed by altimeter, *J. Phys. Oceanogr.*, **43**, 2322–2333, doi:10.1175/JPO-D-13-083.1.
Young, I.R. and Donelan, M.A. (2018). On the determination of global ocean wind and wave climate from satellite observations, *Remote Sens. Environ.*, **215**, 228–241, doi:10.1016/j.rse.2018.06.006.
Young, I.R., Sanina, E. and Babanin, A.V. (2017). Calibration and cross validation of a global wind and wave database of altimeter, radiometer, and scatterometer measurements, *J. Atmos. Ocean. Technol.*, **34**, 1285–1306.
Young, I.R. and Ribal, A. (2019). Multiplatform evaluation of global trends in wind speed and wave height, *Science*, **364**(6440), 548–555.
Zieger, S., Vinoth, J. and Young, I.R. (2009). Joint calibration of multiplatform altimeter measurements of wind speed and wave height over the past 20 years. *J. Atmos. Ocean. Technol.*, **26**, 2549–2564.

Index

A

air–sea coupling, 36
air–sea exchanges, 49
air–sea interaction, 23, 26
Akhmediev Breather, 115, 119, 121, 124
Alber equation, 78
aleatory uncertainty, 309
altimeter, 322
amplitude, 7
amplitude dispersion, 108, 113
amplitude focusing, 71
annual maximum, 276, 279
antenna, 327
asymmetry, 62, 64, 66–67, 70, 82
asymptotic limit, 69
atmospheric boundary layer, 39, 85
atmospheric input, 19
atmospheric stability, 184
atmospheric surface layer, 23

B

B-spline, 291
Babanin–Chalikov model, 90
bathymetry, 183
Bayesian inference, 311
bedforms, 184
Benilov instability, 93, 95
Benilov mechanism, 92, 94
Benilov theory, 89–90, 94
Benjamin–Feir (BF) instability, 112–114, 119, 126, 141, 145, 212, 230, 233
Benjamin–Feir Index (BFI), 77, 130–131
Benney's equation, 108–109, 111
block maxima, 281
bootstrapping, 311
bottom friction, 170
bound waves, 140, 212, 225–226, 236
boundary conditions, 208
boundary layers, 332
Boussinesq approximation, 25
Boussinesq equations, 207
breaking, 173
breaking criteria, 66
breaking events, 48
breaking in progress, 55, 58
breaking onset, 54–55, 57, 65, 83
breaking parameterization, 241, 243–244
breaking probability, 83–84
breaking severity, 83
breaking waves, 123, 125
breaking-in-progress, 83
buoy data, 322, 326
buoyancy forces, 57
Businger, J. A., 25

C

calibrated, 326
calibration, 322
capillary-gravity waves, 105
Cartesian coordinates, 208–209, 219–220, 222
CFOSat, 353
Chalikov–Sheinin Model, 60
climate change, 274, 313

columnar energy, 241, 251
concertina effect, 168
conditional extremes, 301
conformal coordinates, 207, 209, 211, 213, 242, 260
conformal mapping, 213
Coriolis forcing, 27
coupling, 192
covariate, 272–273, 288–291
crest elevations, 274
critical layer, 31
cross-validation, 290
cumulative distribution function, 275
cumulative effect, 82–83, 94
currents, 84, 166–167, 184

D

damping, 84
data assimilation, 194
design criteria, 275
detuned resonance, 104, 110, 112, 117, 123, 129–130, 141
developing breaking, 54, 83
developing stage, 58
DIA, 150
diagnostic tail, 177
diffraction, 167
directional accuracy, 188
directional bandwidth, 79
directional spectrum, 4, 104
directional spreading, 17–18, 292
directional wave fields, 79
directionality, 75, 272
Discrete Interaction Approximation (DIA), 108, 111
dispersion relation, 105, 106, 224–226, 260
dissipation, 19, 48, 52, 55, 82–83, 85, 94, 206, 211, 238–240, 244–248, 250, 252–254, 256, 263
distance to breaking, 71
dominant period, 187
Donelan, M. A., 26, 29–31, 33–34
downshift, 147

downshifting, 84, 103, 115, 126, 146, 239, 248–250, 256–257
downward acceleration, 58
drag coefficient, 32, 39
Draupner Wave, 133
Dynamic Cascade, 117
dynamic criteria, 69
dynamical downscaling, 313
Dysthe equation, 118–119

E

eddy viscosity, 24
Ekman balance, 27–28
energy, 3
energy conservation, 211
energy flux, 211
energy input, 206, 211, 237, 239, 247, 250, 252, 254–255
energy transformation, 246
engineering applications, 208
environmental contour, 295
epistemic uncertainty, 309
equilibrium condition, 32
equilibrium spectrum, 103
ERA-5, 321
ERA-Interim, 321
Eulerian current, 30
exact repeat mission, 322
extreme, 328–339
extreme events, 271
extreme value analysis, 273
extreme value estimates, 338
extreme wave conditions, 271

F

fair weather bias, 332, 343
feedback effects, 26
fetch, 9, 19
fetch law, 103, 147
fetch-limited, 11
fetch-limited growth, 9
finite difference method, 207
finite element method, 207
finite volume method, 207
footprint, 322

forcing, 84
forecast skill, 190
form drag, 32
four-wave resonant interaction, 106
four-wave interactions, 169
four-wave resonance, 103–104, 129, 141
four-wave resonant interaction, 112
Fourier presentation, 206
Fourier transform, 7, 211, 215, 220, 223, 232, 261
Fréchet, 282
freak waves, 104, 119, 128–129, 133, 135, 137, 141–142, 149, 206, 213, 229
free waves, 117, 119, 124, 140
frequency, 1, 67–68
frequency grid, 177
frequency of occurrence, 85
frequency width, 187
friction velocity, 12
front steepness, 56, 58

G

Garden Sprinkler Effect, 181
Gaussian, 285
Gaussian process, 104, 129
general circulation model (GCM), 313
generalized extreme value (GEV), 281–282
generalized Pareto distribution (GPD), 282–283, 339
geostrophic flow, 26
GOMOS hindcast, 275
GOW2, 321
gravity-capillary wave, 35
ground tracks, 322
Gumbel, 281, 301

H

Hasselmann integral, 111, 149–150
high wind speeds, 332
high-order scheme (HOS) model, 217

High-Order Spectral Method (HOSM), 124, 135, 139–140, 150
hurricanes, 271

I

I-FORM, 295
ice, 167
imminent breaking, 83
incipient breaking, 54–55, 58, 65, 67
individual wave heights, 273–274
initial growth, 168, 174
initial steepness, 71, 77
instability, 55, 76, 78–79, 89, 94
integral invariants, 211, 242
inter-annual variability, 274
intermittent turbulence, 88
irregular grid, 178

J

Jeffreys' sheltering, 30
joint criteria, 292
JONSWAP, 9, 13, 36, 78, 80, 233, 245, 259

K

kinematic boundary condition, 242
kinematics, 166
kurtosis, 129–131, 136–138, 141, 149

L

Lagrangian, 28
Lagrangian frame of reference, 26
Lagrangian–Eulerian formulation, 207–208
Langmuir circulation, 51, 85
Langmuir turbulence, 51, 92–94
Laplace, 301
Laplace equation, 208–210, 214–217, 219, 222–223, 262
large eddy simulation, see LES
Law of the Wall, 24
LES, 207, 260, 263
limiting steepness, 48, 56–57, 71–72, 74, 85

linear equations, 205
linear focusing, 53, 84
linear superposition, 48, 59, 74
linear wave theory, 1
lognormal, 297

M

maximum likelihood, 290
maximum wave height, 286
mean currents, 49
mean steepness, 71, 75, 78
measurement error, 189
micro-breakers, 53
Miles, J. W., 29, 31
mixed layer, 85
mixed layer depth, 88
model resolution, 177, 178, 194
modulated wave train, 114–115, 123–126
modulational index, 77
modulational instability, 53, 60, 62, 66, 71, 74, 79, 81–82, 84–85, 93, 110, 112, 131, 136
momentum, 49, 50
momentum transfer, 26, 29
momentum/energy exchange, 57
monochromatic wave trains, 78
MOST, 25–26, 34, 36–37, 39
moving particle, 207
mud, 167, 170, 183–184
multipole technique, 216

N

Navier–Stokes equations, 25, 207
negative wind input, 169
non-breaking dissipation, 168, 173
nonlinear evolution, 71
nonlinear instabilities, 48
nonlinear interactions, 19, 35, 49, 169, 176
Nonlinear Schrödinger Equation (NLS), 85, 115, 117–121, 126, 145–146
nonlinear source term, 111
nonlinear wave trains, 71

numerical error, 180
numerical instability, 218, 240, 242–243, 260–261
numerical model, 206, 215, 218, 261
numerical scheme validation, 210
numerics, 180

O

Obukov length, 25
ocean interface, 85
ocean mixing, 93–94
oceanic boundary layer, 23
one-dimensional wave trains, 59, 76
optical methods, 28
orbital motion, 89

P

90th percentile, 351
P-spline, 291
peak frequency, 9
peak wave period, 292
peaks over threshold (POT), 283, 339
periodicity, 206, 209, 220, 261
phase, 7
phase error, 190
phase speed, 1
phase-averaged models, 8, 321
phase-resolving modeling, 8, 205
Phillips, 29–30, 35
Phillips mechanism, 95
Phillips resonance, 168
Phillips spectrum, 75
Pierson–Moskowitz spectrum, 13
plunging breakers, 58
Poisson, 280
posterior predictive density, 312
potential flow, 89
potential theory, 86
pressure gradients, 24
pressure slope correlation, 33–34
probability density function, 5, 56, 273, 275
probability distribution, 56, 76
probability distribution function, 339, 348

probability of failure, 295
progressive waves, 56

Q

quasi-monochromatic, 59
quasi-monochromatic waves trains, 93
quasi-resonance, 110, 117

R

3% rule, 27–28
radar cross-section, 323
radiometer, 324
rates of breaking occurrence, 81
Rayleigh distribution, 129, 285
rear steepness, 56
recalibration, 173–174
recurrence, 112, 114–115
reflection, 173
refraction, 166
residual breaking, 54
residual stage, 58–59
resolution, 194
resonance condition, 106–107, 109, 151
resonance detuning, 109, 113, 151
resonances, 76
resonant forcing, 108–109
resonant interaction, 29, 150
resonant triad, 106
response-based methods, 293
return level, 277
return period, 281
return period value, 277, 280
return values, 281
Reynolds averaged, 25
Reynolds number, 88
rheology, 171
Richardson number, 25
rogue, 72
roughness, 24
roughness penalty, 290

S

sample size, 309
SAR, 321

SAR-mode altimeters, 353
satellite remote sensing, 321
scattering, 172
scatterometer, 327
Schrödinger equation, 206
sea ice, 171–172, 182, 184
seasonal variations, 334
seasonality, 272
second-order, 76
severity, 53, 85, 94
shallow-water waves, 205
sheltering, 29
shoaling, 40, 166
short-term variability, 273, 284
sideband waves, 112–113
sigma-coordinate, 221
significant wave, 6
significant wave height, 273, 323
site-pooling, 311
skewness, 60, 62, 64, 66–67, 70, 82
slanting fetch, 38
smoothed particle hydrodynamics, 207
source terms, 19, 167, 176
source/sink terms, 49
Southern Ocean, 336
spectral bandwidth, 76
spectral density, 164
spectral evolution, 12
spectral methods, 207
spectral moments, 187
spectral narrowing, 133, 135
spectral peak, 27
spectral resolution, 211, 228
spectral tail, 187, 191–192, 254
spectrum, 1, 3, 19, 231–232, 236, 239–240
spilling, 58
spline, 289
standing waves, 56, 173
statistical downscaling, 314
statistical properties, 212, 225, 228, 264
steepness, 47, 66–67, 84
Stokes drift, 26, 28–29, 40, 51, 92, 94

Stokes limiting steepness, 71
Stokes shear, 92, 94
stokes waves, 112, 211–212, 223–224
storm peak, 284
storm tracks, 339
storm-based approach, 286
stress veering, 41
stretching coefficient, 223
subsiding breaking, 54, 83
subsiding stage, 58
superposition, 81–82, 93
surf breaking, 169
surface currents, 85
surface elevation, 215, 230, 236
surface water waves, 47
surface waves, 49
surface-following coordinates, 209, 213
swath, 325
swell, 19, 36–37, 168, 177, 190
swell waves, 36
SWIM package, 353

T

Tayfun distribution, 129–130
template, 24, 104
third-generation wave models, 111, 128, 136, 150
third-order effects, 76
three dimensions, 87
three-dimensional dynamics, 67
three-dimensional structure, 80
three-dimensional turbulence, 90, 94
three-dimensional vorticity, 89
threshold, 81
threshold exceedances, 282
threshold steepness, 80
threshold values, 70
transient events, 76
trend, 348
triad interactions, 169
tropical cyclones, 271, 344
turbulence, 85, 89
turbulent eddy viscosity, 32
turbulent stress, 32
two-dimensional wavy surfaces, 93

U

ultimate steepness, 48
uncertainty, 273
under-sample, 332
unidirectional wave trains, 76
unstructured grid, 178
upper ocean, 85
upper-ocean dynamics, 49
upper-ocean mixing, 50

V

validation, 186
variance, 4
vegetation, 170, 183
velocity field, 205, 227–228
VESS, 307
viscous stress, 32
viscous sublayer, 32
void fraction, 57
volumetric dissipation, 88
vorticity, 89

W

WAM, 321
water surface, 7
water waves, 1
wave action balance equation, 164
wave amplitude, 1
wave asymmetry, 60
wave breaking, 47–48, 52–53, 93, 147, 211, 213, 240–241, 246, 248, 255, 260, 263
wave celerity, 59
wave climate, 334
wave coherent pressure, 33
wave-coupled processes, 49
wave fields, 75
wave focusing, 74
wave frequency, 28
wave groups, 56, 126, 141–142, 144, 149
wave height, 3, 230–231, 234–235
wave length, 1

wave number, 1
wave orbital motion, 48, 85, 88, 93–95
wave orbital velocity, 27, 92
wave period, 1–2, 186–187
wave spectrum, 9, 52, 248
wave steepness, 76, 78, 81
wave superposition, 76, 85
wave trains, 59, 66, 71, 76, 84
wave trains with continuous wave spectra, 93
wave trains with full spectrum, 77
wave turbulence, 86
wave-based Reynolds number, 86
wave-breaking onset, 71
wave-breaking turbulence, 51
wave-induced mixing, 85
wave-induced pressure fluctuations, 29
wave-induced processes, 50
wave-induced turbulence, 88, 92–94
wave-orbital turbulence, 51
wavelength scale, 86
Weibull, 277, 282

whitecapping, 52, 54, 57–58, 168, 176
whitecaps, 59
wind, 84–85
wind energy input, 52
wind forcing, 70–71, 93
wind input, 32, 48, 167, 173, 176
wind speed, 9, 323
wind stress, 50
wind–wave coupling, 30, 35–37
wind–wave growth, 32, 38, 41
wind–wave interaction, 36
wind–wave momentum, 37
wind–wave momentum coupling, 40
wind–wave momentum flux, 37
wind-forced currents, 26
wind-generated waves, 47
wind-input, 34
windsea, 33, 36, 125, 133

Z

Zakharov's equation, 78, 109–111, 118–119, 139–140, 145, 151
zonal structure, 339

CPSIA information can be obtained
at www.ICGtesting.com
Printed in the USA
BVHW010826220320
575293BV00006B/1

9 789811 208669